Topics in Current Genetics 21

Series Editor: *Stefan Hohmann*

For further volumes:
http://www.springer.com/series/4735

Esther Lubzens • Joan Cerdà • Melody Clark

Dormancy and Resistance in Harsh Environments

Springer

Editors
Esther Lubzens
Israel Oceanographic and Limnological Research
31080 Haifa
Israel
elubzens@hotmail.com

Joan Cerdà
Institut de Recerca i Tecnologia Agroalimentaries (IRTA)
Passeig marítim 37-49
08003 Barcelona
Spain
joan.cerda@irta.cat; jcerda@icm.csic.es

Melody S. Clark
Division of Ecosystems
British Antarctic Survey
High Cross, Madingley Road
Cambridge CB3 0ET
UK
mscl@bas.ac.uk

The cover illustration depicts pseudohyphal filaments of the ascomycete Saccharomyces cerevisiae that enable this organism to forage for nutrients. Pseudohyphal filaments were induced here in a wild-type haploid MATa 1278b strain by an unknown readily diffusible factor provided by growth in confrontation with an isogenic petite yeast strain in a sealed petri dish for two weeks and photographed at 100X magnification (provided by Xuewen Pan and Joseph Heitman).

ISBN 978-3-642-12421-1 e-ISBN 978-3-642-12422-8
DOI 10.1007/978-3-642-12422-8
Springer Heidelberg Dordrecht London New York

Library of Congress Control Number: 2010933088

Cover design: WMXDesign GmbH, Heidelberg, Germany

Printed on acid-free paper

Springer is part of Springer Science+Business Media (www.springer.com)

Topics in Current Genetics publishes review articles of wide interest in volumes that center around a specific topic in genetics, genomics as well as cell, molecular and developmental biology. Particular emphasis is placed on the comparison of several model organisms. Volume editors are invited by the series editor for special topics, but further suggestions for volume topics are highly welcomed.

Each volume is edited by one or several acknowledged leaders in the field, who ensure the highest standard of content and presentation. All contributions are peer-reviewed.

All volumes of Topics in Current Genetics are part of the Springer eBook Collection. The collection includes online access to more than 3,500 newly released books, book series volumes and reference works each year. In addition to the traditional print version, this new, state-of-the-art format of book publications gives every book a global readership and a better visibility.

Editorial office:
Topics in Current Genetics
Series Editor: Stefan Hohmann
Cell and Molecular Biology
Göteborg University
Box 462
40530 Göteborg, Sweden
FAX: +46 31 7862599
E-mail: editor@topics-current-genetics.se

Contents

Contributors

James V. Anderson Plant Science Research Unit, USDA-Agricultural Research Service, 1605 Albrecht Blvd, Fargo, ND, 58105-5674, USA, james.anderson@ars.usda.gov

William Bolingue UMR 1191 Molecular Seed Physiology INRA, 16 Bd Lavoisier, 49045, Angers, France, william.bolingue@angers.inra.fr

Julia Buitink UMR 1191 Molecular Seed Physiology INRA, 16 Bd Lavoisier, 49045, Angers, France, julia.buitink@angers.inra.fr

Cecilia Geijer Department of Cell and Molecular Biology/Microbiology, University of Gothenburg, Box 462, 40530, Gothenburg, Sweden

Joan Cerdà Laboratory of Institut de Recerca i Tecnologia Agroalimentàries (IRTA)-Institut de Ciències del Mar, Consejo Superior de Investigaciones Científicas (CSIC), 08003 Barcelona, Spain, joan.cerda@irta.cat; jcerda@icm.csic.es

Wun S. Chao Plant Science Research Unit, USDA-Agricultural Research Service, 1605 Albrecht Blvd, Fargo, ND 58105-5674, USA, wun.chao@ars.usda.gov

Melody S. Clark Division of Ecosystems, British Antarctic Survey, High Cross, Madingley Road, Cambridge CB3 0ET, UK, mscl@bas.ac.uk

Daphna Joseph-Strauss The Laboratory of Cellular and Molecular Biology, NIDDK, NIH, Bethesda, MD 20892, USA

Luc De Meester Laboratory of Aquatic Ecology and Evolutionary Biology, Katholieke Universiteit Leuven, Ch. Deberiotstraat 32, 3000, Leuven, Belgium

Nadav Y. Denekamp Department of Marine Biology, Israel Oceanographic and Limnological Research, Haifa, Israel

William E. Finch-Savage Warwick HRI, Warwick University, Wellesbourne, Warwick, CV35 9EF, UK, bill.finch-savage@warwick.ac.uk

Michael E. Foley Plant Science Research Unit, USDA-Agricultural Research Service, 1605 Albrecht Blvd, Fargo, ND 58105-5674, USA, michael.foley@ars.usda.gov

Barry Fuller Division of Surgery & Interventional Sciences, Royal Free & University College Medical School, UCL, London, NW3 2QG, UK, bfuller@medsch.ucl.ac.uk

Giora Simchen Department of Genetics, the Hebrew University of Jerusalem, Jerusalem, 91904, Israel

Gordana Grubor-Lajšić University of Novi-Sad, Faculty of Sciences, Trg Dositeja Obradovića 3, 21000, Novi Sad, Republic of Serbia

Edgardo Guibert Fac. de Ciencias Bioquímicas y Farmacéuticas, Universidad Nacional de Rosario, S2002LRK, Rosario, Argentina

Ora Hadas Israel Oceanographic and Limnological Research, Kinneret Limnological Laboratory, P.O. Box 447, Migdal, 14950, Israel

Atsushi Hagiwara Graduate School of Science and Technology, Nagasaki University, Nagasaki, Japan

Gerhard Heldmaier Faculty of Biology, Marburg University, 35032, Marburg, Germany

Henk W.M. Hilhorst Lab of Plant Physiology, Wageningen University, Droevendaalsesteeg 1, 6708 PB, Wageningen, The Netherlands, henk.hilhorst@wur.nl

Dirk K. Hincha Max-Planck-Institute of Molecular Plant Physiology, Potsdam, Germany

David P. Horvath Plant Science Research Unit, USDA-Agricultural Research Service, 1605 Albrecht Blvd, Fargo, ND 58105-5674, USA, david.horvath@ars.usda.gov

Ruth N. Kaplan-Levy Israel Oceanographic and Limnological Research, Kinneret Limnological Laboratory, P.O. Box 447, Migdal 14950, Israel

Olivier Leprince UMR 1191 Physiologie Moléculaire des Semences, Université d'Angers/Agrocampus-Ouest/INRA, 49045 Angers, France

Gerhard Leubner-Metzger Institut fuer Biologie II, Botanik/Pflanzenphysiologie, Albert-Ludwigs-Universitaet Freiburg, Schaenzlestr. 1, 79104 Freiburg i. Br., Germany, gerhard.leubner@biologie.uni-freiburg.de

Esther Lubzens Department of Marine Biology, Israel Oceanographic and Limnological Research, Haifa, Israel, esther@ocean.org.il

David Macherel UMR 1191 Physiologie Moléculaire des Semences, Université d'Angers/Agrocampus-Ouest/INRA, 49045 Angers, France

Thomas H. MacRae Department of Biology, Dalhousie University, Halifax, Canada NS B3H 4J1, tmacrae@dal.ca

Naama Barkai Departments of Molecular Genetics and Physics of Complex system, Weizmann Institute of Science, Rehovot, 76100, Israel

Jason E. Podrabsky Department of Biology, Portland State University, P.O. Box 751, 97207-0751, Portland, OR, USA, podrabsj@pdx.edu

Jelena Purać Biological Sciences Division, British Antarctic Survey, Natural Environment Research Council, High Cross, Madingley Road, Cambridge, CB3 0ET, UK
University of Novi-Sad, Faculty of Sciences, Trg Dositeja Obradovića 3, 21000, Novi Sad, Republic of Serbia

Zhijun Qiu Department of Biology, Dalhousie University, Halifax, Canada NS B3H 4J1, podrabsj@pdx.edu

Richard Reinhardt Molecular Genetics, Max Planck Institute, Berlin-Dahlem, Germany

Mark H. Rider Hormone and Metabolic Research Institute, de Duve Institute and Université catholique de Louvain, Avenue Hippocrate 75, 1200, Brussels, Belgium

Joaquín Rodríguez Fac. de Ciencias Bioquímicas y Farmacéuticas, Universidad Nacional de Rosario, S2002LRK, Rosario, Argentina

Jacqueline Rücker Department of Fresh Water Conservation, Brandenburg University of Technology – Cottbus, Seestrasse 45, Bad Saarow 15526, Germany

Ralph O. Schill Department of Zoology, Biological Institute, University of Stuttgart, Stuttgart, Germany, ralph.schill@bio.uni-stuttgart.de

Stefan Hohmann Department of Cell and Molecular Biology/Microbiology, University of Gothenburg, Box 462, 40530, Gothenburg, Sweden, stefan.hohmann@gu.se

Kenneth B. Storey Institute of Biochemistry, Carleton University, 1125 Colonel By Drive, Ottawa, ON, Canada K1S 5B6, kenneth_storey@carleton.ca

Koushirou Suga Faculty of Fisheries, Nagasaki University, Nagasaki, Japan

Assaf Sukenik Israel Oceanographic and Limnological Research, Kinneret Limnological Laboratory, P.O. Box 447, Migdal 14950, Israel, assaf@ocean.org.il

Michael L. Summers Department of Biology, California State University, Northridge, 18111 Nordhoff St, Northridge, CA 91330-8303, USA

Michael A.S. Thorne Biological Sciences Division, British Antarctic Survey, Natural Environment Research Council, High Cross, Madingley Road, Cambridge CB3 0ET, UK

Angèle Tingaud-Sequeira Laboratory of Institut de Recerca i Tecnologia Agroalimentàries (IRTA)-Institut de Ciències del Mar, Consejo Superior de Investigaciones Científicas (CSIC), 08003 Barcelona, Spain
Génomique et Physiologie des Poissons, Université Bordeaux 1, UMR NuAGe, 33405 Talence, France

Alan Tunnacliffe Department of Chemical Engineering and Biotechnology, University of Cambridge, Cambridge, UK, at10004@biotech.cam.ac.uk

Evelyne Vanvlasselaer Laboratory of Aquatic Ecology and Evolutionary Biology, Katholieke Universiteit Leuven, Ch. Deberiotstraat 32, 3000 Leuven, Belgium

M. Roger Worland Biological Sciences Division, British Antarctic Survey, Natural Environment Research Council, High Cross, Madingley Road, Cambridge, CB3 0ET, UK

Chapter 1
Introduction

Esther Lubzens, Joan Cerdà, and Melody S. Clark

Abstract Dormancy is a general term describing a state whereby metabolism and normal progression of life activities and development are dramatically reduced or brought to a halt. Dormancy facilitates the survival of organisms during environmental conditions that cannot support the regular course of life. Many organisms have evolved the capacity to enter dormancy and exit from it. Common pathways have been identified in several forms, in spite of the diversity and complexity in the survival strategies during dormancy in organisms displaying it. Revealing the mode of regulation of these pathways and the occurrence of other functions associated with dormancy are some of the great future challenges. This may open the way for revealing how cells, tissues or entire organisms can be made dormant or be revived effectively and lead to preserving cells, tissues and organs at ambient temperatures. These aspirations motivated the publication of this book.

1.1 Dormancy

Adaptation to extreme and harsh environments has allowed organisms to populate almost all regions of the planet, from arid deserts through salt encrusted marshes to super-heated hydrothermal vents and polar ice sheets. This has enabled organisms to colonize unique niches with possible reduced competition on resources. Under some conditions, survival is a question of "to be or not to be." This wide array of

E. Lubzens (✉)
Department of Marine Biology, Israel Oceanographic and Limnological Research, Haifa, Israel
e-mail: esther@ocean.org.il

J. Cerdà
Laboratory of Institut de Recerca i Tecnologia Agroalimentàries (IRTA)-Institut de Ciències del Mar, Consejo Superior de Investigaciones Científicas (CSIC), 08003 Barcelona, Spain

M.S. Clark
Division of Ecosystems, British Antarctic Survey, High Cross, Madingley Road, Cambridge CB3 0ET, UK

E. Lubzens et al. (eds.), *Dormancy and Resistance in Harsh Environments*,
Topics in Current Genetics 21,
DOI 10.1007/978-3-642-12422-8_1,

adaptations for survival in adverse conditions has raised the curiosity of scientists for centuries, as many are associated with an apparent cessation of life activity.

> The problem as to whether an organism can be revived after the complete cessation of all life processes, that is, whether life under certain conditions may be a discontinuous process, has, apart from its great scientific interest, a fascination of its own. In fact it is one of the oldest problems which preoccupied man when he began to think of life, death, and immortality

This quote is taken from the monumental treatise published by Keilin (1959). At the time, it provided an historical and state-of-the-art overview on "The problem of anabiosis or latent life: history and current concept." As biologists, we are aware of many terms associated with "anabiosis or latent life" such as hibernation, diapause, quiescence, aestivation, cryptobiosis, abiosis, etc., although they are more often referred to generically, these days, as dormancy. However, whatever the exact terminology or mechanisms involved, they all describe biological phenomena deviating from "active" or "normal" metabolism.

During dormancy normal progression of life activities or development is dramatically reduced or brought to a halt and they are resumed when conditions become supportive to life. The response can be a direct one whereby an organism enters a dormant state in response to an external cue such as lack of water by undergoing desiccation (e.g., tartigrades, bdelloid rotifers), drought tolerance (e.g., embryos of annual fishes), low temperature (e.g., Arctic springtail or hibernating mammals) or through developmentally programmed arrest whereby the information is stored until the appropriate developmental stage is attained and the organism enters into a state of suppressed metabolism (e.g., yeast spores, plant seeds, rotifer resting egg in rotifers, ephippia of waterfleas or *Daphnia*, and cysts of brine shrimp or *Artemia*). Preserving life in the desiccated form is a well-known phenomenon, spectacularly demonstrated by revival of plant seeds after centuries of storage (~1,300 years for a sacred lotus seed (Shen-Miller et al. 1995) and ~2,000 years for that of a date seed (Sallon et al. 2008)). However, there are several examples of survival in a dormant form without undergoing desiccation. These include aquatic organisms producing resting stages such as akinetes of cyanobacteria, resting eggs of rotifers or copepods, or ephippia of daphnids. It also includes annual fishes, where water is maintained within the diapausing embryos and hibernating mammals, where metabolism is reduced to maintain essential life support systems.

Survival in a state of reduced metabolism raises many questions, such as:

- How is metabolism reduced at the cellular level? Is it stopped?
- If metabolism still occurs, what is its nature and how are the potentially toxic waste products removed?
- How do cells maintain their structural integrity without decomposition by protein degrading enzymes such as proteases or bacteria?
- Is the immune response system functional?
- Is there a turnover of structural and functional proteins?
- How are functional units maintained such as ribosomes, so that life functions can be resumed sometimes within minutes?

- How do organisms respond to external stimuli during dormancy?
- How is dormancy regulated?

The complexity of the strategies employed, mechanisms involved, and the sheer variety of different organisms that utilize dormancy provide an almost endless list of potential research topics. In this book, we focus on current achievements in this field with some emphasis on demonstrating the potential of molecular tools in discovering common and divergent physiological pathways, for survival in a dormant form, or for tolerating harsh environments.

The genome sequencing of model organisms such as *Arabidopsis thaliana*, baker's yeast the nematode *Caenorhabditis elegans*, and the subsequent functional analyses, has contributed greatly to unveiling the genes associated in these organisms with dormant processes. These results have provided the impetus and tools to conduct similar investigations in a wide variety of "nonmodel" organisms. This is illustrated by significant molecular contributions summarized in the chapters on dormant cells of cyanobacteria (Chap. 2), rotifer resting eggs (Chap. 7), tardigrades (Chap. 8), Arctic springtails (Chap. 9), brine shrimp cysts (Chap. 10) and waterfleas (Chap. 11). The functional aspects of endurance in harsh environments of buds of perennial plants (Chap. 5), embryos of annual fishes (Chap. 12), and mammals (Chap. 13) provide an additional scope to the wide range of strategies for survival.

In the midst of such complexity, common pathways have been identified in several dormant forms and these include:

- Protection against reactive oxygen species (ROS) and detoxification. ROS are toxic in all life stages, but they are especially problematic for dormant forms. In plant seeds, desiccation causes loss of control mechanisms that maintain low ROS concentrations, thus the antioxidant activity has great importance.
- Maintaining the native folded conformation of proteins: Changes in osmotic pressure, pH, or temperature as well as desiccation, all challenge protein conformation and may cause the formation of cytotoxic protein aggregates. Small molecular weight heat shock proteins (sHSP) and the late embryogenesis abundant (LEA) proteins are proposed as assisting in protecting protein structure. They are demonstrably more abundant in dormant forms.
- Drying is often associated with the formation of an intracellular glassy state with the specific involvement of carbohydrates such as trehalose or other sugars as cryoprotectants.
- Lipid and fatty acid metabolism: lipid metabolism was found to be associated with hibernation in mammals and the nematode dauer stage.
- Indirect evidence suggests a role for aquaporins as they may assist in the passage of water and small soluble molecules across cell membranes.

Revealing how these and other functions associated with dormancy are regulated is one of the great future challenges. It is linked to the recognition that "dormancy" or "cryptobiosis" can be of relevance to preservation of cells and tissues (Chap. 14). This was recognized by Keilin (1959), but gained specific attention with the increasing use of trehalose in cell preservation, spectacularly demonstrated by dry storage of blood components (Wolkers et al. 2002; Crowe et al. 2003; Brumfiel 2004) and cryopreservation of human oocytes (Eroglu et al. 2002 and a review of Crowe 2007). Preserving organs and tissues for transplantation is regularly practiced in medicine today and could potentially benefit greatly from studies on nonmodel organisms, many of which are included in this book.

Preserving ones body and reviving it in future generations is a dream of mankind. In the present book, it is shown that many organisms have evolved the capacity to enter dormant stages and exit from it. This is often associated with desiccation, in which they survive for long periods and then reactivate within minutes or hours. So far, however, we have failed to learn how cells, tissues, or entire organisms can be made dormant or be effectively revived, at ambient temperatures. Attempting to learn that lesson was a vision of a European project named "SLEEPING BEAUTY" (NEST – Adventure # 012674) funded by the European Commission, and this book is the outcome of this project.

We are grateful to a large number of anonymous reviewers for their comments. We also like to thank all the participants of the workshop: "Sleeping beauties: Dormancy and resistance in harsh environments – Molecular, proteomic and metabolomic aspects" that took place at the Max Planck Institute – Molecular Genomics (May 18–21, 2008), for sharing their research, thoughts and aspirations in this book.

References

Brumfiel J (2004) Cell Biology: Just add water. Nature 428:14–15

Crowe JH (2007) Trehalose as a "chemical chaperone" fact and fantasy. In: Csermely P, Vigh L (eds) Molecular aspects of the stress response; Chaperones, membranes and networks. Landes Bioscience and Springer Sciences Business Media, Austin, TX, pp 143–158

Crowe JH, Tablin F, Wolkers WF, Gousset K, Tsvetkova NM, Ricker J (2003) Stabilization of membranes in human platelets freeze-dried with trehalose. Chem Phys of Lipids 122:41–52

Eroglu A, Toner M, Toth TL (2002) Beneficial effect of microinjected trehalose on the cryosurvival of human oocytes. Fertil Steril 77:152–158

Keilin D (1959) The Leeuwenhoek Lecture: The problem of anabiosis or latent life: History and current concept. Proc R Soc Lond B 150:149–191

Sallon S, Solowey E, Cohen Y, Korchinsky R, Egli M, Woodhatch I, Simchoni O, Kislev M (2008) Germination, genetics, and growth of an ancient date seed. Science 320:1464

Shen-Miller J, Mudgett MB, Schopf JW, Clarke S, Berger R (1995) Exceptional seed longevity and robust growth: ancient sacred lotus from China. Am J Bot 82:1367–1380

Wolkers WF, Tablin F, Crowe JH (2002) Fron anhydrobiosis to freeze-drying of eukaryotic cells. Comp Biochem Physiol A 131:535–543

Chapter 2
Akinetes: Dormant Cells of Cyanobacteria

Ruth N. Kaplan-Levy, Ora Hadas, Michael L. Summers, Jacqueline Rücker, and Assaf Sukenik

Abstract Cyanobacteria are an ancient and morphologically diverse group of photosynthetic prokaryotes, which were the first to evolve oxygenic photosynthesis. Cyanobacteria are widely distributed in diversed environments. In the case of members of the orders Nostocales and Stigonematales, their persistence and success were attributed to their ability to form specialized cells: heterocysts, capable of fixing atmospheric nitrogen and spore-like cells, the akinetes. This review focuses on akinetes of Nostocales, emphasizing environmental triggers and cellular responses involved in differentiation, maturation, dormancy, and germination of these resting cells. Morphological and structural changes, variation in akinete composition, and metabolism are summarized. Special attention is given to the genetic regulation of the differentiation process in an attempt to close gaps in our understanding of the dormancy phenomenon in cyanobacteria and to identify open questions for future research.

2.1 Introduction

The cyanobacteria comprise a very diverse group of photoautotrophic oxygenic prokaryotic organisms. They are found all over the world: in seas, soils, glaciers, deserts, and hot springs, but most species reside in freshwater in both benthic and

R.N. Kaplan-Levy, O. Hadas, and A. Sukenik (✉)
Israel Oceanographic and Limnological Research, Kinneret Limnological Laboratory, P.O. Box 447, Migdal 14950, Israel
e-mail: assaf@ocean.org.il

M.L. Summers
Department of Biology, California State University Northridge, 18111 Nordhoff St, Northridge, CA 91330-8303, USA

J. Rücker
Department of Fresh Water Conservation, Brandenburg University of Technology – Cottbus, Seestrasse 45, Bad Saarow 15526, Germany

E. Lubzens et al. (eds.), *Dormancy and Resistance in Harsh Environments*,
Topics in Current Genetics 21,
DOI 10.1007/978-3-642-12422-8_2, © Springer-Verlag Berlin Heidelberg 2010

pelagic habitats (van den Hoek et al. 1998; Adhikary 1996). In open freshwater environments, they can become extremely dominant forming dense blooms. Based on their life strategy, pelagic cyanobacteria can be classified into the following categories (1) species that lack specialized resting cells, for example, members of the orders Chroococcales and Oscillatoriales and (2) species that form specialized resting cells, for example, members of orders Nostocales and Stigonematales. The ability to form resting cells enables these species to survive harsh environmental conditions while dormant in bottom sediments. As environmental conditions improve, vegetative cells germinate from the resting spores and float due to newly formed gas vesicles, thus assisting in dispersal throughout the water column. Dormancy and floating features of these species are responsible for their domination in many water bodies. Within a short period of time, they bloom and influence the phytoplankton composition in a seasonally repetitive pattern. This annual life cycle of planktonic Nostocales is illustrated in Fig. 2.1, using *Aphanizomenon ovalisporum* as the representative Nostocales species.

The resting cells of Nostocales and Stigonematales species are called akinetes (from the Greek "akinetos" – motionless). These are spore-like, thick-walled, nonmotile cells that differentiate from vegetative cells and serve a perennating role. Akinetes are larger and have a thicker wall than vegetative cells and contain large amounts of food reserves and DNA. The akinete shape differs among species

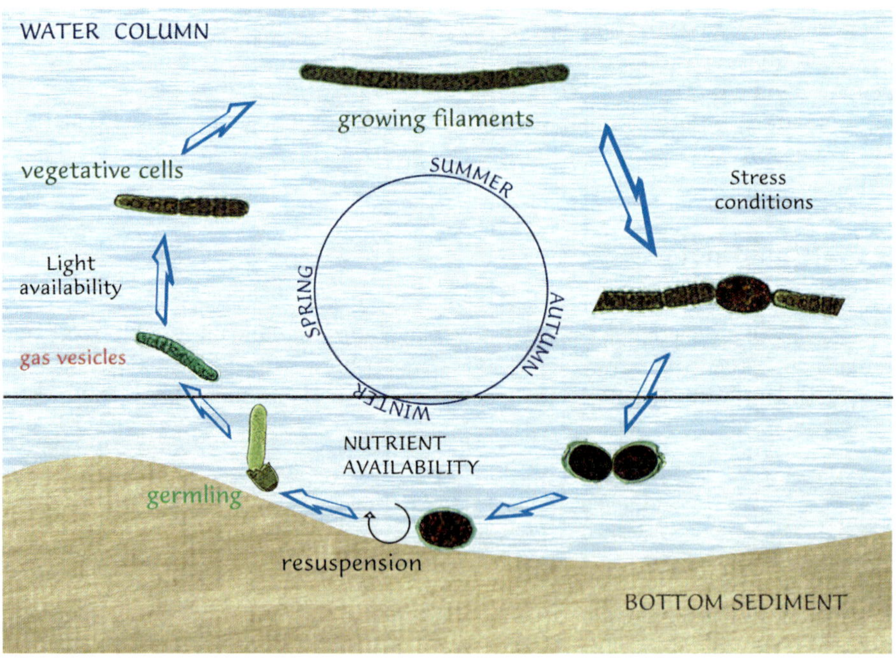

Fig. 2.1 Life cycle of the cyanobacterium *Aphanizomenon ovalisporum* (Nostocales). Adopted from Hense and Beckmann (2006)

from sphere to oblate spheroid and their distribution and position within a filament (trichome) is used as a taxonomic feature. Differentiation, maturation, dormancy, and germination of akinetes in Nostocales share some features with other prokaryotes (endospore in *Bacillus* – see Errington 2003) and some eukaryotes (spore in yeast – see Hohmann et al. 2010, cyst in Protists – see Corliss 2001). Here we present an updated overview on physiological, ecological, and molecular aspects of cyanobacterial akinetes. The reader is referred to earlier review papers of value by Nichols and Adams (1982), Herdman (1987), and by Adams and Duggan (1999).

2.2 Structure, Composition, and Metabolism of Akinetes

Akinetes are larger (sometimes by up to tenfold) than vegetative cells or nitrogen-fixing cells – heterocysts (Adams and Duggan 1999). Akinetes are surrounded by a thickened cell wall and a multilayered extracellular envelope (Nichols and Adams 1982; Herdman 1987, 1988), composed of glucose-rich carbohydrate and amino compounds as shown for *Anabaena cylindrica* by Cardemil and Wolk (1976, 1979). During differentiation, akinetes accumulate both glycogen and granules of cyanophycin (Simon 1987).

The position of akinetes along the trichome varies among cyanobacterial species and strains, where in some cases heterocysts were reported to influence their location (Wolk 1966). Akinetes develop immediately adjacent to heterocysts in *A. cylindrica* but several cells away from the heterocyst in *Anabaena circinalis* and in some other planktonic species (Fay et al. 1984; Li et al. 1997). In most cases, akinetes develop in strings, showing a gradient of decreasing maturity away from the first to develop. Adams and Duggan (1999) explained the akinete placement in relation to heterocysts by the need to accumulate large amounts of cyanophycin.

Akinetes undergo various metabolic and morphological changes during their development and maturation. Metabolic activities of akinetes such as CO_2 fixation showed reduced rates in *A. cylindrica* and *Nostoc* PCC 7524 (Fay 1969a; Sutherland et al. 1985a; Rao et al. 1987; Rai et al. 1985), whereas the rate of respiration was often elevated (Yamamoto 1976; Herdman 1987), presumably in relation to the maturation process but lost in older akinetes (Chauvat et al. 1982). Isolated akinetes of *Nostoc spongiaeforma* respired in the dark, evolved oxygen in the light and retained residual capability to synthesize proteins and lipids (Thiel and Wolk 1983). While developing akinetes of *A. cylindrica* are metabolically active, they have significantly decreased activity as they mature (Fay 1969a). Mature akinetes of *A. cylindrica* were reported to have little chlorophyll and no functional photosystem I (PSI) (Fay 1969b). However, the pigment content of akinetes of a different isolate of *A. cylindrica* was similar to that of the vegetative cells (Wolk and Simon 1969). Akinetes of *Anabaena doliolum* lost both chlorophyll and phycocyanin following incubation in the dark for several weeks (Singh and Sunita 1974). In vivo fluorescence measurements of *A. variabilis* akinetes suggested that the akinetes lacked a functional Photosystem II (PSII), although the reaction center chlorophyll was present (Bjorn et al. 1983). Using transmission electron microscopy and immunocytological labeling, the 32 kDa – PsbA protein (D1 polypeptide) of PSII was detected in akinetes (and other cell types) of the cyanobionts within leaf cavities of *Azolla caroliniana* Willd (Braun-Howland and Nierzwicki-Bauer 1990). In a recent study, Sukenik et al. (2007) demonstrated changes in the photosynthetic activities of individual vegetative cells and akinetes in trichomes of *A. ovalisporum* during akinete formation, whereas mature isolated akinetes retained only residual photosynthetic capacity. In mature akinetes of *A. ovalisporum*, the phycobilisome antenna was reduced in size and apparently detached from the reaction centers. Similarly, the disappearance of phycocyanin from *A. cylindrica* akinetes was reported by Fay (1969b) in accordance with observations on diminishing photosynthetic activity in isolated akinetes (Fay 1969a). The stoichiometric ratio of PSI to PSII in *A. ovalisporum* akinetes remained more or less the same as in vegetative cells and the cellular content of PsbA (D1) and PsaC proteins per cell volume remained fairly stable (Sukenik et al. 2007). Furthermore, preliminary immunoblotting experiments indicated the presence of the RubisCO large subunit in *A. ovalisporum* akinetes (Sukenik unpublished). Thus, it was concluded that in *A. ovalisporum* the reduction in the phycobilisome pool in mature akinetes is targeted at minimizing absorption of light energy to diminish potential damage to reaction centers during dormancy. The presence of reaction center complexes in mature akinetes ensures a prompt recruitment of photosynthesis upon germination, to energize essential cellular processes (Sukenik et al. 2009).

Akinetes accumulate both glycogen and cyanophycin, a nonribosomally produced reserve polymer composed of an aspartate backbone with arginine side groups (Simon 1987). In akinetes of *Nostoc* PCC 7524, the mean cellular content of cyanophycin was eightfold higher than in vegetative cells (Sutherland et al. 1979). However, accumulation of cyanophycin was not specific for akinete development, vegetative cells also accumulated glycogen and cyanophycin when entering the stationary phase (Herdman 1987). Incubation of *A. cylindrica* with the arginine analogue, canavanine (Nichols and Adams 1982), and mutation of the arginine biosynthesis gene, *argL*, in *Nostoc ellipsosporum* (Leganés et al. 1998), resulted in the production of akinetes lacking cyanophycin, suggesting that cyanophycin accumulation is not essential for the formation of akinetes. The mean cellular content of RNA, DNA, and protein was similar in vegetative cells and akinetes of *Nostoc* PCC 7524 (Sutherland et al. 1979), whereas the akinetes of *A. cylindrica* contained the same amount of RNA, but more than twice as much DNA, and ten times as much protein as vegetative cells (Simon 1977). These high values are probably a consequence of the increased size of the *A. cylindrica* akinetes, which were up to ten times the volume of the vegetative cells (Fay 1969b). In *A. ovalisporum*, DAPI staining demonstrated the accumulation of nucleic acids in developed akinetes. The intensity and localization of the DAPI signal indicate a homogeneous dispersion of nucleic acids in the entire akinete volume and the absence of polyphosphate bodies (Sukenik et al. 2009). Polyphosphate bodies were rare in mature akinetes of *Nostoc* PCC 7524, although they were commonly present in vegetative cells during akinete differentiation (Sutherland et al 1979).

2.3 Factors that Influence Akinete Differentiation

Various environmental factors were reported as triggers for differentiation of akinetes in different cyanobacterial species and strains (Table 2.1). The major, although not the only, trigger for akinete development is light intensities (Adams and Duggan 1999). For example, in *Nostoc* PCC 7524 cultivated in the presence of excess inorganic nutrients, akinetes differentiated as light availability was reduced by 90% or more of the incident light, due to the culture self shading (Sutherland et al 1979). However, high light intensities triggered the formation of akinete in *Cylindrospermopsis raciborskii* (Moore et al. 2005).

Light quality also plays a role in the control of akinete formation. In *Gloeotrichia*, akinete differentiation was stimulated by green rather than white light. As green light is the dominant spectral component during bloom conditions, this could also explain observations by Rother and Fay (1977) that akinete differentiation in natural populations is frequently associated with the development of surface blooms. Similar observations were recently reported by Thompson et al. (2009) for the toxic cyanobacterium *A. circinalis*, red or green irradiance were much more effective for akinete production than blue light. For cells grown under a predominantly red, white, or green irradiance, even short exposures to blue light substantially reduced the number of akinetes, suggesting that blue light inhibits akinete formation.

Limitation of phosphate has been implicated as a trigger for akinete development (Nichols and Adams 1982; Herdman 1987, 1988) and increasing numbers of akinetes were found during phosphorus deficiency (Sinclair and Whitton 1977). In *A. circinalis*, phosphate limitation appeared to be the major trigger, whereas limitations for N, inorganic C, iron, trace elements, or light had no effect on the development of akinetes (van Dok and Hart 1996). In *N. punctiforme*, akinetes were induced within 2 weeks starvation for phosphate (Meeks et al. 2002). However, phosphorus was required to allow full development of akinetes in *C. raciborskii* (Moore et al. 2003, 2005) and in *A. circinalis* (Fay et al. 1984).

In addition to phosphate, other nutrients and abiotic conditions are also known to affect the formation of akinetes. Deficiencies in Mg, Ca, Fe, and S, for example, led to a decrease in the number of akinetes in *Gloeotrichia ghosei*, while in a range of planktonic *Anabaena* isolates, temperature was important for triggering akinete differentiation (Li et al. 1997). In *A. doliolum* (Rao et al. 1987) and *Anabaena torulosa* (Sarma and Khattar 1993) a critical C:N ratio appeared to be important. In *C. raciborskii*, the formation of akinetes was triggered by an initial temperature shock, by the frequency of temperature fluctuations, and by high light intensity (Moore et al. 2005). Recently it was reported that deprivation of potassium ions (K^+) triggered the formation of akinetes in the cyanobacterium *A. ovalisporum*. A burst of akinete formation was observed within 1–2 weeks after the induction (K^+ depletion) was imposed (Sukenik et al. 2009). K^+-deficiency was found to induce akinete formation also in *Nostoc spongiaeforme* and in *N. punctiforme* (Sukenik and Summers unpublished). K^+-deficiency stimulus seems to induce a secondary signal, apparently related to cellular osmo-regulation and desiccation

Table 2.1 Summary of environmental conditions and factors that influence differentiation and germination of akinetes in various cyanobacterial species (Order Nostocales)

Environmental variable	Cyanobacterial species	Reference	Observation
A. Differentiation of akinetes			
Light intensity[1]	*Anabaena circinalis*	Fay et al. (1984)	Light limitation induced akinete formation
	Nostoc PCC 7524	Sutherland et al. (1979)	
	Anabaena cylindrica	Nichols et al. (1980), Fay (1969a)	
	Aphanizomenon flos-aquae	Rother and Fay (1977)	
	Cylindrospermopsis raciborskii	Moore et al. (2005)	Increase in light intensity resulted in an increase in akinete concentration
Light quality	*Gloeotrichia echinulata*	Wyman and Fay (1986)	Akinete differentiation was stimulated by green light
	Anabaena circinalis	Thompson et al. (2009)	Red light enhanced differentiation, short exposure to blue light inhibited akinete production
Phosphate[2]	*Anabaena cylindrica*	Nichols and Adams (1982)	Phosphate limitation has been implicated as a trigger
	Cylindrospermum licheniforme	Fisher and Wolk (1976)	
	Anabaena circinalis	van Dok and Hart (1996)	
	Gleotrichia ghosei	Sinclair and Whitton (1977)	
	Anabaena circinalis, *Nostoc* PCC 7524	Fay et al. (1984)	Phosphate stimulated akinete differentiation
	Nostoc PCC 7524	Sutherland et al. (1979)	Akinetes were never produced in the absence of phosphate
Temperature[3]	*Anabaena* spp.	Li et al. (1997)	Important for triggering of akinete differentiation in a range of strains
	Cylindrospermopsis raciborskii	Moore et al. (2005)	Maximum akinete concentrations were observed in cultures that experienced multiple diurnal temperature fluctuations with a magnitude of 10°C (25°C–15°C)
C:N ratio	*Anabaena doliolum*.	Rao et al. (1987)	Critical C:N ratio appeared to be important
	Anabaena torulosa.	Sarma and Khattar (1993)	
K^+ ion[4]	*Aphanizomenon ovalisporum*	Sukenik et al. (2007)	Depletion of potassium from the medium triggered akinete formation
	Nostoc spongiaeforme, *Nostoc punctiforme*	Sukenik and Summers (unpubl.)	
B. Germination of akinetes			
Light[5]	*Nodularia spumigena*	Huber (1985)	Low light was sufficient for germination
	Anabaena cylindrica	Yamamoto (1976)	Light was essential, germination did not occur in dark
	Anabaena circinalis	van Dok and Hart (1997)	

(*continued*)

Table 2.1 (continued)

Environmental variable	Cyanobacterial species	Reference	Observation
	Anabaena, Aphanizomenon	Karlsson-Elfgren and Brunberg (2004)	
	Gleotrichia echinulata	Karlsson-Elfgren et al. (2004)	
Light quality[5]	*Cylindrospermopsis raciborskii*	Wiedner et al. (2007)	Red light supported germination
	Nodularia spumigena	Huber (1985)	
	Anabaena variabilis	Braune (1979)	
	Anabaena doliolum, Fischrella mucicola	Kaushik and Kumar (1970)	Germination occurred also in non-photosynthetic light
Phosphate[5]	*Anabaena circinalis*	Thompson et al. (2009)	Phosphate was required for germination
	Anabaena circinalis	van Dok and Hart (1997)	
	Nodularia spumigena	Huber (1985)	
Temperature[5]	*Anabaena circinalis*	Fay (1988)	Incubation in high temperatures (37-45 °C) imposed reduction in germination rate
	Anabaenopsis arnoldii	Reddy (1983), Pandey and Talpasayi (1981)	High germination rate at around optimal temperature for growth
	Nostoc spumigena, Anabaena vaginicola	Rai and Pandey (1981)	
	Aphanizomenon ovalisporum	Hadas (unpubl.)	
	Cylindrospermopsis raciborskii	Wiedner et al. (2007)	
Sediment mixing and resuspension[6]	*Gleotrichia echinulata*	Ståhl-Delbanco and Hansson (2002), Karlsson-Elfgren et al. (2004)	Germination is enhanced by mixing of bottom sediment
	Anabaena, Aphanizomenon	Karlsson-Elfgren and Brunberg (2004)	
	Aphanizomenon ovalisporum	Hadas et al. (1999)	
	Anabaena circinalis	Baker and Bellifemine (2000)	
	Anabaena sp., *A. solitaria*, and *A. lemmermannii*	Rengefors et al. (2004)	
	Anabaena cylindrica	Yamamoto (1976)	
Oxygen	*Nostoc PCC 7524*	Chauvat et al. (1982)	Oxygen was essential for germination
	Anabaena circinalis	Kezhi et al. (1985)	

[1]The response to light intensity is species dependent
[2]In some species phosphate deficiency triggers akinete formation while in others a basal level of phosphate is required for akinete development
[3]Different temperature optima for different species. Temperature fluctuations play a role in some species
[4]K^+ -deficiency may be involved in secondary internal signals
[5]In most cases, conditions that support growth of vegetative cultures are required for germination
[6]Observations are mainly from lakes and water reservoirs

that leads to the induction of akinete formation. Adams and Duggan (1999) postulated that the diverse stimuli, reported to affect akinete formation, induce a common physiological trigger – perhaps decreased cell division or low energy – which results in akinete development. Argueta and Summers (2005) speculated that a metabolic imbalance triggers akinete formation as a *zwf* mutant of *N. spongiae-forme*, lacking the first enzyme of the oxidative pentose phosphate pathway, formed functional akinetes during dark incubation in the presence of fructose. The collective observations on the environmental stimuli that trigger akinete formation in different cyanobacterial species and strains (Table 2.1) are mostly consistent with cellular energy limitation and cessation of cell division as primary signals.

2.4 Factors Influencing Akinete Germination

Germination of akinetes is a complex coordinated metabolic process triggered by various ambient conditions such as temperature, increased light availability (day length and penetration to sediments), and by sediment resuspension induced by turbulence in proximity to the bottom sediments (Reynolds 1972; Karlsson-Elfgren et al. 2004) as specified in Table 2.1. *Light* – This was identified as a significant factor triggering germination of akinetes. In *A. cylindrica*, germination was dependent on light intensity and did not take place in the dark or in the presence of DCMU (Yamamoto 1976). However, in *Nodularia spumigena* very low light intensities (0.5 μmol photon m^{-2} s^{-1}) were enough to initiate germination. Akinetes were not able to germinate in the dark (Huber 1985; Rengefors et al. 2004) even under heterotrophic conditions; however, supply of suitable organic carbon may result in germination (van Dok and Hart 1997). The most active spectral range for germination was between 620 and 630 nm, coinciding with the maximum light absorption by C-phycocyanin (Nichols and Adams 1982). Light and phosphate were required for germination of *A. circinalis* (van Dok and Hart 1997) and *N. spumigena* (Huber 1985). Light was an important trigger for the recruitment of *Gloeotrichia echinulata* in Lake Erken but not as important for *Anabaena and Aphanizomenon* in Lake Limmaren, Sweden. In all species, light was correlated to the scale of recruitment via germination (Karlsson-Elfgren and Brunberg 2004; Karlsson-Elfgren et al. 2004). Dilution of an akinete-containing culture with a fresh medium stimulated germination, apparently due to increased light intensity (Herdman 1988; Adams and Duggan 1999). The process of germination may be photoperiodic (day-length) dependent and germination would occur only after maturation perioda was completed (Karlsson-Elfgren et al. 2004). *Temperature* – The tolerance of akinetes to temperature extremes vary among species. *A. fertilissima* akinetes when pretreated at high (37–45°C) or low (0–7°C) temperatures for 48 h showed no effect on germination, while a reduced germination rate was observed in akinetes of *Anabaenopsis arnoldii*, *N. spumigena*, and *A. vaginicola* when incubated in extreme temperatures (Reddy 1983; Pandey and Talpasayi 1981; Rai and Pandey 1981). Unlike bacterial and fungal spores, germination of akinetes of *A. cylindrica* was not stimulated by heat shock (Yamamoto 1976). Akinetes of *A. ovalisporum* isolated from Lake Kinneret (Israel) and grown in cultures, germinated within a temperature range of 18–25°C but germination yield was low and unsynchronized (Hadas unpublished). Based on a field study in a shallow lake in northern Germany, Wiedner et al. (2007) suggested a germination temperature for *C. raciborskii* of 15–17°C, but Tingwey et al. (personal communication) found germination down to 13°C in an experimental set up with sediment from the same lake. *Sediment mixing and resuspension* – In *G. echinulata*, the process of recruitment from bottom sediments via akinete germination was influenced by high temperature and light, and significantly enhanced by mixing of bottom sediment imposed by bioturbation and physical processes (Ståhl-Delbanco and Hansson 2002; Karlsson-Elfgren et al. 2004). In Lake Kinneret, the benthic boundary layer and the sediment water interface are subject to turbulence processes, whereas sediments in the littoral zone are resuspended due to wave breaks, thus possibly affecting recruitment of akinetes. Bottom sediments collected from Lake Kinneret and incubated under control conditions in N-free BG11 medium yielded many filaments of *A. ovalisporum*, pointing to the role of akinetes in the establishment of a new population (Hadas et al. 1999). It is possible that shallow wetlands, shallow lakes, and littoral zones of deep lakes provide a conducive environment for germination of akinetes due to continuous resuspension of akinetes from the sediments and their exposure to an appropriate level of light (Baker and Bellifemine 2000; Rengefors et al. 2004) and oxygen, which are crucial for germination (Fay 1988). *Salinity* – When *N. spumigena* akinetes were pretreated at low (Pandey and Talpasayi 1981) or high concentrations of sodium chloride (Huber 1985), germination rate was reduced. The appearance of *A. circinalis* germlings increased with increased salinity up to 2.5 g l^{-1} with 26.9% germination and decreased to 0.2% at 5 g l^{-1}. No germination was observed at 10 g l^{-1} salinity (Baker and Bellifemine 2000). *Nutrients* – Addition of organic compounds such as sucrose and a supply of oxygen increased the efficiency of germination in *Nostoc* PCC 7524. Under these conditions all akinetes germinated, although slowly, indicating that successful germination required respiration and cyclic photophosphorylation (Chauvat et al. 1982). Germination of akinetes of *A. cylindrica* was completely inhibited by DCMU (Yamamoto 1976). Accumulated cyanophycin served as a source of nitrogen required for protein synthesis in the early stages of germination in *A. variabilis* (Braune and Doehler 1996). Degradation of cyanophycin during germination was observed in *Cylindrospermum* (Miller and Lang 1968), *A. flos-aquae* (Wildman et al. 1975), *A. cylindrica* (Fay 1969a), and *Nostoc* PCC 6720 (Skill and Smith 1987), whereas in *Nostoc* PCC 7524 other intracellular storage compounds were consumed (Sutherland et al. 1985a). The involvement of hydrolytic enzymes that degrade cyanophycin during germination was postulated (Braune 1979).

The environmental stimuli that trigger akinete germination in different cyanobacterial species (Table 2.1) generally correspond to the conditions that support growth of vegetative cultures. In addition, sediment mixing and resuspension play an important role in germination as they relocate the akinetes from the bottom sediment into the water column and photic zone.

2.5 The Germination Process

Germination of akinetes begins with cell division that occurs inside the akinete's envelope as described for *Nostoc* PCC 2574 (Sutherland et al. 1985b; Herdman 1988) and other cyanobacterial species (Moore et al 2004; Hori 2003; Baker and Bellifemine 2000; Braune 1980). Expansion of the cells results in an increase in turgor pressure which, consequently leads to a disruption of the envelope and emergence of the germling from the akinete's envelope. The open envelope may remain associated with the developing filament for some time. Morphological changes of the germling eventually leads to a fully developed young trichome (Moore et al. 2004). The germination process of *A. ovalisporum* akinetes, demonstrated in Fig. 2.2, begins with reorganization of cellular material (Fig. 2.2b) followed by elongation and division of the spore-like cell and opening of the akinete envelope on either terminal side of its slightly longer axis (Fig. 2.2c–g). It is unclear whether the envelope disruption is assisted by enzymatic activity or if it occurs merely due to the increased internal pressure resulting from the expanded cells. Finally, the germling, comprised of several cells, emerges from the akinete (Fig. 2.2 h). In some strains such as *N. punctiforme*, during germination the entire akinete wall may dissolve and, hence, not be microscopically visible (Adams and Duggan 1999; Meeks et al. 2002). Akinetes of *A. ovalisporum* isolated from Lake Kinneret did not germinate synchronously and the germination frequency was low.

Germination of *Cyanospira* akinetes was accompanied by de novo synthesis of proteins which took place prior to the first cell division (Sili et al. 1994). In *A. circinalis*, photosynthetic activity provided the energy for akinete germination (Kezhi et al. 1985), but the rate of germination was determined by the respiratory oxygen uptake of the akinetes, in a temperature-dependent manner (Fay 1988).

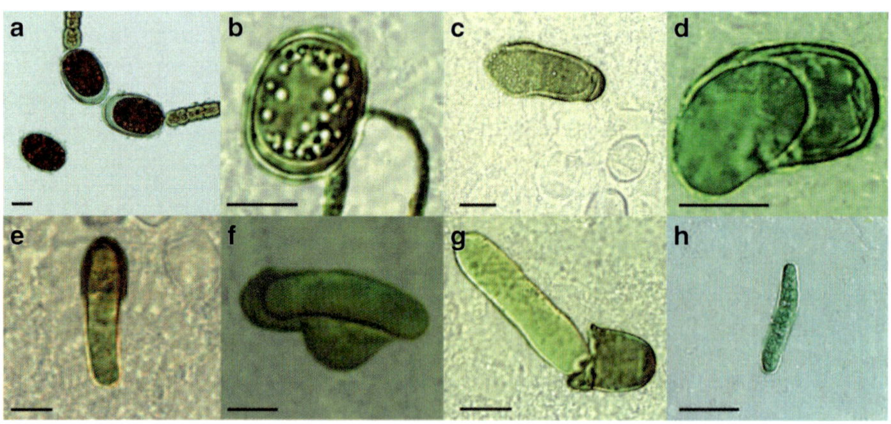

Fig. 2.2 Germination stages in akinetes of *A. ovalisporum*. (**a**) A free akinete and two connected akinetes within a filament of vegetative cells; (**b**) A matured granulated akinete; (**c**) The envelope of the akinete is opened (ruptured); (**d**)–(**g**) A germling emerging from the envelope; (**h**) A young filament; *Black horizontal scale bar* indicates 10 μm

An additional stage (although not an obligatory one) in the germination process is the development of gas vacuoles that support successful flotation of germlings and trichomes in the water column as depicted in Fig. 2.1. In *G. echinulata*, the newly formed filaments float 2–4 days after germination (Karlsson-Elfgren et al. 2004).

2.6 Ecological Functions of Akinetes

Variable harsh conditions imposed in laboratory studies showed that akinetes are resistant to low temperatures and desiccation, but not to heat, with the exception of *A. cylindrica* which germinate after drying at 60°C or under sunlight (Hori et al. 2003). Extended survival was reported for akinetes of *A. cylindrica* surviving in the dark and dry state for 5 years, whereas vegetative cells survived no longer than 2 weeks under similar conditions (Yamamoto 1975). Akinetes of *Nostoc* PCC 7524 survived in the dark at 4°C for 15 months, whereas vegetative cells lost viability within 7 days (Sutherland et al. 1979). Akinetes of *Aphanizomenon* and *Anabaena*, 18 and 64 years old, respectively, found in the sediment of Rostherne Mere (England) were viable and successfully germinated (Livingstone and Jaworski 1980). Thus, akinetes do not only have a temporary resting function, but may also ensure the long-term survival of a species giving it an ecological advantage. The term temporary resting means the overwintering and survival through dry periods. In temperate climatic zones, where the vegetative cells die in autumn, akinetes are a key factor in the annual life cycle of Nostocales (Fig. 2.1). A good example is the life cycle regulations of *C. raciborskii* in North German lakes where the time of germination was temperature mediated but further growth was mainly controlled by underwater light supply (Wiedner et al. 2007). Using a simple mathematical model it was demonstrated that temperature is the most important variable determining the population size: the earlier the germination took place in spring a larger population was recorded the next summer. *C. raciborskii* population size determines the annual input of akinetes to the sediment (Rücker et al. 2009 submitted). Consequently, interannual variations in pelagic populations were reflected by a varying number of akinetes deposited in the sediment, representing different inoculum sizes for the proceeding growing season. Although the akinete "seed bank" in the sediments of lakes is important for the recolonization of the pelagic zone, the contribution of akinetes toward the bloom success of next year's population of Nostocales seems to be rather small: 0.62% in Green Lake, OR (Barbiero and Welch 1992), 8% in Agency Lake, OR (Barbiero and Kann 1994), and 0.003–0.05% in Lake Limmaren, Sweden (Karlsson-Elfgren and Brunberg 2004). However, small deposits of akinetes may be sufficient for later colonization. For instance, *C. raciborskii* population size in a shallow German lake was more dependent on abiotic conditions after germination than on the inoculum size (Rücker et al. 2009 submitted).

Besides their role in survival, akinetes have the ability to serve as dispersal units. The most dramatic change in geographic distribution could be observed for the originally tropical cyanobacterium *C. raciborskii*, which spread from tropical to temperate regions on all continents except Antarctica during recent decades (Padisák 1997). Two hypotheses have been put forward to explain these changes in biogeography (a) the species spread to temperate regions due to increasing water temperatures associated with climate change and (b) selected ecotypes with lower temperature and light requirements have spread northward. Wiedner et al. (2007) assumed that an earlier rise in water temperature associated with climate change has promoted the species expansion. Transport of akinetes by migratory birds as a possible means of dispersal (Padisák 1998) may increase the chances of akinete-producing strains to be spread. The possible role of akinetes as a prerequisite for spreading of *Anabaena bergii* and *Aphanizomenon aphanizomenoides*, which invaded lakes of northern Germany was also hypothesized by Stüken et al. (2006)

The robust shells of akinetes are useful microfossil indicators, which may contribute to the reconstruction of earlier phytoplankton composition and trophic state of water bodies (van Geel 1986; van Geel et al. 1994). The invasion of *C. raciborskii* to north German lakes in the last 10–20 years could be proved by the detection of akinete shells in the upper part of sediment cores of two shallow lakes (Rücker et al. unpublished data). Since akinetes may stay viable in deeper sediment layers for a long time, providing an interesting tool for studying genetic variability of ancient Nostocales populations, or perhaps even physiological studies if they could be germinated and induced to grow in the laboratory.

2.7 Genes Involved in Akinete Differentiation

While the formation of akinetes presents a relatively simple model for cellular differentiation, the elucidation of the molecular mechanism regulating and involved in this process lagged until recently. Many studies were carried out using filamentous cyanobacteria to decipher the differentiation of nitrogen-fixing cells, heterocysts, from photosynthetically active vegetative cells but only few attempts focused on the differentiation of the dormant forms (Meeks et al. 2002). Heterocysts were used as a preferred model, mainly for their simple cell differentiation triggered by deprivation of fixed nitrogen. The advanced data accumulated on heterocyst formation suggest that these cells and akinetes share some commonalities in the molecular pathway of cell differentiation (Zhang et al. 2006). Four genes were found to be involved in both differentiating cells. One of these genes is *hepA*. A mutation in this gene resulted in alterations of akinete and heterocysts envelopes in *Anabaena variabilis* (Leganés 1994). This gene encodes for an ABC transporter required for the deposition of polysaccharides in the envelope of both cell types. The second gene, also implicated in polysaccharide synthesis is *devR*. *devR* encodes for a response regulator of a two-component system. When this gene was overexpressed in *Nostoc punctiforme*, an increase in akinete differentiating cells was observed (Campbell et al. 1996). The third gene found to be involved in both heterocysts and akinetes differentiation is *hetR*, which encodes for a DNA-binding protease. This gene when mutated by a transposon insertion in *N. ellipsosporum* resulted in a failure of cells to differentiate either to heterocysts or akinetes. Further analysis using the luciferase reporter gene, showed that *hetR* was expressed in akinetes (Leganés et al. 1994). In *Nostoc punctiforme*, a *hetR* mutant was capable of producing akinete-like cells. These akinetes-like cells lacked the granular characteristics found in the wild type. Both types of akinetes, however, mutant and wild type, had similar viability upon low-temperature treatment following phosphate starvation, when compared to vegetative cells. Therefore, it was suggested that although involved in the process, *hetR* is not essential for akinete differentiation (Wong and Meeks 2002). Another gene affecting akinete and heterocyst development in *Nostoc ellipsosporum* is *argL*, which encodes for an *N*-acetylglutamate semialdehyde dehydrogenase, an enzyme involved in L-arginine biosynthesis. A mutation caused by a transposon insertion in *argL* of *N. elipsosporum* resulted in smaller than wild type akinetes, which lack cyanophycin granules and failed to germinate (Leganés et al. 1998).

The study of akinete differentiation at the molecular level has been limited by the asynchronous development and restricted number of akinetes formed within a filament and by the lack of a marker gene for developing or mature akinetes. The first akinete marker was identified in *Anabaena variabilis* (Zhou and Wolk 2002) representing a breakthrough in the study dormant cells development in cyanobacteria. Separation of total protein extract by SDS-PAGE showed the presence of a 43-kDa protein in akinetes. This protein was designated AvaK. *avaK* was highly expressed in akinetes but to a small degree in vegetative cells as was demonstrated by GFP fusion in this strain (Zhou and Wolk 2002) and in *N. punctiforme* (Argueta et al. 2004). The deduced protein sequence of AvaK shows the existence of a PRC barrel domain in its N-terminal region, a domain implicated in RNA metabolism (Anantharaman and Aravind 2002); however, the function of this gene remains unknown. Synchronized differentiation of akinetes was reported in the *zwf* mutant of *Nostoc punctiforme* (Argueta and Summers 2005), which lacks the activity of glucose-6-phosphate dehydrogenase, the first enzyme of the oxidative pentose phosphate pathway (Summers et al. 1995). In this mutant, vegetative cells differentiate into akinetes synchronously, following dark incubation of cultures with fructose as an external carbon source. It was, therefore, chosen as a preferred strain for studies of akinete development (Argueta and Summers 2005).

The identification of an akinete marker, together with a reliable system that gives synchronous akinete differentiation allowed the application of high-throughput technology to study akinete formation in cyanobacteria. Argueta et al. (2006) reported the detection of three novel genes involved in akinete differentiation. These genes were detected by differential display and confirmed with quantitative RT-PCR and promoter fusions to a reporter gene (GFP) to demonstrate cell-type-specific gene expression. The genes were designated (a) *aet* (Npun_F0062) an akinete expressed transporter encoding an ABC transporter with high similarity to the *E. coli* MsbA a lipopolysaccharide transporter. (b) *aapN* (Npun_F5999) – an akinete aminopeptidase belonging to the M28 peptidase family. (c) *hap* (Npun_R4070) a hormogonium/akinete-expressed protease homologous to the β-subunit group of the M16 zinc-dependent proteases complex.

Sequencing of the *N. punctiforme* genome allowed the production of an open reading frame (ORF) microarray. This was then used to compare the global gene expression of *N. punctiforme* cultures with heterocyst and during the differentiation of hormogonia and *zwf* akinetes (Campbell et al. 2007). In that study, a single time point, 3 days into the akinete differentiation process was tested. During that time window, 255 genes were up-regulated, 41% of which encoded for characterized proteins. The global gene expression 3 days after induction, showed an increase in transcript levels of four transcription regulators: two transcription factors members of the Crp family and two sigma factors. There was an increase in genes involved in cell envelope metabolism, such as *amiC*, encoding for an enzyme that biodegrades peptidoglycan linker bonds. The expression of *nblA* gene that encodes for a phycobilisome degradation protein increased as well (Campbell et al. 2007). It is postulated that the increase in the expression of *nblA* facilitates the degradation of phycobilisome antenna in maturating akinetes as reported by Sukenik et al (2007). The *avaK* orthologous gene encoding the akinete marker was up-regulated as well in a 3-days-old akinete induced culture of *N. punctiforme* (Campbell et al. 2007). *A. ovalisporum* genes orthologous to *avaK*, *aet*, and *nblA* were highly expressed in isolated akinetes as compared to their expression level in vegetative cells of an exponentially grown culture as shown by a semiquantitative RT-PCR experiment (Fig. 2.3). These results are consistent with the expression of *avaK* in *A. variabilis* (Zhou and Wolk 2002), with *N. punctiforme* differential display results for *aet* (Argueta et al. 2006), and with microarray results from *zwf* akinetes for *nblA* and *avaK* (Campbell et al. 2007).

Transcript levels of *patA* and the CHF class protease – *hetF* genes increased during heterocyst differentiation. Transcript levels of both genes were also increased in the akinete-forming culture (Campbell et al. 2007), suggesting a common regulatory pathway for differentiation of these two cell types. During heterocyst development, the expression pattern of the cell differentiation regulatory protein *hetR* was

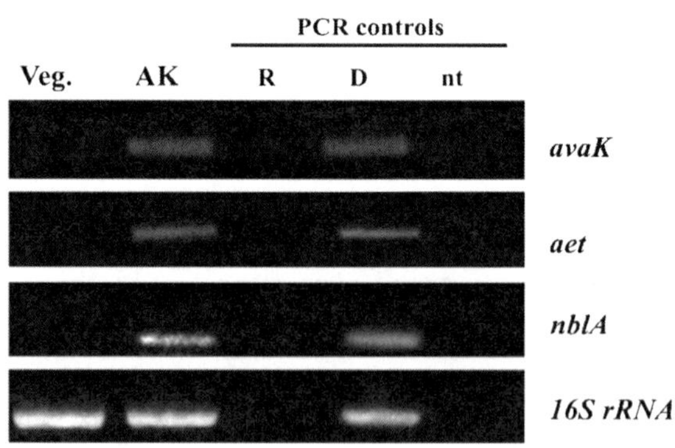

Fig. 2.3 SQ-RT-PCR (semiquantitative reverse transcriptase-PCR) of akinete marker genes in *Aphanizomenon ovalisporum* using specific primers for the *A. ovalisporum* orthologous genes to *avaK*, *aet*, and *nblA*. RNA was extracted from an exponentially grown culture lacking akinetes (Veg.) and from isolated akinetes (AK). Negative controls contained only RNA as template (R) or no template (nt). The positive control contained genomic DNA (D) as template in the PCR reaction

similar to that of *ntcA*, suggesting a mutual dependency in the expression of these two genes (Muro-Pastor et al. 2002). HetF was found to be essential for proper regulation of HetR both in the transcription and posttranslational level (Risser and Callahan 2008). The gene *hetF* is constitutively expressed in both vegetative cells and heterocysts (Wong and Meeks 2001). In heterocysts, PatA facilitates HetF activity to regulate the levels of HetR in an unknown manner (Risser and Callahan 2008). Interestingly, in *A. ovalisporum* cultures induced to form akinetes, the expression of *patA* was observed only after 3 weeks of induction, and its transcript was preferentially found in mature akinetes (Kaplan-Levy unpublished). *devR* encodes a small protein similar to the receiver domain of two-component regulatory systems that was implicated in heterocyst cell envelope formation and required for normal nitrogen fixation in *N. punctiforme* (Campbell et al. 1996). The presence of a complementing *devR* gene on a multicopy plasmid resulted in induction of akinete formation under noninducing conditions, implicating it in a phosphorelay system involved directly, or indirectly, through crosstalk, with development of heterocysts and akinetes. In akinete induced cultures of *A. ovalisporum*, the transcript levels of *devR* increased in a similar manner to that of *hetR* and *hetF*, with high levels in the isolated akinetes (Kaplan-Levy unpublished). It is suggested that the HetR regulatory pathway is involved also in the akinete differentiation process. However, unlike in heterocysts differentiation, we postulate that this pathway is activated at later stages of akinete differentiation, leading to akinete maturation. In heterocysts, the HetR pathway leads to activation of several processes (a) the synthesis of a polysaccharide envelope a process in which DevR is implicated; (b) deposition of a glycolipid layer (possibly via expression of *aet*); (c) cell division is stopped; and (d) cessation of oxygenic photosynthetic activity (Zhao and Wolk 2007). The first three processes are also essential for the formation and maturation of akinetes.

2.8 Similarity of Akinetes to Dormant Forms of Other Prokaryotes

Other types of prokaryotes form specialized differentiated resting cells in response to nutritional stress. These cells display less metabolic activity than their vegetative counterparts and do not divide. As is found for akinetes, differentiated resting cells are commonly more resistant to environmental stress, and exhibit an altered morphology relative to vegetative cells.

Endospores, so termed due to spore formation within an existing cell, are among the most resistant dormant cells. They are commonly found among the gram-positive *Bacillus*, *Clostridia*, and the thermophilic genus *Thermoactinomyces* (Cross 1968). Endospore development begins with an asymmetric septation within a single cell. The larger compartment, destined to become the “mother cell,” engulfs the smaller cell destined to become the endospore. Each cell type contributes materials to the endospore envelope to create a thickened multilayered protective envelope, while the nucleoid of the endospore is condensed and protected by interactions with newly synthesized proteins and compounds. Lysis of the mother cell releases the mature endospore, which is resistant to boiling, radiation, and chemical attack (Setlow 2000). Akinetes do not undergo internal septation and engulfment, instead create the dormant form by deposition of additional protective layers around an existing cell. Endospores exhibit no detectable metabolism or ATP, a characteristic that also separates them from akinetes, and spores of streptomyces and myxobacteria (Setlow 2000). The timing and gene regulation involved in septation, engulfment, and deposition of endospore envelope layers between the mother cell and developing endospore has been extensively studied, and used as a model for comparison with other prokaryotic developmental systems. It is controlled by the sequential action of different compartment-specific sigma factors and signaling by two-component regulatory systems (Kroos 2007).

Another type of dormant cells are cysts, such as those formed by *Azotobacter* and *Rhodospirillum*. In *Azotobacter vinelandii*, the differentiating cell accumulates poly-β-hydroxybutyrate (PHB) and forms a large sphere surrounded by a thick multilayered covering consisting of an inner layer containing carbohydrates and lipids, and an outer layer composed of lipopolysaccharides and lipoproteins (Pope and Wyss 1970). Like akinetes, *A. vinelandii* cysts are minimally resistant to heat, are resistant to desiccation, and can be observed to germinate from ruptured cyst envelopes (Socilifsky and Wyss 1962). Akinetes also accumulate storage material, albeit in the form of glycogen and cyanophycin. Cysts of the anoxygenic photosynthetic bacterium *Rhodospirillum centenum* contain multiple cells per cyst, but show many similarities to cyst formation in *A. vinelandii* (Berleman and Bauer 2004).

Streptomyces and myxobacteria also contain well-studied examples of bacteria that differentiate into spores. Streptomyces are the most complex type of gram-positive actinomycetes that grow as a mycelium of branching hyphal filaments. The best studied is *Streptomyces coelicolor* that produces a series of aerial spores from long hyphae growing up from the colony upon nutrient depletion, similar to the sporulation and dispersal strategy used by molds (Wildermuth 1970). Although superficial similarity exists between *S. coelicolor* spore formation and that of akinetes in strains exhibiting contiguous stretches of maturing akinetes within a filament, cyanobacterial akinete formation does not physically resemble this process.

In gram-negative myxobacteria such as *Myxococcus xanthus*, large numbers of spores are formed within fruiting bodies. Fruiting body formation occurs on solid substrates when large numbers of motile myxobacteria sense a nutritional downshift (Dworkin 1996). By comparison, cyanobacterial akinetes form individually within non-motile filaments and are not enclosed in a larger structure. Any cell–cell signaling would be limited to adjacent cells and those in close proximity on other filaments. In myxobacteria, only a small proportion of cells is destined to become spores in a fruiting body, whereas in some cyanobacteria as differentiation proceeds down a filament, all the vegetative cells can eventually convert to akinetes (e.g., *Nostoc* strains). In other strains, differentiation into akinetes occurs not progressively along a filament but simultaneously along long sequences of cells (Sarma and Khattar 1993). Akinetes are similar to spores of *Streptomyces* and *Myxococcus*, but unlike endospores they are not resistance to extreme heat (Setlow 2000).

Both *M. xanthus* and *Streptomyces* spores contain large amounts of the disaccharide trehalose, which has been implicated in their desiccation protection (Cruze-Martin et al. 1989; McBride and Ensign 1987). Trehalose and sucrose have been shown to be induced in cyanobacteria found in desert crusts (Hershkovitz et al. 1991). Extracellular polysaccharides of desiccation-resistant cyanobacteria in combination with trehalose or sucrose have been shown to stabilize membrane structure (Hill et al. 1997), which could account for an alternative mechanism of desiccation resistance for these sugars, in addition to their role as "chemical chaperones" within the cytoplasm (Crowe et al. 1998). A smaller but significant induction of sucrose by desiccation stress was found in *Anabaena* 7120, although very little trehalose accumulation was observed (Higo et al. 2006), indicating a plausible explanation for the range of desiccation resistance observed among cyanobacteria. The increased amount of polysaccharides in the envelope of akinetes (Cardemil and Wolk 1981; Wolk et al. 1994) could play a role similar to the external polysaccharides of desiccation-resistant cyanobacterial species; however, the presence and role of these polymers remain largely unexplored.

As in *Bacillus* endospore formation, regulation of sporulation in streptomyces and myxobacteria has been linked to regulatory cascades that include interactions between sigma factors and members of two-component regulatory systems (Chater 2000; Kroos 2007). Research into the regulation of akinete formation is still in its infancy. However, the identification of two alternative sigma factors and a subset of two-component regulatory systems up-regulated in *zwf* akinetes (Campbell et al. 2007) provide hints that similar regulatory cascades may be involved in akinete formation.

2.9 Conclusions and Future Prospects

Differentiation of vegetative cells to dormant forms (akinetes) in cyanobacteria that belong to the orders Nostocales and Stigonematales, and the role of akinetes in the life history and the success of cyanobacteria in nature have been studied since 1856 (see Herdman 1987). Structural changes and metabolic variations during akinete formation and maturation were described for a wide range of species and strains. Nevertheless, understanding of mechanisms that trigger akinete formation via the conversion of a vegetative cell within a filament into a resting cell was hindered until recently. Identification of environmental triggers that induce akinete formation, the availability of mutants that form akinetes under well-defined conditions and above all, development of advanced genomic tools that have been implemented to study the regulation and differentiation of cyanobacteria, have allowed progress in understanding the akinete differentiation process. Significant progress in analysis of mechanisms that lead to akinete maturation has been made recently. Further identification of signal perception and transduction as well as characterization of cellular processes and metabolic activities leading to the formation of akinetes are expected in the near future.

Numerous genes, including regulatory genes that are involved specifically in maturation have been identified, and microarray experiments have demonstrated that many genes are activated at different times during akinete induction and maturation. The involvement of regulatory cascade and transcription factors, primarily associated with the formation of heterocysts, were also identified in the akinete induction process. These findings clearly support an early notion (Wolk et al. 1994) that heterocysts may have evolved from akinetes.

Implementation of various molecular techniques and data from fully sequenced genomes of several Nostocales species ensure a rapid advancement toward a better understanding of the dormancy phenomenon in cyanobacteria. Akinete transcriptomic, proteomic, and metabolomic data is rapidly accumulating and the mechanism of dormancy in cyanobacteria is emerging as a heterogeneous process, as has been found in other prokaryotes, protists, and higher organisms. However, many questions are yet to be resolved: How are external signals that initiate germination perceived by an akinete and how are they processed to resume a fully active dividing vegetative cell? What are the factors that determine which vegetative cells along a filament will differentiate into akinetes? How are cellular and regulatory processes integrated into the environmental phenomenon of seasonal repetitive blooms? And finally, can we learn from akinete formation and dormancy processes about long-term preservation of eukaryotic cells under permissive temperatures and other environmental conditions?

Acknowledgment Our work was supported by a EU-NEST program project No 12674 "sleeping beauty" (AS, OH), by BMBF/MOST program project No WT803/2316 (OH AS JR), and by US-NIH SCORE grant 5S06GM048680 (MLS). We thank Dr. R. Reinhardt and Dr. M. Kube from Max Planck Institute for Molecular Genetics, Berlin for their guidance and support with the molecular analysis of *A. ovalisporum* genome. We wish to thank two anonymous reviewers who contributed to the improvement of an earlier version by their constructive comments and suggestions.

References

Adams DG, Duggan PS (1999) Heterocyst and akinete differentiation in cyanobacteria. New Phytol 144:3–33

Adhikary SP (1996) Ecology of freshwater and terrestrial cyanobacteria. J Sci Ind Res 55:753–762

Anantharaman V, Aravind L (2002) The PRC-barrel: a widespread, conserved domain shared by photosynthetic reaction center subunits and proteins of RNA metabolism. Genome Biol 3: research0061.0061–research0061.0069

Argueta C, Summers ML (2005) Characterization of a model system for the study of *Nostoc punctiforme* akinetes. Arch Microbiol 183:338–346

Argueta C, Yuksek K, Summers M (2004) Construction and use of GFP reporter vectors for analysis of cell-type-specific gene expression in *Nostoc punctiforme*. J Microbiol Methods 59:181–188

Argueta C, Yuksek K, Patel R, Summers ML (2006) Identification of *Nostoc punctiforme* akinete-expressed genes using differential display. Mol Microbiol 61:748–757

Baker PD, Bellifemine D (2000) Environmental influences on akinete germination of *Anabaena circinalis* and implications for management of cyanobacterial blooms. Hydrobiologia 427: 65–73

Barbiero RP, Kann J (1994) The importance of benthic recruitment to the population development of *Aphanizomenon flos-aquae* and internal loading in a shallow lake. J Plankton Res 16: 1581–1588

Barbiero RP, Welch EB (1992) Contribution of benthic blue-green algal recruitment to lake populations and phosphorous translocation. Freshwater Biol 27:249–260

Berleman JE, Bauer CE (2004) Characterization of cyst cell formation in the purple photosynthetic bacterium *Rhodospirillum centenum*. Microbiology 150:383–390

Bjorn GS, Braune W, Bjorn LO (1983) Photochromic pigments in akinetes and pigment characteristics of akinetes in comparison with vegetative cells of *Anabaena variabilis*. Physiol Plant 59:493–500

Braune W (1979) C-Phycocyanin: The main photoreceptor in the light-dependent germination process of *Anabaena* akinetes. Arch Microbiol 122:289–296

Braune W (1980) Structural aspects of akinete germination in the cyanobacterium *Anabaena variabilis*. Arch Microbiol 126:257–262

Braune W, Doehler G (1996) ^{15}N-uptake, influenced by UV-B radiation, and pattern of amino-acid pools during akinete germination in *Anabaena variabilis* (Cyanobacteria). J Basic Microbiol 36:219–227

Braun-Howland EB, Nierzwicki-Bauer SA (1990) Occurrence of the 32-kDa Q_B-binding protein of photosystem II in vegetative cells, heterocysts and akinetes of *Azolla caroliniana* cyanobionts. Planta 180:361–371

Campbell EL, Hagen KD, Cohen MF, Summers ML, Meeks JC (1996) The *devR* gene product is characteristic of receivers of two-component regulatory systems and is essential for heterocyst development in the filamentous cyanobacterium *Nostoc* sp. strain ATCC 29133. J Bacteriol 178:2037–2043

Campbell EL, Summers ML, Christman H, Martin ME, Meeks JC (2007) Global gene expression patterns of *Nostoc punctiforme* in steady-state dinitrogen-grown heterocyst -containing cultures and at single time points during the differentiation of akinetes and hormogonia. J Bacteriol 189:5247–5256

Cardemil L, Wolk CP (1976) The polysaccharides from heterocyst and spore envelopes of a blue-green alga. Methylation analysis and structure of the backbones. J Biol Chem 251:2967–2975

Cardemil L, Wolk CP (1979) The polysaccharides from heterocyst and spore envelopes of a blue-green alga. Structure of the basic repeating unit. J Biol Chem 254:736–741

Cardemil L, Wolk CP (1981) Polysaccharides from the envelopes of heterocysts and spores from the blue-green algae *Anabaena variabilis* and *Cylindrospermum licheniforme*. J Phycol 17:234–240

Chater KF (2000) Developmental decisions during sporulation in the areal mycelium in *Streptomyces*. In: Brun YV, Shimkets LJ (eds) Prokaryotic Development. American Society for Microbiology, Washington, DC, pp 33–48

Chauvat F, Corre B, Herdman M, Joset-Espardellier F (1982) Energetic and metabolic requirements for the germination of akinetes of the cyanobacterium *Nostoc* PCC 7524. Arch Microbiol 133:44–49

Corliss J (2001) Protozoan cysts and spores. Encyclopedia of life sciences. Wileyd, Chichester

Cross T (1968) The thermophilic actinomycetes. J Appl Bacteriol 31:36–53

Crowe JH, Carpenter JF, Crowe LM (1998) The role of vitrification in anhydrobiosis. Annu Rev Physiol 60:73–103

Cruze-Martin MA, Diaz A, Manzanal MB, Hardisson C (1989) Role of trehalose in the spores of *Streptomyces*. FEMS Microbiol Ecol 35:49–54

Dworkin M (1996) Recent advances in the social and developmental biology of the myxobacteria. Microbiol Rev 60:70–120

Errington J (2003) Regulation of endospore formation in *Bacillus subtilis*. Nat Rev Microbiol 1:117–125

Fay P (1969a) Metabolic activities of isolated spores of *Anabaena cylindrica*. J Exp Bot 20:100–109

Fay P (1969b) Cell differentiation and pigment composition in *Anabaena cylindrica*. Arch Mikrobiol 67:62–70
Fay P (1988) Viability of akinetes of the planktonic cyanobacterium *Anabaena circinalis*. Proc R Soc Lond B 234:283–301
Fay P, Lynn JA, Majer SC (1984) Akinete development in the planktonic blue-green alga *Anabaena circinalis*. Br Phycol J 19:163–174
Fisher RW, Wolk CP (1976) Substance stimulating the differentiation of spores of the blue-green alga *Cylindrospermum licheniforme*. Nature 259:394–395
Hadas O, Pinkas R, Delphine E, Vardi A, Kaplan A, Sukenik A (1999) Limnological and ecophysiological aspects of *Aphanizomenon ovalisporum* bloom in Lake Kinneret, Israel. J Plankton Res 21:1439–1453
Hense I, Beckmann A (2006) Towards a model of cyanobacteria life cycle – effects of growing and resting stages on bloom formation of N_2-fixing species. Ecol Model 195:205–218
Herdman M (1987) Akinetes: structure and function. In: Fay P, van Baalen C (eds) The cyanobacteria. Elsevier, Amsterdam, pp 227–250
Herdman M (1988) Cellular differentiation: akinetes. Methods Enzymol 167:222–232
Hershkovitz N, Oren A, Cohen Y (1991) Accumulation of trehalose and sucrose in cyanobacteria exposed to matric water stress. Appl Environ Microbiol 57:645–648
Higo A, Katoh H, Ohmori K, Ikeuchi M, Ohmori M (2006) The role of a gene cluster for trehalose metabolism in dehydration tolerance of the filamentous cyanobacterium *Anabaena* sp. PCC 7120. Microbiology 152:979–987
Hill DR, Thomas EK, Helm RF, Potts M, Crowe LM, Crowe JH (1997) Extracellular polysaccharide of *Nostoc commune* (Cyanobacteria) inhibits fusion of membrane vesicles during desiccation. J Appl Phycol 9:237–248
Hohmann S et al (2010) *Saccharomyces cerevisiae* spore germination. In: Lubzens E, Cerda J, Clark M (eds) Dormancy and resistance in harsh environments. Springer, Heidelberg
Hori K, Ji O, Tanji Y, Unno H (2003) Formation, sedimentation and germination properties of *Anabaena* akinetes. Biochem Eng J 14:67–73
Huber AL (1985) Factors affecting the germination of akinetes of *Nodularia spumigena* (cyanobacteriaceae). Appl Environ Microbiol 49:73–78
Karlsson-Elfgren I, Brunberg A-K (2004) The importance of shallow sediments in the recruitment of *Anabaena* and *Aphanizomenon* (Cyanophyceae). J Phycol 40:831–836
Karlsson-Elfgren I, Rengefors K, Gustafsson S (2004) Factors regulating recruitment from the sediment to the water column in the bloom-forming cyanobacterium *Gleotrichia echinulata*. Freshwater Biol 49:265–273
Kaushik M, Kumar HD (1970) The effect of light on growth and development of two nitrogen fixing blue-green algae. Arch Microbiol 74:52–57
Kezhi B, Guoliang W, Cheng C (1985) Studies on the mechanism of light-dependent germination of akinetes of blue-green algae. Hydrobiologia 123:89–91
Kroos L (2007) The *Bacillus* and *Myxococcus* developmental networks and their transcriptional regulators. Annu Rev Genet 41:13–39
Leganés F (1994) Genetic evidence that *hepA* gene is involved in the normal deposition of the envelope of both heterocysts and akinetes in *Anabaena variabilis* ATCC 29413. FEMS Microbiol Lett 123:63–67
Leganés F, Fernández-Piñas F, Wolk CP (1994) Two mutations that block heterocyst differentiation have different effects on akinete differentiation in *Nostoc ellipsosporum*. Mol Microbiol 12:679–684
Leganés F, Fernández-Piñas F, Wolk CP (1998) A transposition-induced mutant of *Nostoc ellipsosporum* implicates an arginine-biosynthetic gene in the formation of cyanophycin granules and of functional heterocysts and akinetes. Microbiol (Reading) 144: 1799–1805
Li R, Watanabe M, Watanabe MM (1997) Akinete formation in planktonic *Anabaena* spp (Cyanobacteria) by treatment with low temperature. J Phycol 33:576–584

Livingstone D, Jaworski GHM (1980) The viability of akinetes of blue-green algae recovered from the sediments of Rostherne Mere. Br Phycol J 15:357–364
McBride MJ, Ensign JC (1987) Effects of intracellular trehalose content of *Streptomyces griseus* spores. J Bacteriol 169:4995–5001
Meeks JC, Campbell EL, Summers ML, Wong FCY (2002) Cellular differentiation in the cyanobacterium *Nostoc punctiforme*. Arch Microbiol 178:395–403
Miller MM, Lang NJ (1968) The fine structure of akinete formation and germination in *Cylindrospermum*. Arch Microbiol 60:303–313
Moore D, O'Donohue M, Shaw G, Critchley C (2003) Potential triggers for akinete differentiation in an Australian strain of the cyanobacterium *Cylindrospermopsis raciborskii* (AWT 205/1). Hydrobiologia 506–509:175–180
Moore D, McGregor G, Shaw G (2004) Morphological changes during akinete germination in *Cylindrospermopsis raciborskii* (Nostocales, cyanobacteria). J Phycol 40:1098–1105
Moore D, O'Donohue M, Garnett C, Critchley C, Shaw G (2005) Factors affecting akinete differentiation in *Cylindrospermopsis raciborskii* (Nostocales, Cyanobacteria). Freshwater Biol 50:345–352
Muro-Pastor AM, Valladares A, Flores E, Herrero A (2002) Mutual dependence of the expression of the cell differentiation regulatory protein HetR and the global nitrogen regulator NtcA during heterocyst development. Mol Microbiol 44:1377–1385
Nichols JM, Adams DG (1982) Akinetes. In: Carr NG, Whitton BA (eds) The biology of cyanobacteria. Blackwell, Oxford, pp 387–412
Nichols JM, Adams DG, Carr NG (1980) Effect of canavanine and other amino acid analogues on akinete formation in the cyanobacterium *Anabaena cylindrica*. Arch Microbiol 127:67–75
Padisák J (1997) *Cylindrospermopsis raciborskii* (Woloszynska) Seenayya et Subba Raju, an expanding, highly adaptive cyanobacterium: worldwide distribution and review of its ecology. Arch Hydrobiol 107:563–593
Padisák J (1998) Sudden and gradual responses of phytoplankton to global climate change: case studies from two large, shallow lakes (Balaton, Hungary, and the Neusiedlersee, Austria/Hungary). In: George D, Jones J, Puncochar P, Reynolds C, Sutcliffe D (eds) Management of lakes and reservoirs during global change. Kluwer, London, pp 111–125
Pandey RK, Talpasayi ERS (1981) Factors affecting germination of spores in a blue-green alga *Nodularia spumigena*. Acta Bot Ind 9:35–42
Pope LM, Wyss O (1970) Outer layers of the *Azotobacter vinelandii* cyst. J Bacteriol 102:234–239
Rai AK, Pandey GP (1981) Influence of environmental stress on the germination of *Anabaena vaginicola* akinetes. Ann Bot (London) 48:361–370
Rai AN, Rao VV, Singh HV (1985) The biology of cyanobacterial akinetes. J Plant Sci Res 1:1–20
Rao VV, Ghosh R, Singh HN (1987) Diazotrophic regulation of akinete development in the cyanobacterium *Anabaena doliolum*. New Phytol 106:161–168
Reddy PM (1983) Effects of temperature pre-treatment, desiccation and aging on the viability of spores of halophilic blue-green algae. Hydrobiol 106:235–240
Rengefors K, Gustafsson S, Ståhl-Delbanco A (2004) Factors regulating the recruitment of cyanobacterial and eukaryotic phytoplankton from littoral and profundal sediments. Aquat Microb Ecol 36:213–226
Reynolds CS (1972) Growth, gas vacuolation and buoyancy in a natural population of a planktonic blue-green alga. Freshwat Biol 2:87–106
Risser DD, Callahan SM (2008) HetF and PatA control levels of HetR in *Anabaena* sp. strain PCC 7120. J Bacteriol 190:7645–7654
Rother JA, Fay P (1977) Sporulation and the development of planktonic blue-green algae in two salopian meres. Proc R Soc Lond B 196:317–332
Rücker J, Tingwey E, Wiedner C, Anu C, Nixdorf B (2009) Impact of the inoculum size on the population of Nostocales cyanobacteria in temperate lakes. J Plankton Res 31:1151–1159
Sarma TA, Khattar JIS (1993) Akinete differentiation in phototrophic, photoheterotrophic and chemoheterotrophic conditions in *Anabaena torulosa*. Folia Microbiol 38:335–340

Setlow P (2000) Resistance of bacterial spores. In: Storez G, Hengge-Atonis R (eds) Bacterial stress responses. American Society for Microbiology, Washington, DC, pp 217–230
Sili C, Ena A, Materassi R, Vincenzini M (1994) Germination of desiccated aged akinetes of alkaliphilic cyanobacteria. Arch Microbiol 162:20–25
Simon RD (1977) Macromolecular composition of spores from the filamentous cyanobacterium *Anabaena cylindrica*. J Bacteriol 129:1154–1155
Simon RD (1987) Inclusion bodies in the cyanobacteria: cyanophycin, polyphosphate, and polyhedral bodies. In: Fay P, van Baalen C (eds) The Cyanobacteria: current research. Elsevier/ North Holland Biomedical Press, Amsterdam, pp 199–226
Sinclair C, Whitton BA (1977) Influence of nitrogen source on morphology of *Rivulariaceae* (cyanophyta). J Phycol 13:335–340
Singh HN, Sunita K (1974) Biochemical study of spore germination in blue-green alga *Anabaena doliolum*. J Exp Bot 25:837–845
Skill SC, Smith RJ (1987) Synchronous akinete germination and heterocyst differentiation in *Anabaena* PCC 7937 and *Nostoc* PCC 6720. J Gen Microbiol 133:299–304
Socilifsky MD, Wyss O (1962) Resistance of the *Azotobacter* cyst. J Bacteriol 84:119–124
Ståhl-Delbanco A, Hansson LA (2002) Effects of bioturbation on recruitment of algal cells from the "seed bank" of lake sediments. Limnol Oceanogr 47:1836–1843
Stüken A, Rücker J, Endrulat T, Preussel K, Hemm M, Nixdorf B, Karsten U, Wiedner C (2006) Distribution of three alien cyanobacterial species (Nostocales) in northeast Germany: *Cylindrospermopsis raciborskii*, *Anabaena bergii* and *Aphanizomenon aphanizomenoides*. Phycologia 45:696–703
Sukenik A, Hadas O, Stojkovic S, Malinsky-Rushansky N, Viner-Motzini Y, Beardall J (2009) Fluorescence approaches reveal variations in cellular composition during formation of akinetes in the cyanobacterium Aphanizomenon ovalisporum. Eur J Phycol 44:309–317
Sukenik A, Beardall J, Hadas O (2007) Photosynthetic characterization of developing and mature akinetes of *Aphanizomenon ovalisporum* (cyanoprokaryota). J Phycol 43:780–788
Summers ML, Wallis JG, Campbell EL, Meeks JC (1995) Genetic evidence of a major role for glucose-6-phosphate dehydrogenase in nitrogen fixation and dark growth of the cyanobacterium *Nostoc* sp. strain ATCC 29133. J Bacteriol 177:6184–6194
Sutherland JM, Herdman M, Stewart WDP (1979) Akinetes of the cyanobacterium *Nostoc* PCC 7524: macromolecular composition, structure and control of differentiation. J Gen Microbiol 115:273–287
Sutherland JM, Reaston J, Stewart WDP, Herdman M (1985a) Akinetes of cyanobacterium *Nostoc* PCC 7524 macromolecular and biochemical changes during synchronous germination. J Gen Microbiol 131:2855–2864
Sutherland JM, Stewart WDP, Herdman M (1985b) Akinetes of the cyanobacterium *Nostoc* PCC 7524 morphological changes during synchronous germination. Arch Microbiol 142:269–274
Thiel T, Wolk CP (1983) Metabolic activities of isolated akinetes of the cyanobacterium *Nostoc spongiaeforme*. J Bacteriol 156:369–374
Thompson PA, Jameson I, Blackburn SI (2009) The influence of light quality on akinete formation and germination in the toxic cyanobacterium *Anabaena circinalis*. Harmful Algae 8:504–512
van den Hoek C, Mann DG, Jahns HM (1998) Algae: An introduction to phycology. Cambridge University Press, Cambridge
van Dok W, Hart BT (1996) Akinete differentiation in *Anabaena circinalis* (Cyanophyta). J Phycol 32:557–565
van Dok W, Hart BT (1997) Akinete germination in *Anabaena circinalis* (Cyanophyta). J Phycol 33:12–17
van Geel B (ed) (1986) Application of fungal and algal remains and other microfossils in palynological analyses. Wiley, Chichester, pp 497–505
van Geel B, Mur LR, Ralska-Jasiewiczowa M, Goslar T (1994) Fossil akinetes of *Aphanizomenon* and *Anabaena* as indicators for medieval phosphate-eutrophication of Lake Gosciaz (Central Poland). Rev Palaeobot Palynol 83:97–105

Wiedner C, Rücker J, Brüggemann R, Nixdorf B (2007) Climate change affects timing and size of populations of an invasive cyanobacterium in temperate regions. Oecologia 152:473–484

Wildermuth H (1970) Development and organization of the aerial mycelium in *Streptomyces coelicolor*. J Gen Microbiol 60:43–50

Wildman RB, Loescher JH, Winger CL (1975) Development and germination of akinetes of *Aphanizomenon-flos-aquae*. J Phycol 11:96–104

Wolk CP (1966) Evidence of a role of heterocysts in the sporulation of a blue-green alga. Am J Bot 53:260–262

Wolk CP, Simon RD (1969) Pigments and lipids of heterocysts. Planta 86:92–97

Wolk CP, Ernst A, Elhai J (1994) Heterocyst metabolism and development. In: Bryant DA (ed) The molecular biology of cyanobacteria. Kluwer, The Netherlands, pp 769–823

Wong FCY, Meeks JC (2001) The *hetF* gene product is essential to heterocyst differentiation and affects HetR function in the cyanobacterium *Nostoc punctiforme*. J Bacteriol 183:2654–2661

Wong FCY, Meeks JC (2002) Establishment of a functional symbiosis between the cyanobacterium *Nostoc punctiforme* and the bryophyte *Anthoceros punctatus* requires genes involved in nitrogen control and initiation of heterocyst differentiation. Microbiology 148:315–323

Wyman M, Fay P (1986) Interaction between light quality and nitrogen availability in the differentiation of akinetes in the planktonic cyanobacterium *Gloeotrichia echinulata*. Br Phycol J 21:147–154

Yamamoto Y (1975) Effect of desiccation on the germination of akinetes of *Anabaena cylindrica*. Plant Cell Physiol 16:749–752

Yamamoto Y (1976) Effect of some physical and chemical factors on the germination of akinetes of *Anabaena cylindrica*. J Gen App Microbiol 22:311–323

Zhang C-C, Laurent S, Sakr S, Peng L, Bédu S (2006) Heterocyst differentiation and pattern formation in cyanobacteria: a chorus of signals. Mol Microbiol 59:367–375

Zhao J, Wolk CP (2007) Developmental biology of heterocysts. In: Whitworth DE (ed) Myxobacteria: multicellularity and differentiation. ASM Press, Herndon, VA, pp 397–418

Zhou R, Wolk CP (2002) Identification of an akinete marker gene in *Anabaena variabilis*. J Bacteriol 184:2529–2532

Chapter 3
Saccharomyces cerevisiae Spore Germination

Cecilia Geijer, Daphna Joseph-Strauss, Giora Simchen, Naama Barkai, and Stefan Hohmann

Abstract *Saccharomyces cerevisiae* spore germination is the process in which dormant spores resume growth. Upon exposure to glucose and other essential nutrients, the spore gradually loses its spore characteristics and starts acquiring properties of a vegetative cell. Translation and transcription are initiated early in the germination process. Global gene expression analysis has revealed that germination can be divided into two stages prior to the first cell cycle. During the first stage, the transcriptional programme resembles the general response of yeast cells to glucose. During the second stage, the spores sense and respond also to other nutrients than glucose. In addition, genes involved in conjugation are upregulated in germinating spores and mating is initiated before the first mitotic cell cycle. Here, we review the current understanding of the cellular rearrangements and the genes and proteins involved in germination.

C. Geijer and S. Hohmann (✉)
Department of Cell and Molecular Biology/Microbiology, University of Gothenburg, Box 462 40530, Gothenburg, Sweden
e-mail: stefan.hohmann@gu.se

D. Joseph-Strauss
The Laboratory of Cellular and Molecular Biology, NIDDK, NIH, Bethesda, MD 20892, USA

G. Simchen
Department of Genetics, the Hebrew University of Jerusalem, Jerusalem 91904, Israel

N. Barkai
Departments of Molecular Genetics and Physics of Complex system, Weizmann Institute of Science, Rehovot 76100, Israel

E. Lubzens et al. (eds.), *Dormancy and Resistance in Harsh Environments*, Topics in Current Genetics 21,
DOI 10.1007/978-3-642-12422-8_3,

3.1 *Saccharomyces cerevisiae* as a Model Organism for Dormancy

Most eukaryotic cells spend the majority of their lives in a resting state. This is true for unicellular organisms as well as cells in multicellular organisms. Basic cellular processes are conserved among eukaryotes and studying the molecular and genetic mechanisms of the establishment of and exit from dormancy in baker's yeast will broaden the understanding of these shared mechanisms.

Upon starvation, *S. cerevisiae* can form two different types of dormant cells: quiescent or stationary phase cells and spores. When conditions become favourable again, cells exit from stationary phase and spores germinate. How stationary phase cells respond to nutrient addition and resume growth has been intensively studied (Gray et al. 2004). On the other hand, the mechanisms controlling spore germination are still relatively unexplored.

As sporulation has been the subject of a number of excellent reviews (Mitchell 1994; Vershon and Pierce 2000; Engebrecht 2003; Neiman 2005), focus of this chapter will be on yeast spore germination.

3.2 Yeast Sporulation

The sporulation process is initiated when a diploid **a**/α cell is faced with starvation (Fig. 3.1). There are three nutritional conditions that need to be met for sporulation to initiate. First, the growth media must lack a fermentable carbon source, such as glucose or fructose. Second, a non-fermentable carbon source (such as acetate) must be present, since energy is needed for the cell to complete the sporulation process. Third, at least one essential nutrient such as nitrogen, phosphate, sulphate or methionine must be lacking (Freese et al. 1982).

Sporulation in yeast consists of two overlapping processes: meiosis and spore morphogenesis. When transferred to sporulation medium, cells exit the mitotic cell cycle at G_1. They go through the meiotic prophase where DNA is synthesised, meiotic recombination occurs and the synaptonemal complex forms. Prophase is followed by meiosis I (chromosome segregation) and meiosis II (chromatid separation). At the end of meiosis II, four so-called forespore membranes are formed around each haploid nucleus within the mother cell cytoplasm. Spore walls are subsequently formed around the forespores, which is then followed by spore maturation (Mitchell 1994). Starting from one diploid cell the end product is one ascus with four ascospores, two of mating type **a** and two of mating type α. Ascus formation is a feature that gives the entire group of fungi its name: ascomycetes.

The spore wall consists of four layers (compared to the vegetative cell wall with only two layers). The inner two layers mainly consist of mannan and beta-1,3-glucans, respectively. Those are followed by a layer of chitosan and finally a thin layer consisting for the most part of dityrosine molecules. The outer two layers

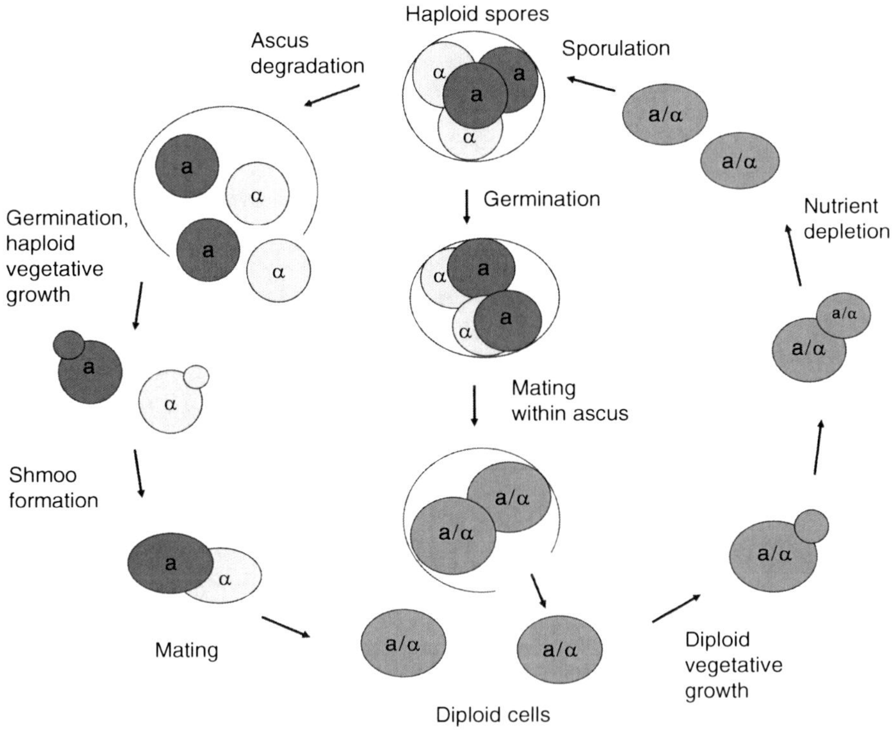

Fig. 3.1 Life cycle of *Saccharomyces cerevisiae*. A diploid cell experiencing starvation sporulates and forms an ascus containing four spores, two of mating type **a** and two of mating type α. Upon exposure to carbon and nutrients, the spores germinate. Either the germinated spores resume vegetative growth, form shmoos and mate with a haploid cell of the opposite mating type, or mating occurs within the ascus before the first mitotic cell cycle is initiated

(the outer spore coat) give the spore much of its characteristic resistance to environmental stresses such as heat, desiccation, chemicals and UV-irradiation (Briza et al. 1988).

3.3 The Dormant Spore

Spores kept in solution at 4°C are able to survive for months and years, and the mRNA they contain is stable under these conditions (Brengues et al. 2002). Among mRNAs found in spores are those transcribed from sporulation-specific genes and genes needed for utilisation of non-fermentable carbon sources, amino acid biosynthesis and stress-induced genes (Joseph-Strauss et al. 2007). Spores incubated in water at 30°C show a metabolic rate of about 5% of vegetative cells, incorporating both ^{3}[H]uracil and ^{35}[S]methionine at a steady state (Brengues et al. 2002). This is contradictory to the general view of a yeast spore being completely dormant. It has been shown that spores incubated for weeks at 30°C lose viability (Brengues et al. 2002), possibly because they are metabolically active and eventually deplete all sources for energy.

mRNAs in dormant spores are capped and polyadenylated and both the heterogeneous size and molecular weight of polyadenylated RNA in spores are comparable to RNA in a vegetative cell. The mRNA in spores is associated with both ribosomes and polysomes, but there seem to be a greater fraction of mono-ribosomes in spores than in vegetative cells. Also, spore polysomes contain fewer ribosomes per polysome as compared to vegetative cells (Harper et al. 1980; Brengues et al. 2002). This suggests that those mRNAs are translated into a protein at a lower rate than in vegetative cells.

3.4 Germination

S. cerevisiae spore germination is the multi-step process in which spores break dormancy, re-enter the mitotic cell cycle and resume vegetative growth (Fig. 3.1). During the germination process, the spore slowly loses its spore characteristics and starts acquiring properties of a vegetative cell. Spore germination resembles in many ways exit of stationary phase for vegetative cells (Gray et al. 2004). At the same time, it has unique features such as uncoating of the spore cell wall.

3.4.1 Nutritional Requirements for Germination

Spores germinate most readily in rich growth medium together with the fermentable carbon source glucose. The rate of germination in synthetic medium is somewhat slower than in rich medium, indicating that growth factors in the medium support the process. The presence of a carbon source is essential for germination to take place as spores incubated in synthetic medium without addition of a carbon source do not germinate. In fact, glucose alone can induce germination whereas oxygen and nitrogen are dispensable for the early germination events (Palleroni 1961; Herman and Rine 1997). Germination can also start upon addition of a non-fermentable carbon source in the presence of oxygen, but then at a slower pace and frequency than for glucose-treated spores (Donnini et al. 1986). It has been shown that addition of glucose to spores induces a rapid spike in the cAMP level, similar to what is seen for vegetative cells starved for glucose. Interestingly, addition of nitrogen (in the presence of glucose) to spores also triggers an increase in cAMP. This is not seen for nitrogen-starved vegetative cells (Thevelein 1984). It is not known exactly how the dormant spores sense the presence of nutrients. Whether the spore detects the essential nutrients simultaneously or if glucose is sensed first and this initial signal then enables the cell to recognise other nutrients still needs to be determined.

The glucose needs to be taken up and metabolised; its mere presence is not enough for germination to take place. Addition to the medium of the non-metabolisable glucose analog 2-deoxyglucose is not enough to initiate germination (Herman and Rine 1997). Consistently, a *gal1* mutant (a mutant unable to convert galactose to glucose) cannot germinate on galactose-containing medium. Spores become committed to germinate already after a short exposure to glucose-containing media: when spores incubated with glucose for 1 h are shifted to just water they continue to germinate (Herman and Rine 1997). Glucose alone is sufficient to stimulate spore wall uncoating and swelling, but cannot induce mitotic cell cycle events (Herman and Rine 1997; Joseph-Strauss et al. 2007).

3.4.2 Germination: The Process

Upon germination, the first noticeable morphological change is a decrease in electron density, which indicates a swelling of the tightly packed spore. Viewing a non-germinating spore with a scanning electron microscope, the outer surface looks smooth. Throughout germination, the spore wall surface becomes more and more uneven with continuous ridges over the entire surface (Rousseau et al. 1972). The outer two spore layers are locally degraded to allow swelling, whereas the inner two layers stay intact. These inner layers will serve as the cell wall of the out-growing vegetative cell (Hashimoto et al. 1958). Next, the germinating spore becomes pear-shaped, and further out-growth is seen as an elongation at one end of the cell. The outgrowth part of the cell has a smoother surface than the older part. Elongation continues until the initial bud formation takes place several hours into germination (Rousseau et al. 1972; Kono et al. 2005). There is a rapid decrease in optical density by the spore population at the onset of germination, followed by an increase in optical density again about 2 h into germination. Cell weight decreases about 10% during the first hour and then increases at a steady state for the rest of the developmental process (Rousseau et al. 1972).

Germination is composed of both a polarised growth phase and a depolarised growth phase. When germination begins, the long axis increases rapidly. Later, both the long and the short axis increase at approximately the same rate until the first bud appears (Kono et al. 2005). The actin cytoskeleton has been shown to play an important role during germination. Treatment of germinating spores with an actin polymerisation inhibitor results in a severe delay of germination progression, whereas germinating spores expressing a constitutively active Rho1 (GTP-binding protein involved in establishment of cell polarity) show an aberrantly elongated morphology (Kono et al. 2005).

While the dormant spore is resistant to a range of environmental stresses, the germinating spore gradually loses resistance to those conditions. Zymolyase, a cocktail of cell-wall degrading enzymes, breaks down the wall of vegetative cells but not of spores. Acquisition of Zymolyase sensitivity is defined as an early germination event. It occurs much earlier than bud emergence, and also before spore elongation and either before or in parallel to spore swelling (Herman and Rine 1997). Also, loss of heat shock resistance is an early event of germination (Joseph-Strauss et al. 2007).

3.4.3 Macromolecular Synthesis and Degradation

Proteins are synthesised very early on in germination, closely followed by or in parallel to synthesis of RNA. In contrast, DNA synthesis starts first some hours after addition of germinant (Rousseau and Halvorson 1973; Armstrong et al. 1984; Joseph-Strauss et al. 2007). As dormant spores contain mRNA associated with ribosomes, the initial protein synthesis is likely dependent on, at least in part, pre-existing mRNAs in the spore (Rousseau and Halvorson 1973). Intriguingly, late in the sporulation process genes encoding ribosomal proteins are upregulated (Primig et al. 2000). Possibly these ribosomal proteins are required for spore dormancy maintenance and/or germination.

Dormant spores contain mRNAs that are stable even when spores are kept for months in water at 4°C and at least for days at 30°C. Upon germination, many of these mRNAs are rapidly degraded (Brengues et al. 2002; Joseph-Strauss et al. 2007).

3.4.4 Trehalose Breakdown

Dormant spores contain high amounts of the stress metabolite trehalose. Trehalose serves as a storage carbohydrate and has a protective role during dormancy. Likely, trehalose functions as a carbon source for energy during the dormant stage (Barton et al. 1982). A drop in adenosine 5′-triphosphate (ATP) and the resulting drop in intracellular pH are suggested to trigger the slow mobilisation of trehalose in dormant spores (Thevelein et al. 1982).

Spore germination is characterised by a rapid breakdown of the stored trehalose. The spore content of trehalose decreases with the loss of optical density seen during the first hours of germination. The addition of glucose alone is sufficient to stimulate trehalose breakdown, but after some time in only glucose the spores stop germinating and start to re-accumulate trehalose. In the presence of both glucose and nitrogen, the trehalose is completely mobilised and the spores are able to germinate (Thevelein et al. 1982). Addition of acetate as a carbon source supports the initiation of germination but does not induce breakdown and mobilisation of stored trehalose (Donnini et al. 1988). Trehalase, the enzyme that catalyses breakdown of trehalose to glucose, is stored in an inactive form in the dormant spore. Within the first minutes after addition of germination medium, a rapid increase in trehalase activity is observed. Initial activation of trehalase cannot be inhibited by the protein synthesis inhibitor cycloheximide. In germinating spores, most of the energy comes from breakdown of the exogenous supplied carbon source with only a small contribution deriving from the stored trehalose. In rich medium, the energy contribution from the medium has been shown to be at least 70% of the total energy demand (Thevelein et al. 1982).

3.4.5 Global Changes in Gene Expression During Germination

The initiation of germination is followed by a rapid and intensive change in gene expression pattern. Although the importance of these changes during early stages of spore germination is not well understood, studying the global transcription programme has provided a comprehensive insight into spore germination and the progression throughout this multi-step process (Joseph-Strauss et al. 2007).

Consistent with the immediate induction of protein synthesis in the germinating spores, genes that are related to protein translation are strongly and rapidly upregulated. In contrast, genes that are associated with stress are expressed in spores and downregulated immediately upon the initiation of germination, as spores are released from the stress. The shift to glucose metabolism is reflected in the immediate and strong downregulation of genes that are associated with growth on a non-optimal carbon source (e.g. genes encoding enzymes in gluconeogenesis). Further, genes associated with the proteosome, oxidative phosphorylation and the TCA cycle are expressed in dormant spores and their expression is downregulated when germination starts (Fig. 3.2) (Joseph-Strauss et al. 2007).

3.4.5.1 Two Stages of Germination Prior to the First Mitotic Cell Cycle

Genome-wide expression analysis of the response of spores to different components of the germination medium suggests that spore germination, prior to the first cell cycle, can be divided into two major stages. The first stage starts immediately upon addition of germinant and continues for 1.5–2 h. During this early stage, the spores respond mainly to glucose in the growth medium, and glucose alone is sufficient to fully induce the transcriptional response. The transcriptional programme principally recapitulates the general response of yeast cells to glucose and reflects the shift to glucose metabolism, the release from stress and the initiation of growth. Spores incubated in growth medium without glucose show almost no change in gene expression pattern relative to resting spores (Joseph-Strauss et al. 2007).

At the transition between the first and the second stage the germinating **a** and α spores start to sense each other's presence, and they respond by strong induction of genes involved in conjugation (Fig. 3.2). The second stage of germination starts about 2 h following the induction of germination and continues until the germinating spores enter the first mitotic cell cycle. Only during the second stage the germinating spores are able to readily sense and respond to components in the environment

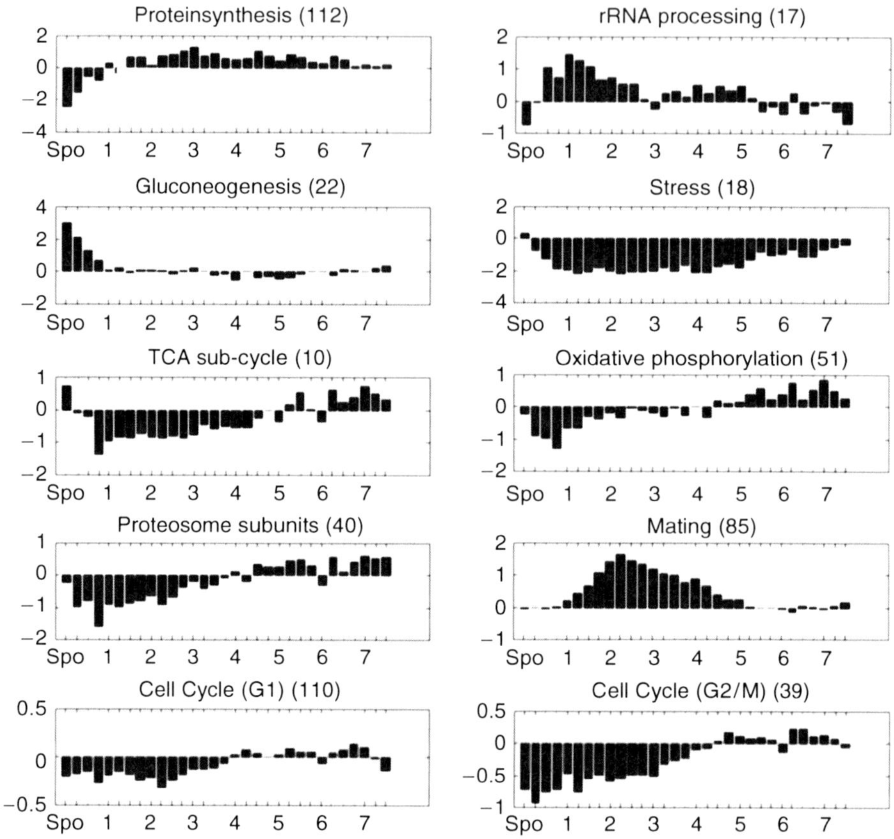

Fig. 3.2 The general transcriptional response during the first 7 h after germination initiation. Upon addition of rich growth medium to dormant spores, genes associated with protein synthesis (rRNA processing and ribosomal proteins) are rapidly upregulated. Genes associated with proteosome subunits are expressed in spores and downregulated as germination begins. Genes encoding proteins involved in stress and growth on non-optimal carbon sources (gluconeogenesis, TCA cycle and oxidative phosphorylation) are also seen to be downregulated. Further, genes associated with mating are upregulated after a few hours in germination medium, followed by cell cycle related genes. The figure shows the average expression of genes in specific modules, and the number of genes in each module is shown in parentheses. *Y*-axis is the log2 values of expression relative to the expression in vegetative cells, *X*-axis is time in hours. (For further details, see original figure text in Joseph-Strauss et al. 2007)

other than glucose. While gene expression during rich medium-induced germination continues to change gradually, gene expression in spores incubated in glucose is relatively static during this phase. The spores in glucose begin to respond to the lack of nitrogen, as indicated by the upregulation of genes that are related to the utilisation of alternative nitrogen sources. Also, genes involved in protein synthesis are downregulated in glucose-treated spores, reflecting the arrest of growth (Joseph-Strauss et al. 2007).

3.4.5.2 Germination and the General Response of Yeast Cells to Glucose

Comparison of the gene expression programme in germination to the general response of yeast cells to glucose has revealed many similarities. The overall correlation in gene expression between germination, exit from stationary phase and the response to glucose of cells grown on a non-fermentable carbon source is relatively high (Joseph-Strauss et al. 2007; Radonjic et al. 2005; Wang et al. 2004). However, the germination-specific transcriptional response for genes related to the cell cycle and to DNA replication reflects the existence of specific processes that are taking place during germination. The germinating spores utilise components of the mitotic machinery but through a unique pattern of regulation that is germination specific. During the mitotic cell cycle, DNA replication is coordinated with a phase of polarised growth (Lew and Reed 1993). In spore germination, these processes are separated and the germinating spores go through a phase of polarised growth in the absence of DNA replication. One sub-group of cell cycle genes is induced in the first phase of germination, concomitant with a phase of polar growth. This group of genes is enriched in genes related to cytoskeleton organisation and polar budding. A second sub-group is enriched in genes associated with DNA replication. These genes are downregulated during early germination and are upregulated at a later stage concomitant to DNA replication (Joseph-Strauss et al. 2007).

3.4.6 Conjugation Between Germinating Spores of Opposite Mating Type

As stated above, genes involved in conjugation are upregulated during germination. Mating between germinating spores has been shown to occur before mitotic cell cycle initiation and budding (Taxis et al. 2005; Joseph-Strauss et al. 2007), and it is very likely that spores in nature mate within the ascus with a sibling spore of opposite mating type (Fig. 3.1). Intra-tetrad mating has proved advantageous compared to germination followed by cell division, mating type switching (homothallic strains) and subsequent mating of mother and daughter. This as intra-tetrad mating rescues haploid genomes harbouring lethal mutations and preserves heterozygosity that confers a fitness advantage (Taxis et al. 2005).

In a population of dividing haploid cells, exposure to mating pheromone produced by cells of the opposite mating type causes cell cycle arrest (Bardwell 2004). A temporary arrest in the progress towards the first cell cycle is also seen in germinating spores, but it may be prevented by the use of diploids whose spores are unable to mate with each other (Joseph-Strauss 2006). Two different mutant diploids were used for this purpose: (1) a diploid *MAT*α/*MAT*α strain carrying an extra *MAT**a*** gene inserted near *LYS2*, producing viable spores that are either *MAT*α or *MAT*α/*MAT**a***; (2) a *sir2*/*sir2* diploid, homozygous at the mating-type locus *MAT*, producing *sir2* spores that expresses mating-type information of both α and **a** cassettes from the *MAT*, *HML* and *HMR* loci (derepressed due to the *sir2* mutation).

Spore germination of these mating-defective mutants was considerably faster than germination of spores from wild-type cells. Thus, the early expression of mating genes during spore germination leads to a temporary G_1 arrest and therefore the germinating spore embarks on the cell cycle programme only later, after having recovered from the mating response (Joseph-Strauss 2006).

3.4.7 *Specific Proteins Required for Germination*

Synthesis of new proteins begins within minutes after spores are reintroduced with nutrients, and translation has been demonstrated to be essential for germination initiation (Rousseau and Halvorson 1973). Addition of the translation inhibitor cycloheximide to the germination reaction results in a complete inhibition of germination. Consistently, mutants with temperature-sensitive translational machinery cannot germinate when incubated in glucose-containing medium at the non-permissive temperature (Herman and Rine 1997). Also, the Ras signalling pathway has proved essential for germination to take place. Spores of mutants defective in the Ras signalling pathway do not lose resistance to the cell-wall degrading enzymes in Zymolyase when incubated in rich medium containing glucose as wild-type spores do. Over-expression of Ras2 accelerates germination in minimal medium containing galactose, but it does not bypass the requirement for a carbon source to trigger germination (Herman and Rine 1997).

Interestingly, early events in germination such as spore uncoating and acquisition of Zymolyase sensitivity do not require many key regulators of cell cycle progression. Cdc28 and Cdc37 are two proteins needed for passage through the Start checkpoint in late G_1. When temperature sensitive mutants for these proteins were tested for their ability to germinate, neither protein proved essential for acquisition of Zymolyase sensitivity, spore swelling or elongation. Also, other cell cycle proteins such as Cdc4 and Cdc34 (in G_1/S) and Cdc7 (in S phase) are dispensable for these early germination events (Herman and Rine 1997).

In a global screen for mutants with germination and post-germination growth defects, different types of mutants were identified. Mutants defective in meiosis (likely because these mutants form inviable spores), aneuploid strains and strains with lethal recessive mutations (the heterozygous mother is phenotypically wild type) were among those mutants picked up by the screen. Also, potentially interesting mutants defective in vesicle transport, nutrient signalling/cell-wall integrity and carbon metabolism were identified but not further analysed (Deutschbauer et al. 2002). Two proteins involved in *N*-glycosylation, Rhk1 and Alg7, have been reported to be important for germination (Deutschbauer et al. 2002; Kukuruzinska and Lennon 1995). It is not determined whether mutants in these genes are impaired in the early germination events (spore uncoating, swelling and elongation) or in following rounds of mitotic divisions or both. Ubc1, an Ubiquitin-conjugating enzyme that mediates degradation of short-lived and abnormal proteins, and Cmk1, a calmodulin-dependent protein kinase, have also been reported to play a role in germination (Seufert et al. 1990; Pausch et al. 1991). Both *ubc1* and *cmk1* spores acquire Zymolyase sensitivity with wild-type kinetics, indicating that the germination defects are mainly due to impaired post-germination growth (Herman and Rine 1997).

3.4.8 Future Challenges

Most genes and pathways involved in germination are still unidentified as it has proved rather difficult to recover mutants defective in this essential developmental process. There are several reasons for this. The lack of high-throughput assays that can distinguish between inviability of spores, true germination mutants and mutants defective in post-germination growth makes screening complicated. Also, spores of both **a** and α mating type are formed during sporulation, and germinating spores can mate and become diploids even before the first bud appears. This is detrimental when screening for mutants defective in post-germination growth as mutant spores close to a neighbouring wild-type spore can be rescued by mating, forming a heterozygous diploid. Also, the resulting mix of cell types in the germinating culture makes analysing of data difficult. Using a micro-manipulator to separate spores from each other before germination is possible but both tedious and time consuming.

Future challenges in the field of yeast germination will involve identification of proteins and pathways important for the different stages of resumption of growth as well as characterisation of proteomic and metabolomic changes associated with this multi-step process.

References

Armstrong RL, West TP, Magee PT (1984) Protein synthesis in germinating *Saccharomyces cerevisiae* ascospores. Can J Microbiol 30:345–352

Bardwell L (2004) A walk-through of the yeast mating pheromone response pathway. Peptides 25:1465–1476

Barton JK, Den Hollander JA, Hopfield JJ, Shulman RG (1982) 13C nuclear magnetic resonance study of trehalose mobilization in yeast spores. J Bacteriol 151:177–185

Brengues M, Pintard L, Lapeyre B (2002) mRNA decay is rapidly induced after spore germination of *Saccharomyces cerevisiae*. J Biol Chem 277:40505–40512

Briza P, Ellinger A, Winkler G, Breitenbach M (1988) Chemical composition of the yeast ascospore wall. The second outer layer consists of chitosan. J Biol Chem 263:11569–11574

Deutschbauer AM, Williams RM, Chu AM, Davis RW (2002) Parallel phenotypic analysis of sporulation and postgermination growth in *Saccharomyces cerevisiae*. Proc Natl Acad Sci USA 99:15530–15535

Donnini C, Artoni N, Marmiroli N (1986) Germination conditions that require mitochondrial function in *Saccharomyces cerevisiae*: utilization of acetate and galactose. J Bacteriol 168:1250–1253

Donnini C, Puglisi PP, Vecli A, Marmiroli N (1988) Germination of *Saccharomyces cerevisiae* ascospores without trehalose mobilization as revealed by in vivo 13C nuclear magnetic resonance spectroscopy. J Bacteriol 170:3789–3791

Engebrecht J (2003) Cell signaling in yeast sporulation. Biochem Biophys Res Commun 27 (306):325–328

Freese EB, Chu MI, Freese E (1982) Initiation of yeast sporulation of partial carbon, nitrogen, or phosphate deprivation. J Bacteriol 149:840–851

Gray JV, Petsko GA, Johnston GC, Ringe D, Singer RA, Werner-Washburne M (2004) "Sleeping beauty": quiescence in *Saccharomyces cerevisiae*. Microbiol Mol Biol Rev 68:187–206

Harper JF, Clancy MJ, Magee PT (1980) Properties of polyadenylate-associated ribonucleic acid from *Saccharomyces cerevisiae* ascospores. J Bacteriol 143:958–965

Hashimoto T, Conti SF, Naylor HB (1958) Fine structure of microorganisms: III. Electron microscopy of resting and germinating ascospores of Saccharomyces cerevisiae. J Bacteriol 76:406–416

Herman PK, Rine J (1997) Yeast spore germination: a requirement for Ras protein activity during re-entry into the cell cycle. EMBO J 16:6171–6181

Joseph-Strauss D (2006) Spore germination in *Saccharomyces cerevisiae*: global gene expression patterns and cell cycle landmarks. Thesis submitted to the Weizmann Institute of Science, Rehovot, Israel

Joseph-Strauss D, Zenvirth D, Simchen G, Barkai N (2007) Spore germination in *Saccharomyces cerevisiae*: global gene expression patterns and cell cycle landmarks. Genome Biol 8:R241

Kono K, Matsunaga R, Hirata A, Suzuki G, Abe M, Ohya Y (2005) Involvement of actin and polarisome in morphological change during spore germination of *Saccharomyces cerevisiae*. Yeast 22:129–139

Kukuruzinska MA, Lennon K (1995) Diminished activity of the first N-glycosylation enzyme, dolichol-P-dependent N-acetylglucosamine-1-P transferase (GPT), gives rise to mutant phenotypes in yeast. Biochim Biophys Acta 1247:51–59

Lew DJ, Reed SI (1993) Morphogenesis in the yeast cell cycle: regulation by Cdc28 and cyclins. J Cell Biol 120:1305–1320

Mitchell AP (1994) Control of meiotic gene expression in *Saccharomyces cerevisiae*. Microbiol Rev 58:56–70

Neiman AM (2005) Ascospore formation in the yeast *Saccharomyces cerevisiae*. Microbiol Mol Biol Rev 69:565–584

Palleroni NJ (1961) The nutritional requirements for the germination of yeast spores. Phyton 16:117–128

Pausch MH, Kaim D, Kunisawa R, Admon A, Thorner J (1991) Multiple Ca2+/calmodulin-dependent protein kinase genes in a unicellular eukaryote. EMBO J 10:1511–1522

Primig M, Williams RM, Winzeler EA, Tevzadze GG, Conway AR, Hwang SY, Davis RW, Esposito RE (2000) The core meiotic transcriptome in budding yeasts. Nat Genet 26: 415–423

Radonjic M, Andrau JC, Lijnzaad P, Kemmeren P, Kockelkorn TT, van Leenen D, van Berkum NL, Holstege FC (2005) Genome-wide analyses reveal RNA polymerase II located upstream of genes poised for rapid response upon *S. cerevisiae* stationary phase exit. Mol Cell 18: 171–183

Rousseau P, Halvorson HO (1973) Macromolecular synthesis during the germination of *Saccharomyces cerevisiae* spores. J Bacteriol 113:1289–1295

Rousseau P, Halvorson HO, Bulla LA Jr, St Julian G (1972) Germination and outgrowth of single spores of *Saccharomyces cerevisiae* viewed by scanning electron and phase-contrast microscopy. J Bacteriol 109:1232–1238

Seufert W, McGrath JP, Jentsch S (1990) UBC1 encodes a novel member of an essential subfamily of yeast ubiquitin-conjugating enzymes involved in protein degradation. EMBO J 9: 4535–4541

Taxis C, Keller P, Kavagiou Z, Jensen LJ, Colombelli J, Bork P, Stelzer EH, Knop M (2005) Spore number control and breeding in *Saccharomyces cerevisiae*: a key role for a self-organizing system. J Cell Biol 171:627–640

Thevelein JM (1984) Cyclic-AMP content and trehalase activation in vegetative cells and ascospores of yeast. Arch Microbiol 138:64–67

Thevelein JM, den Hollander JA, Shulman RG (1982) Changes in the activity and properties of trehalase during early germination of yeast ascospores: correlation with trehalose breakdown as studied by in vivo 13C NMR. Proc Natl Acad Sci U S A 79:3503–3507

Vershon AK, Pierce M (2000) Transcriptional regulation of meiosis in yeast. Curr Opin Cell Biol 12:334–339

Wang Y, Pierce M, Schneper L, Güldal CG, Zhang X, Tavazoie S, Broach JR (2004) Ras and Gpa2 mediate one branch of a redundant glucose signaling pathway in yeast. PLoS Biol 2:E128

Chapter 4
Dormancy in Plant Seeds

Henk W.M. Hilhorst, William E. Finch-Savage, Julia Buitink, William Bolingue and Gerhard Leubner-Metzger

Abstract Seed dormancy has been studied intensely over the past decades and, at present, knowledge of this plant trait is at the forefront of plant biology. The main model species is *Arabidopsis thaliana*, an annual weed, possessing nondeep physiological dormancy. This overview presents the state-of-the-art of seed dormancy research, focusing mainly on physiological and molecular-genetic aspects in this species. It has become clear that, like in many other organisms, the dormancy and stress responses are tightly associated in seeds. The plant hormones abscisic acid and gibberellins play a pivotal role in the acquisition of developmental arrest or repression of metabolic inactivity, respectively. Some attention is given to the overlapping dormancy and stress responses, commonly studied in many other organisms but only marginally in seeds.

H.W.M. Hilhorst (✉)
Lab of Plant Physiology, Wageningen University, Droevendaalsesteeg 1, 6708 PB, Wageningen, The Netherlands
e-mail: henk.hilhorst@wur.nl

W.E. Finch-Savage
Warwick HRI, Warwick University, Wellesbourne, Warwick CV35 9EF, UK
e-mail: bill.finch-savage@warwick.ac.uk

J. Buitink and W. Bolingue
UMR 1191 Molecular Seed Physiology, INRA, 16 Bd Lavoisier, 49045, Angers, France
e-mail: julia.buitink@angers.inra.fr; william.bolingue@angers.inra.fr

G. Leubner-Metzger
Institut fuer Biologie II, Botanik/Pflanzenphysiologie, Albert-Ludwigs-Universitaet Freiburg, Schaenzlestr. 1, D-79104, Freiburg i. Br., Germany
e-mail: gerhard.leubner@biologie.uni-freiburg.de

E. Lubzens et al. (eds.), *Dormancy and Resistance in Harsh Environments*, Topics in Current Genetics 21,
DOI 10.1007/978-3-642-12422-8_4,

4.1 Introduction

Seeds are the principal propagules of the majority of higher plants. They ensure dispersal of the species in space and time and, by their adaptation potential, make an important contribution to the introduction and survival of species. Seed forms and shapes are highly varied, in line with specific environmental requirements for dispersal and establishment. Most seeds consist of an embryo, surrounded by one or more covering layers. The covering layers usually consist of a living endosperm of one to several cell layers and a testa, which is mostly dead tissue (Fig. 4.1a, b). Most seeds can withstand desiccation to water contents as low as 2–3% and this gives seeds the ability to survive for long periods under adverse conditions.

Seed germination and dormancy represent key ecological and agronomical traits that determine plant establishment in natural or agricultural ecosystems. Seeds are mostly shed from the mother plant in a dry state in which the seed tissues (embryo, covering layers) are preserved at low water content. Seed germination commences with the uptake of water by the dry seed, followed by embryo expansion growth. Germination is completed when the radicle has protruded through the surrounding covering layers. Seed germination depends on the interaction of the seed with the environment, and occurs under favourable conditions with the key environmental factors: water availability, appropriate temperature and in some cases light.

Germination timing is a plant trait with the highest selection pressure by the environment and has, during seed evolution, led to a connected second key trait: seed dormancy. This can be defined as the (temporary) incapacity of a viable imbibed seed to germinate under favourable conditions. Primary dormancy (PD)

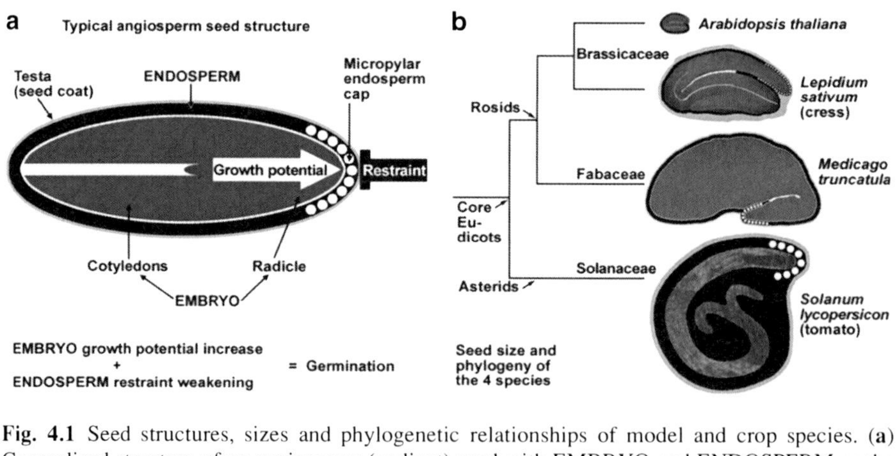

Fig. 4.1 Seed structures, sizes and phylogenetic relationships of model and crop species. (**a**) Generalised structure of an angiosperm (eudicot) seed with EMBRYO and ENDOSPERM as the two important seed components. The direction of embryo growth (*arrow* 'Growth potential') that results in germination (rupture of the endosperm) and repressive function of endosperm (*block* 'restrain') are shown. (**b**) Phylogenetic relationship and seed size comparison for *Arabidopsis*, *Lepidium*, *Medicago*, and tomato. The four species represent important model and crop plants. Figure reproduced with permission from The Seed Biology Place (http://www.seedbiology.de)

refers to the type of dormancy that occurs prior to dispersal as part of the seed developmental program, whereas secondary dormancy (SD) refers to the acquisition of dormancy in a mature seed after imbibition as a result of the lack of proper conditions for germination (Amen 1968).

Dormancy may be located in the embryo or imposed by the tissues that surround the embryo. In a number of species, both the embryo and the tissues enclosing it impose dormancy. The completion of germination (radicle protrusion) is the net result of the opposing forces: the "thrust" of the embryo and the restraints by the surrounding tissues. In the case of embryo dormancy, the properties of the embryo are of principal importance. In coat-imposed dormancy, the properties of the covering tissues are the determinants, including mechanical, chemical and permeability features, all of which may interfere with the successful completion of germination. For example, many seeds possess a seed coat that poses a mechanical restraint to embryonic growth and that may also contain chemical inhibitors, such as phenolic compounds, that prevent embryo growth (mechanical and chemical dormancy). Endosperm tissue may restrict embryo growth until the thick endosperm cell walls are degraded by hydrolytic enzymes that can be induced by factors (e.g. plant hormones) derived from the embryo (physiological/mechanical dormancy). Both embryo and coat-imposed dormancy are common and there does not seem to be a preference for a specific category or type of dormancy among plant families or genera (Baskin and Baskin 1998). Among the several different types and classes of dormancy, the study of physiological dormancy has received most attention. This class of dormancy is caused by metabolic blocks in the seed and is essentially reversible. This enables the seed (in the soil) to go through several successive cycles of dormancy break and induction until the conditions for germination and seedling establishment are optimal (Hilhorst 2007). Here, we will give an update on the progress in dormancy research of physiological dormancy and mainly in *Arabidopsis thaliana*. For a full account of the other dormancy types, the reader is referred to several excellent reviews (Baskin and Baskin 1998, 2004).

4.1.1 Embryo–Endosperm Interaction as a Mechanistic Model for Germination

The mature seeds of most angiosperm species are endospermic, that is have retained a more or less abundant endosperm layer (Finch-Savage and Leubner-Metzger 2006; Holdsworth et al. 2008a). In typical seeds, the embryo is surrounded by two covering layers ('coats', Fig. 4.1): the endosperm (living cells in most species) and the testa (seed coat, dead cells). On the mechanistic level, successful seed germination/breaking of dormancy depends simply on the net sum between two opposing forces:

- The embryo growth potential (mainly associated with the radicle) must increase to allow radicle extension growth and protrusion of the covering layers (EMBRYO = promotive).

- The restraint of the covering layers (testa, endosperm) must be weakened and weakening of the micropylar endosperm cap covering the radicle is of utmost importance (endosperm cap weakening, ENDOSPERM = repressive).

Radicle extension growth, 'coat' dormancy release and endosperm cap weakening are the key processes of seed germination and dormancy break in most species and share known molecular mechanisms of which several are evolutionary conserved.

4.2 Seed Dormancy Research: An Update

4.2.1 Global Analysis

There have been several recent reviews reporting advances in our understanding of dormancy and the control of germination in seeds resulting from large-scale gene expression profiling at both RNA and protein levels (Finch-Savage and Leubner-Metzger 2006; Bradford and Nonogaki 2007; Holdsworth et al. 2008a, b; Catusse et al. 2008a, b; Finkelstein et al. 2008). It is clear from work on both transcriptome (Ogawa et al. 2003; Nakabayashi et al. 2005; Cao et al. 2006; Cadman et al. 2006; Finch-Savage et al. 2007; Carrera et al. 2007, 2008) and proteome (Gallardo et al. 2001; Rajjou et al. 2004; Job et al. 2005; Chibani et al. 2006; Oracz et al. 2007) that there are extensive changes in genome expression involved in the control of cycling through different levels of dormancy and the final transition to the completion of germination. Holdsworth et al. (2008b) conclude from this work that RNA translation and post-translation are the major levels of control for germination completion and that transcriptome changes reflect more the alteration in dormancy status, enhancement of germination potential and effects on post-germination functions related to seedling growth. However, Nakabayashi et al. (2005) have shown that more than half (>12,000) of all genes in *Arabidopsis* have transcripts present in dry mature seeds. Holdsworth et al. (2008b), therefore, also suggest that changes in the transcriptome following seed imbibition indicate a dynamic relationship between these RNAs 'stored' from late seed development and synthesis of new RNAs related to post-imbibition germinating or dormant seed states. A further dynamic is now also thought to exist through changes in the 'dry state', apparently resulting from transcription and protein metabolism, which are manifested as altered dormancy status upon imbibition. These various levels of control are temporally coordinated from seed maturation through dormancy to germination and this provides the flexibility that is required for seeds to respond to the variable environment that surrounds them (Finch-Savage and Leubner-Metzger 2006; Fig. 4.2). In this way, seeds continually change their dormancy status to optimise the timing of germination completion in tune with seasonal cycles to maximise subsequent plant survival and reproduction.

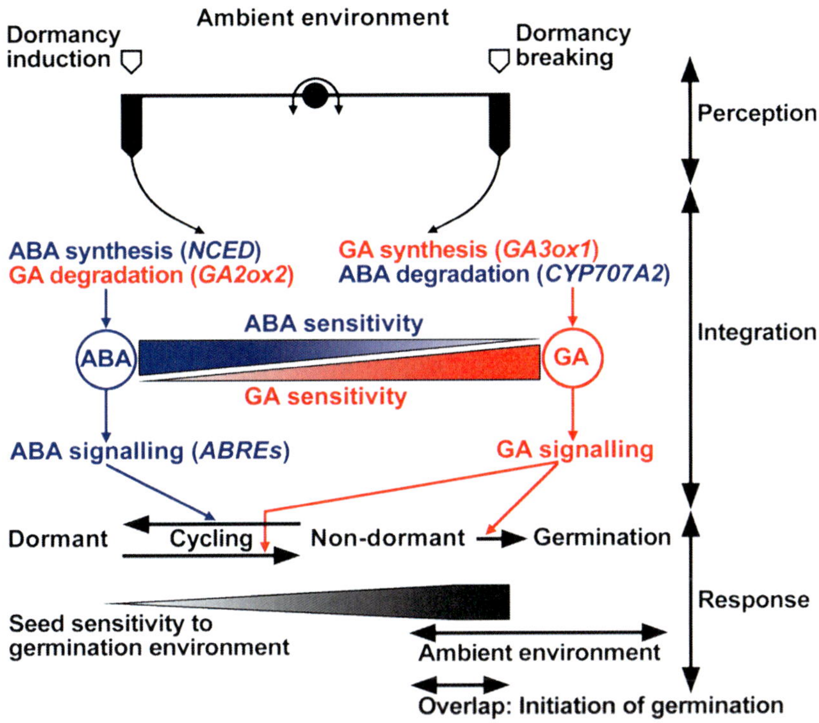

Fig. 4.2 Model for the regulation of dormancy and germination by ABA and GA in response to the environment. According to this model ambient environmental factors (e.g. temperature) affect the ABA/GA balance and the sensitivity to these hormones. ABA synthesis and signalling (GA catabolism) dominates the dormant state, whereas GA synthesis and signalling (ABA catabolism) dominates the transition to germination. The complex interplay between hormone synthesis, degradation and sensitivities in response to ambient environmental conditions can result in dormancy cycling. Change in the depth of dormancy alters the requirements for germination (sensitivity to the germination environment); when these overlap with changing ambient conditions, germination will proceed to completion. Model based on work with *A. thaliana* ecotype Cvi, modified from Cadman et al. (2006). Key target genes are in parenthesis. Figure reproduced from Finch-Savage and Leubner-Metzger (2006) with permission from Elsevier Ltd

4.2.1.1 Similarities and Differences between Physiological States

The analysis of global expression patterns has provided a new opportunity to address some old questions such as: whether seeds imbibed in the dormant state are fundamentally different from those in a non-dormant state; and whether different dormant states such as primary dormancy (PD) and secondary dormancy (SD) are similar. It is clear that specific sets of transcripts have higher abundance in seeds that will complete germination, when compared to seeds that will remain dormant and vice versa (Cadman et al. 2006; Carrera et al. 2007, 2008). From this, characteristic sets of gene transcripts have been assigned to dormant states (D-set) and fully after-ripened states (AR-set). However, principal component analysis has shown that primary dormant seeds that have had relatively short periods of imbibition (24 and 48 h) group separately from seeds that have been imbibed for longer and have entered a 'maintained' primary or secondary dormant state (Cadman et al. 2006). Thus, on the basis of transcript abundance there is little difference between these maintained states during dormancy cycling. However, it is likely that newly imbibed primary dormant seeds are dominated by transcripts remaining from seed development (stored) and thus differ from maintained states. Most published work on dormancy has centred on the former and this may have produced some misconceptions about the dormant state.

In complementary work, transcriptomes were compared from *Arabidopsis* seeds of the deeply dormant Cvi accession, exposed to different dormancy releasing factors (after-ripening, cold, nitrate, light; Finch-Savage et al. 2007). To complete germination, these seeds require more than one of these factors and thus exposure to only one factor or an incorrect combination of factors will result in different depths of dormancy. Principal component analyses of the expression patterns observed grouped physiological states in a way that related to this depth of seed dormancy, rather than the type of environmental exposure (Finch-Savage et al. 2007). This suggests similarity in the response to different environments. Furthermore, opposite changes in transcript abundance of genes in D- and AR-sets were also related to the depth of dormancy and common to different environments. Thus, transcription of these gene sets responds in a quantitative way to specific environmental signals when they are presented to the seeds in the order appropriate to relieve dormancy and facilitate the completion of germination in seasonal conditions that are suitable to sustain subsequent growth.

In addition to these common quantitative changes, environment-specific gene expression patterns during dormancy relief were also found. For example, higher transcript abundance for genes linked to the process of nitrate accumulation and reduction was associated with dormancy relief. Further patterns were consistent with a role for the balance of the plant hormones abscisic acid (ABA) and gibberellins (GAs) in integrating dormancy-relieving environmental signals, which is discussed further below.

4.2.1.2 Genes Associated with Different States

Work at the level of gene expression has confirmed that dormancy is an active state, with complex regulatory networks continuously integrating environmental signals and responding to them by positive maintenance of dormancy through de novo ABA synthesis and/or negative regulation of germination (Fig. 4.1; Cadman et al. 2006; Carrera et al. 2007, 2008; Finch-Savage et al. 2007). Cadman et al. (2006) show that changes in dormancy status are consistent with differential expression of large numbers of transcription factors present in the D- and AR-sets, along with genes encoding histones, which are suggestive of a complete switch in gene expression, resulting from a change in chromatin structure. Genes in the D-set are associated with embryo maturation including storage proteins, heat shock proteins and dehydrins etc. It was also found that ABA-, stress- and dormancy responses significantly overlap at the transcriptome level (Cadman et al. 2006; Finch-Savage et al. 2007). Many of the genes more highly expressed in the dormant states appeared to be related to stress. This may be linked to the synthesis of ABA, which appears essential to the maintenance of dormancy. Under prolonged conditions that are non-permissive to germination, there is an increase of ABA content (Ali-Rachedi et al. 2004). These conditions will, therefore, always lead to co-expression of dormant- and stress-related genes that are controlled by ABA. There is an evolutionary advantage in this co-expression, since dormancy is a mechanism to survive prolonged periods of environmental stress that are unfavourable for growth. In contrast, many genes of the AR-set are associated with the establishment of translation machinery, the potential for cell-wall remodelling and reserve mobilisation in advance of germination completion. The genes represented in this set appear also to, at least partly, anticipate the next likely stage of development, that is radicle extension and subsequent seedling growth.

Results from Rajjou et al. (2004) have shown that chemical inhibition of transcription in non-dormant *Arabidopsis* seeds did not affect the eventual completion of germination, but inhibited further growth of the seedling after radicle protrusion. In contrast, translation inhibitors effectively blocked the completion of germination. This suggests that the transcripts for the completion of germination of non-dormant seeds are pre-formed during their development (stored transcripts), and then translated to enable progress of germination all the way to completion. Thus, transcription is not essential for the completion of germination in these previously unimbibed non-dormant seeds. However, the work of Cadman et al. (2006) shows that dormancy is characterised by an absence of transcripts related to establishing translational machinery, whereas dormancy release is accompanied by transcription of genes associated with the completion of germination (AR-set) including those encoding for proteins involved in translation machinery. They, therefore, hypothesise that an important molecular event in release from the maintained dormant state, when stored transcripts may no longer be available, is establishing the capacity for the translational control of germination completion.

4.2.1.3 The Hormone Balance and Regulation of Dormancy

A dynamic balance of hormone synthesis and catabolism operates that establishes a controlling balance of ABA–GA ratio (Ali-Rachedi et al. 2004; Cadman et al. 2006). This intrinsic balance directs signalling pathways that regulate dormancy level by altering the seed's sensitivity to the ambient germination environment (Fig. 4.2). While the release of primary dormancy in Cvi seeds occurs effectively by after-ripening, stratification or inhibition of ABA biosynthesis, the addition of GA appears less effective and can cause a transient increase in ABA levels (Ali-Rachedi et al. 2004; Finch-Savage et al. 2007). This suggests that in dormant seeds a feedback mechanism exists that maintains a high ABA–GA ratio. However, dormancy release also involves a net shift to increased GA biosynthesis and ABA degradation resulting in a low ABA–GA ratio (Ali-Rachedi et al. 2004; Cadman et al. 2006).

D-set genes had an over-representation of ABA-responsive elements (ABRE) in their promoters, and of genes for transcription factors that bind to the ABRE (Cadman et al. 2006). Such an over-representation of ABRE-containing genes is also evident in stored mRNAs of dry *A. thaliana* seeds (Nakabayashi et al. 2005). ABRE-binding transcription factors appear to be master regulators that mediate ABA responses in seeds including the regulation of dormancy. On the other hand, during imbibition of non-dormant seeds, there are many GA-responsive genes induced, but GA also causes down-regulation of many ABRE-containing genes (Yamaguchi and Kamiya 2002; Ogawa et al. 2003; Yamauchi et al. 2004).

4.2.1.4 Exposure to Dormancy Releasing Environmental Factors

The timing, extent and pattern of seed germination and subsequent seedling emergence within a seed population are determined by a complex interaction of ambient weather conditions, soil, and seed characteristics (Finch-Savage and Leubner-Metzger 2006). The key weather/soil factors for germination and dormancy are:

- Water availability
- Temperature
- Light
- Abiotic stresses

In crops, the rate and extent of seed germination is key to successful seedling establishment, which in turn is the cornerstone of sustainable and profitable crop production. Transitions from the primary dormant to the non-dormant state and from the non-dormant state to germination or the secondary dormant state depend on the ambient environment, which determines both rate and extent of the response. This interactive process can be very complex, but population-based threshold models provide a universal approach to quantifying the array of ecophysiological responses exhibited by seeds (Finch-Savage and Leubner-Metzger 2006). The models use biological time in which the process of germination progresses to completion at different rates according to the ambient conditions. The quantitative effects of temperature (thermal time), water availability (hydrotime), and the combination of both (hydrothermal time), as well as seed after-ripening, dormancy or any abiotic stress can be described by these models. They can be used to simulate and predict the impact of environment on seed germination in field soils.

In fully AR seeds that require only light to germinate and those exposed to light, the transcript expression of AtGA3ox2 increases dramatically (Yamaguchi et al. 1998; Cadman et al. 2006; Finch-Savage et al. 2007) presumably facilitating the final step of the biosynthesis of biologically active GA. Cold release of dormancy is also mediated, at least in part, by promoting GA biosynthesis via enhanced expression of AtGA3ox (Yamaguchi and Kamiya 2002; Oh et al. 2004; Yamauchi et al. 2004; Liu et al. 2005a, b; Penfield et al. 2005) and by promoting ABA catabolism *via* activity of the flowering gene *FLC (FLOWERING LOCUS C*; Chiang et al. 2009).

Dormancy can also be released by AR at rates that are determined by moisture and oil content, seed-covering structures and temperature (e.g. Manz et al. 2005). Bove et al. (2005) provide evidence that Nicotiana seed AR generates a developmental switch at the transcript level that is evident upon imbibition and this is supported by work with *Arabidopsis* Cvi (Cadman et al. 2006). In part, this may result from gene expression in air-dry seeds during after-ripening (Bove et al. 2005; Leubner-Metzger 2005). Carrera et al. (2008) studied changes in gene expression in imbibed non-dormant mutants (aba1 and abi1) and compared them to wild-type seeds with and without AR. This indicated that AR acts as a developmental pathway that can be separated from dormancy of the imbibed seed. This work also showed that exogenous application of ABA did not re-impose the gene expression of seeds that had not been after-ripened, and that seeds of the non-dormant mutants demonstrated changes in genome expression during dry storage that were characteristic of AR. This provided a clear demonstration that ABA is not a major regulator of AR in dry seed (Holdsworth et al. 2008a).

It is also clear that exogenous application of ABA to seeds does not result in seed phenotypes that mimic dormancy at the proteome (Chibani et al. 2006) or transcriptome levels (Carrera et al. 2008). Despite this, in the studies so far carried out where samples are comparable there is little correlation between observations at the transcriptome and proteome levels, for example following imbibition of AR seeds (Cadman et al. 2006 and Chibani et al. 2006 respectively). To date the reason for this is open to speculation and more work is required. However, a proteome study of dormancy relief by AR in sunflower by Oracz et al. (2007) has highlighted the potential importance of reactive oxygen species (ROS) in this process. From this work, they raise the hypothesis that dormancy release involves a change in proteome oxidation, resulting from the accumulation of ROS during AR. ROS accumulation, therefore, appears to be a key signal governing cell activity during AR (Oracz et al. 2007). They suggest that this mechanism may also have relevance for dormancy breaking in the imbibed state.

4.2.1.5 Different Seed Tissues and Sensitivities

Global expression analyses are consistent with the induction and maintenance of the dormant state being characterised by increased ABA biosynthesis and GA degradation and the reverse during dormancy release. In seeds of different species these changes in the two hormones may occur at the same time or at different times and at different sites within the seed. However, the emerging picture is incomplete without considering the influence of the seed coat and the antagonism of different tissues (embryo, endosperm) within the seed and hormone sensitivities. The sensitivities for GA and ABA, their perception by receptors, their interconnected signalling chains, and their developmental regulation are of utmost importance for germination and dormancy (Kucera et al. 2005). Thus, dormancy loss in many seeds is also characterised by a decrease in ABA sensitivity and an increase in GA sensitivity (e.g. Le Page-Degivry et al. 1996; Corbineau et al. 2002; Koornneef et al. 2002; Leubner-Metzger 2002; Ali-Rachedi et al. 2004; Chiwocha et al. 2005).

In endospermic seeds, the endosperm acts as a mechanical barrier to germination and is, therefore, intimately involved in dormancy mechanisms (Kucera et al. 2005; Finch-Savage and Leubner-Metzger 2006). Again the emergence of the radicle through the endosperm is regulated via the ratio of ABA–GA, which controls weakening of the micropylar endosperm in many species (reviewed by Finch-Savage and Leubner-Metzger 2006). Recent evidence suggests that the endosperm may be the primary determinant of seed dormancy in *Arabidopsis* (Bethke et al. 2007). It is anticipated that future genome wide expression studies will have emphasis on the separate analysis of seed tissues (Holdsworth et al. 2008b). Indeed, differential global gene expression patterns have already been demonstrated between different seed tissues of *Arabidopsis* at the level of the transcriptome (Penfield et al. 2006), and of sugar beet at the proteome level (Catusse et al. 2008a, b).

4.2.2 *Specific Analyses: Key Genes and Processes Related to the Hormonal Regulation of Dormancy, After-Ripening and Germination*

The previous sections provide a global overview and introduced the general concept of the antagonistic hormonal interactions like GA–ABA and the importance of the seed tissues for dormancy, after-ripening and germination. The following parts present specific examples for key genes and processes in seeds that are exemplary. They are of course not exclusive, and other important case studies are summarised in several recent reviews (Kucera et al. 2005; Finch-Savage and Leubner-Metzger 2006; Bentsink and Koornneef 2008; Finkelstein et al. 2008; Holdsworth et al. 2008a).

4.2.2.1 ABA: A Positive Regulator of Dormancy Induction and Maintenance, and a Negative Regulator of Germination

In many plant species, endogenous ABA is involved in the induction and perhaps in the maintenance of the dormant state (reviews: Hilhorst 1995; Kucera et al. 2005; Holdsworth et al. 2008a). Mutants with reduced seed ABA biosynthesis exhibit reduced dormancy. Over-expression of genes for ABA biosynthesis can increase seed ABA content and enhance seed dormancy or delay germination (e.g. Grappin et al. 2000; Nambara and Marion-Poll 2003). Enhanced dormancy is also evident in *Arabidopsis cyp707a2* mutants with increased ABA content due to a block of seed ABA catabolism (ABA 8′ hydroxylase, Kushiro et al. 2004; Müller et al. 2006).

Several of the *Arabidopsis* ABA-insensitive (*abi*) response mutants, *abi1* to *abi5* and *abi8*, exhibit, like the ABA-deficient mutants, a marked reduction in seed dormancy (Kucera et al. 2005; Finkelstein et al. 2008; Holdsworth et al. 2008a). The seed responses of strong alleles of the *Arabidopsis ABI3* gene are severe compared to the *abi1*, *abi2* and the ABA-deficient mutant alleles. ABI3 may play a major role in seed and bud dormancy (Rohde et al. 2000; Bassel et al. 2006). The ABA-insensitive *viviparous1* (*vp1*) mutant of maize is characterised by severe seed responses, including reduced sensitivity of germination to exogenous ABA and vivipary. The *Arabidopsis ABI3* and the maize *VP1* are orthologous genes that encode transcription factors of the B3 domain class that are essential for ABA action. VP1/ABI3-like proteins are multifunctional transcription factors that integrate ABA and other regulatory signals of seed maturation and developmental arrest. Post-translational targeting of ABI3 for protein degradation and perhaps also farnesylation of ABI3 are mechanisms to regulate ABI3-mediated ABA signalling (Finkelstein et al. 2008). The interaction of ABI3 with other factors in the network that establishes seed dormancy during seed maturation is summarised by Holdsworth et al. (2008a).

ABA is not only a positive regulator of dormancy induction; it also inhibits seed germination and has been proposed to be a positive regulator of dormancy maintenance. ABA inhibits embryo growth potential and endosperm cap weakening during coffee seed germination (da Silva et al. 2004). A transient rise in ABA content in the embryo was evident early during imbibition. ABA treatment inhibits and fluridone treatment accelerates radicle protrusion of coffee seeds. Vegetation-derived ABA is also of ecological importance in the regulation of seed dormancy and germination. ABA leached from plant litter plays an important role in the germination control of the post-fire annual *Nicotiana attenuata* (Krock et al. 2002; Schwachtje and Baldwin 2004).

Rupture of the testa and the endosperm are distinct and temporally separate events during the germination of many species; such two-step germination with testa rupture subsequently followed by endosperm rupture, is known for Nicotiana spp. (Solanaceae, e.g. Leubner-Metzger 2003), *Lepidium sativum* (cress) and *A. thaliana* (Liu et al. 2005a, b; Müller et al. 2006; Piskurewicz et al. 2008). Addition of ABA to the medium during imbibition resembles the effects of maternal ABA during seed development and residual ABA in mature seeds. In after-ripened seeds, this does not appreciably affect the kinetics of testa rupture, but it delays endosperm rupture and results in the formation of a novel structure, consisting of the enlarged radicle with a sheath of greatly elongated endosperm tissue (Leubner-Metzger and Meins 2000; Leubner-Metzger 2003).

4.2.2.2 Gibberellins Release Coat Dormancy, Promote Germination and Counteract ABA Effects

According to the revised hormone-balance hypothesis for seed dormancy proposed by Karssen and Laçka (1986), ABA and GA act at different times and sites during the 'seed life'. ABA induces dormancy during maturation, and GAs play a key role in dormancy release and in the promotion of germination. GA biosynthesis in developing seeds of many species leads to the accumulation and storage of either bioinactive GA precursors or bioactive GA (Yamaguchi and Kamiya 2002; Kucera et al. 2005). GA biosynthesis in developing seeds appears not to be involved in the establishment of primary dormancy per se, but in other aspects of seed development, including fertilisation, embryo growth, assimilate uptake, fruit growth, and the prevention of seed abortion.

The temporal and spatial expression pattern of GA biosynthesis genes has been investigated during *Arabidopsis* seed germination (Yamaguchi et al. 2001; Ogawa et al. 2003; Yamauchi et al. 2004). Bioactive GAs accumulate just prior to radicle protrusion and appear to occur in two separate locations within the embryo: (1) the early biosynthetic pathway, including the geranylgeranyl diphosphate cyclisation reaction catalysed by ent-copalyl diphosphate synthetase (CPS), in the provascular tissue where *AtCPS1* gene promoter activity is localised, and (2) the late biosynthetic pathway, including the formation of bioactive GA by GA 3-oxidase, in the cortex and endodermis of the root where *AtGA3ox1* and *AtGA3ox2* transcripts accumulate and *AtGA3ox2* gene promoter activity is localised. This implies that intercellular transport of an intermediate of the GA biosynthetic pathway (probably ent-kaurene) is required to produce bioactive GA. Two functions for GA during seed germination have been proposed (reviews: Hilhorst 1995; Bewley 1997a, b; Kucera et al. 2005; Finch-Savage and Leubner-Metzger 2006). First, GA increases the growth potential of the embryo. Second, GA is necessary to overcome the mechanical restraint conferred by the seed-covering layers by weakening of the tissues surrounding the radicle. The localisation of seed GA biosynthesis in the *Arabidopsis* radicle (Yamaguchi et al. 2001) is consistent with the hypothesis that embryonic GA is released and triggers the weakening of seed-covering layers. This is further supported by the finding that at least some GA responsive genes are expressed in non-GA-producing seed tissues (Ogawa et al. 2003). Environmental cues like light and temperature can alter the tissue-specific localisation of GA biosynthesis (Yamauchi et al. 2004). The temporal and spatial pattern of GA biosynthesis and sensitivity are both important for the GA-mediated seed responses. Seed germination of GA-deficient biosynthesis mutants of *Arabidopsis* (e.g. *ga1*) and tomato (e.g. *gib-1*) absolutely depends on the addition of GA to the medium during imbibition (Hilhorst 1995; Kucera et al. 2005). The mechanisms imposing a GA requirement to promote the germination of dormant and non-dormant *Arabidopsis* seeds have been analysed using the GA-deficient mutant *ga1* and the ABA-deficient mutant *aba1*, and is described in Sect. 2.2.5.

Among the GA-response mutants of *Arabidopsis*, some of the GA-insensitive DELLA repressor mutants, including *gai* (GA-insensitive), *rga* (repressor-of-ga1-3), *rgl1* (*rga-like1*), *rgl2* and *rgl3*, have been investigated in detail (e.g. Richards et al. 2001; Kucera et al. 2005; Achard et al. 2008; Piskurewicz et al. 2008). GA signalling causes proteasome-mediated degradation of these repressor proteins which is the mechanism by which many GA responses are mediated. The gain-of-function mutants in these DELLA repressor mutants are characterised by dominant GA-insensitive repression of GA responses leading to a dwarf phenotype, increased GA content and complex seed effects that are consistent with a severely decreased GA-sensitivity of dormancy release and germination. It has been proposed that *RGL1* plays a greater role in seed germination than do *GAI* and *RGA* (Wen and Chang 2002), but *RGL2* has been proposed to be the most important regulator of *Arabidopsis* seed germination in response to GA (Lee et al. 2002; Tyler et al. 2004; Cao et al. 2005). However, two detailed studies demonstrate that the involvement of DELLA repressor degradation in seed germination is complex: A careful time-course analysis of *Arabidopsis* seed germination showed that the *RGL2* mRNA decline occurred after radicle emergence, that is after germination had been completed (Bassel et al. 2004). The work of Piskurewicz et al. (2008) shows by a combination of time course analyses of testa rupture and endosperm rupture, transcript and protein analyses, that RGL2 inhibits *Arabidopsis* seed germination by stimulating ABA synthesis and ABI5 activity. These results support the notion that ABI5 acts as the final common repressor of germination in response to changes in ABA and GA levels.

4.2.2.3 Identification of Dormancy-Specific Genes and Other Key Genes that Control Germination Timing

While a major role for ABA in the establishment and maintenance of seed dormancy is evident, hardly anything is known about its downstream targets and the molecular mechanisms of the induction of dormancy and the release by temperature and after-ripening. Due to the overall importance of ABA in plant development, the ABA-related mutants exhibit pleiotropic phenotypes and are, therefore, not seed- or dormancy-specific. ABA-independent pathways and genes specific for seed dormancy are evident from the *Arabidopsis rdo* (reduced dormancy) and *dog* (delay of germination) mutants (Bentsink and Koornneef 2008; Holdsworth et al. 2008a). Besides a mild pleiotropic phenotype, the *rdo* mutants are ABA-independent, have a strong effect on dormancy, and *rdo2* and *rdo4* mutant seeds are thermoinhibition resistant (Peeters et al. 2002; Tamura et al. 2006). The *RDO4* (*REDUCED DORMANCY4*) = *HUB1* (*HISTONE MONOUBIQUITINATION1*) gene encodes a RING finger protein necessary for monoubiquitination of histone H2B (Liu et al. 2007). The importance of the peroxisome has been highlighted by the observation that the ABC transporter COMATOSE (CTS) controls germination (Carrera et al. 2007; Holdsworth et al. 2008a).

A very promising and successful approach to find specific genes involved in *Arabidopsis* seed dormancy is based on natural genetic variation, as it exists between the ecotype Ler (low dormancy) and the deeply dormant ecotype Cvi (Alonso-Blanco et al. 2003; Koornneef et al. 2004; Bentsink et al. 2006). The substantial influence of environmental effects on the expression of germination characteristics and the involvement of many genes make dormancy a typical quantitative trait. Such traits are becoming more amenable to genetic analysis, because the position of individual quantitative trait loci (QTL) and the relative contribution of these loci can now be determined. QTL analysis for seed dormancy requires permanent mapping populations, such as recombinant inbred lines (RILs), because these allow the testing of a large number of genetically identical seeds, that is seeds from the same RIL, in different environmental conditions. Seven dormancy QTLs, *DOG1* to *DOG7*, have been identified by Alonso-Blanco et al. (2003) and several more by Laserna et al. (2008). Cvi alleles at six loci (*DOG1*, *DOG3–DOG7*) increased dormancy, while Cvi alleles at *DOG2* decreased dormancy, compared to Ler alleles. The cloning of such a dormancy QTL has yet been published only for the case of *DOG1* (Bentsink et al. 2006). *A. thaliana DOG1*, for which the Cvi allele increases the level of seed dormancy, explains 12% of the variance observed in seed dormancy. The *dog1* mutant lacks dormancy, but it does not show any obvious pleiotropic effects and is, therefore, a dormancy-specific mutant. The positional cloning of this major seed dormancy QTL *DOG1* has been reported by Bentsink et al. (2006). With the isolation of *DOG1*, the first seed dormancy gene accounting for genetic variation in natural populations has been identified at the molecular level. The *DOG1* gene encodes a novel protein of unknown mode of action, but it is absolutely required for *Arabidopsis* seed dormancy. *DOG1* transcripts are expressed during seed development, are present in dry fresh (dormant; higher *DOG1* mRNA content) and dry after-ripened (non-dormant; lower *DOG1* mRNA content) seeds, and disappear upon imbibition of fresh and after-ripened seeds. A recent transcriptome analysis with *Arabidopsis* Cvi seeds demonstrated that *DOG1* transcript expression is regulated in a complex manner during dormancy induction and release (Finch-Savage et al. 2007). *DOG1* is not specifically involved in ABA signal transduction; the *dog1* mutant has a normal sensitivity to applied ABA. DOG1 function is, however, clearly related to ABA, it might affect dry seed ABA levels (Bentsink et al. 2006). The *DOG1* Cvi allele is induced by the ABA-mediated sugar signalling pathway, and enhances sugar sensitivity by stimulating *ABI4* expression (Teng et al. 2008).

4.2.2.4 Control of Germination by the Seed Coat: Testa Mutant Studies

Embryo and coat (testa and/or endosperm) dormancy are the components of physiological dormancy, their sum and interaction determine the degree of 'whole-seed' dormancy (Kucera et al. 2005; Finch-Savage and Leubner-Metzger 2006; Bentsink and Koornneef 2008; Holdsworth et al. 2008a). Embryo dormancy is characterised by an intrinsic block within the embryo itself that inhibits extension growth, and therefore excised embryos do not grow. Coat dormancy is characterised by a block to germination that is conferred to the seed by the covering layers ('coats'). 'Coat' is used in a loose sense and can be any embryo-covering structure, for example testa and/or endosperm. Based on this definition, the physiological seed dormancy of *A. thaliana* is due to coat dormancy: testa (Debeaujon and Koornneef 2000) and endosperm (Bethke et al. 2007) confer a (mechanical, chemical, etc.) resistance, which in the dormant state prevents embryo growth. In physiologically dormant seeds the embryo-covering layers can confer mechanical constraint (coat dormancy) that must be overcome by the growth potential of the embryo (Finch-Savage and Leubner-Metzger 2006; Bentsink and Koornneef 2008). For dead seed covering layers, for example the testa, pre-determined breaking points may facilitate tissue rips prior to germination. Enzymes that facilitate testa rupture might be released by the endosperm and/or the radicle. The testa is a maternal tissue and the reduced seed dormancy phenotype is inherited maternally. A series of *A. thaliana* testa mutants show reduced dormancy that is caused by alterations of the testa characteristics (Debeaujon and Koornneef 2000; Koornneef et al. 2002; Rajjou et al. 2004) and highlight the importance of the testa structure as a constraint to radicle emergence. The GA requirement for *A. thaliana* seed germination is determined by testa characteristics, embryonic growth potential and by embryonic ABA.

4.2.2.5 Control of Germination by the Endosperm: Endosperm Dormancy and Endosperm Weakening

Endosperm dormancy requires that the restraint of the embryo-covering layers must be overcome by the growth potential of the embryo (Finch-Savage and Leubner-Metzger 2006; Holdsworth et al. 2008a). Since the endosperm in many species is a living tissue, seed-covering weakening occurs prior to germination and the tissue itself can produce enzymes for this process. The work of Bethke et al. (2007) demonstrates the importance of the endosperm for *Arabidopsis* seed dormancy: when the testas of dormant seeds were removed, the endosperm prevented the germination upon imbibition. Treatments, known to release *Arabidopsis* seed dormancy, induced endosperm rupture and radicle emergence of these 'testa-less' seeds. Excised *Arabidopsis* embryos, even from seeds of the deeply dormant accessions Cvi and C24 (Bethke et al. 2007) or from GA-deficient or-insensitive mutants (e.g. Iuchi et al. 2007), have coat-dormancy; their excised embryos grow and exhibit at least the initial extension growth required for germination. Thus, based on current knowledge, the testa and the endosperm are both major determinants conferring coat dormancy to *Arabidopsis* seeds; the excised embryos grow, but may exhibit reduced growth potential. The contributions of the different tissues to the degree of the 'whole-seed' dormancy are a matter of controversial debate. The small size of *Arabidopsis* seeds is a disadvantage for directly quantifying these tissue-specific processes in order to calculate the degree of the 'whole-seed' dormancy. It is not precisely known if 'dormancy genes' affect only the embryo, only the endosperm, only the testa, or any combination of the three seed components.

The endosperm acts as a mechanical barrier to the germination of seeds in several angiosperm clades (Finch-Savage and Leubner-Metzger 2006). A decline in this mechanical resistance of the micropylar endosperm (the endosperm layer covering the radicle tip) appears to be a prerequisite for radicle protrusion during seed germination. This endosperm weakening can be promoted by GA and, at least in part, inhibited by ABA. Solanaceae species like tomato, tobacco, pepper and Datura have become model species for endosperm weakening.

Direct biomechanical measurement of endosperm weakening by puncture-force experiments with coffee and tomato seeds, have shown that endosperm weakening is biphasic with regard to the ABA inhibition (Finch-Savage and Leubner-Metzger 2006). The first phase is ABA-insensitive and this is followed by the second phase that is inhibited by ABA (Toorop et al. 2000; da Silva et al. 2004). In coffee seeds ABA controls germination by inhibiting both the embryo growth potential and the second step of endosperm weakening (da Silva et al. 2004). Coffee (Rubiaceae) and tomato (Solanaceae) belong to the Asterid clade of angiosperms. Endosperm weakening appears to be a widespread phenomenon and has also been demonstrated for the Rosid clade of angiosperms: in Brassicaceae seeds the endosperm is also a constraint to germination (Müller et al. 2006). In this work, seeds of both *A. thaliana* and its much larger-seeded relative *L. sativum* (garden cress) were studied. Both species belong to the subclade I of the Brassicaceae and are highly similar in seed structure and physiology. Testa rupture and endosperm rupture are separate events and only the latter is inhibited by ABA in after-ripened seeds of both species. Direct biomechanical measurement of the puncture force required to rupture the endosperm showed that the *L. sativum* micropylar endosperm weakened prior to radicle emergence (Müller et al. 2006). ABA delayed the onset and inhibited the rate of endosperm weakening in a dose-dependent manner. An early embryo signal which was required to induce endosperm weakening could be replaced by GA, and that weakening was found to be regulated by the GA–ABA ratio. These results suggest that the control of radicle protrusion in *L. sativum* and probably also *A. thaliana* seeds is mediated, at least in part, by endosperm weakening. In contrast to coffee and tomato, a 'one-phase' ABA-inhibited endosperm weakening is evident in *Lepidium* seeds (Müller et al. 2006). Based on the 'comparative seed biology' approach with *Lepidium* and *Arabidopsis*, one can speculate that during evolution the endospermic Brassicaceae seeds have retained ABA-inhibitable and evolutionary conserved molecular mechanism(s) found in both clades, whereas the ABA-insensitive phase of endosperm weakening was lost.

Ikuma and Thimann (1963) in their 'hatching hypothesis' of seed biology suggested that '... the final step in the germination control process is the production of an enzyme whose action enables the tip of the radicle to penetrate through the coat'. In searching for this 'hatching enzyme', evidence has been uncovered for the contribution of various cell-wall modifying proteins, including endo-β-1,4-mannanases and endo-β-1,3-glucanases (summarised in: Hilhorst 1995; Bewley 1997a; Leubner-Metzger 2003; Kucera et al. 2005; Finch-Savage and Leubner-Metzger 2006; Holdsworth et al. 2008a). Taken together, the current findings support the view that germination control by the seed-covering layers is achieved through the combined or successive action of several cell-wall modifying proteins. One intriguing issue arising from these studies is that there seem to be evolutionary conserved molecular mechanisms as well as species-specific adaptations for endosperm weakening and/or coat dormancy release. Analysis of endosperm-specific transcriptome data sets of germinated *Arabidopsis* seeds, provide information about the expression of genes for cell-wall modifying proteins (Penfield et al. 2006; Holdsworth et al. 2008a). In addition to typical cell-wall polysaccharide hydrolases,

ROS seem to be involved in seed dormancy release and germination and may contribute to endosperm weakening and embryo growth (Bailly 2004; Oracz et al. 2007).

4.3 Dormancy and Harsh Environments

4.3.1 Seed Dormancy and Tolerance in the Dry State

At the final stages of seed maturation, the induction of dormancy and subsequent desiccation on the mother plant results in dry seeds that are extremely tolerant to many types of stress. For instance, dry seeds can survive exposure to extremely high (120°C) or low (liquid nitrogen) temperatures or vacuum (Leprince and Vertucci 1995). The lifespan of seeds in the dry state can be extremely long, ranging from decades to centuries and even millennia. The most remarkable discovery was that on ancient seeds of Sacred Lotus from China; radiocarbon dating showed an age of these seeds of 1,288 ± 271 years while still being capable to germinate (Shen-Miller et al. 1995). The main reason for long-term protection is that the removal of water results in glass formation of the cytoplasm. Glasses are semi-equilibrium solid liquids with an extremely high viscosity (see Buitink and Leprince 2004, for review). Low temperatures and low water contents drive the viscosity to such high values that the cytoplasm will form a glassy state. The high viscosity is thought to be responsible for the decreased ageing rates observed at these low water contents and temperatures. Indeed, cellular viscosity and molecular mobility measurements in the cytoplasm correlate with seed longevity over a wide range of temperatures and water contents (Buitink et al. 2000). Thus, this intracellular glass formation, together with direct interaction between molecules that are imbedded in the glassy matrix through hydrogen bonding will maintain structural integrity lead to optimal preservation of the dormant seeds in the dry state (reviewed in Buitink and Leprince 2004).

Although glass formation in seeds drastically decreases molecular mobility, the molecules in a glass are not completely restricted in their movement, and this can have known repercussions on the survival in the dry state and probably as well on the dormancy status. In time, diffusion will be possible in the dry state, albeit at a rate considerably slower than that in hydrated cytoplasm. Using theoretical considerations coupled to measurements of relaxation times, Walters (2004) demonstrated that mobility is not restricted until at least 70°C below the glass transition temperature. This explains why seeds still age, because deteriorative processes such as lipid oxidation can take place, though at a very slow rate. This could also explain the natural after-ripening process that occurs in 'dry' seeds after harvest, during which seeds escape dormancy (reviewed in Holdsworth et al. 2008a). The processes taking place in dry seeds, that is with a water content below 0.10 g H_20 g/g DW (corrected for lipid content), can not involve true metabolism (i.e. ATP production via electron transport chains) because it has been shown to be arrested at water contents below 0.2 g/g. However, the release of dormancy could well be determined by the diffusion rate of certain molecules released from or diffusing within the glassy cytoplasm. Interestingly, both the rate of after-ripening and aging increase with increasing water content and temperature, as does the molecular mobility of the cytoplasm (J. Buitink, unpublished data).

Another interesting question that remains to be answered is whether dry seeds in a dormant state are more tolerant to stress than non-dormant seeds. Although this has been suggested, we found no experimental evidence in the literature. Seed longevity and seed dormancy seem to be controlled by different genetic factors in rice as well as in *Arabidopsis* seeds (Miura et al. 2002; Clerkx et al. 2004) as suggested by the different chromosomal locations of QTL for these traits. Mutant seeds of *ABI3*, a master-regulator of seed maturation, are affected both in dormancy and longevity (Ooms et al. 1993), but this regulation could involve independent signalling pathways. The only mutation that directly affects both dormancy and longevity is related to the seed testa of *Arabidopsis* (Debeaujon et al. 2000). These testa mutants were shown to take up tetrazolium much more readily than the wild types. This was related to defects in the pigmentation of the endothelium and its neighbouring crushed parenchymatic layers. The degree of seed deterioration was not strictly correlated with dormancy characteristics. Where the increased permeability of the seed coat may result in reduced dormancy, it is most likely the absence of the flavonoids that affect longevity, play a protective role against solute leakage, imbibition damage, and oxidative stress.

4.3.2 Stress Tolerance of Dormant Seeds in the Hydrated State

Although seeds that remain dry are very tolerant, environmental conditions fluctuate in nature, and seeds in soil banks are submitted to hydration and dehydration cycles. Regardless of their dormancy status, seeds can undergo several cycles of hydration and dehydration and, prior to radicle emergence, seeds remain desiccation-tolerant, unless this cycle is repeated too often (Sliwinska and Jendrzejczak 2002). Interestingly, seed mitochondria of desiccation-tolerant, non-germinated pea seeds have a remarkable temperature tolerance in response to both cold and heat stress, when compared to mitochondria isolated from etiolated epicotyls, and contain large amounts of a small heat shock protein, HSP22, and a late embryogenesis abundant (LEA) protein, LEAm (Stupnikova et al. 2006). It has been even shown that re-drying hydrated, even germinated seeds can re-induce desiccation tolerance (Buitink et al. 2003; Faria et al. 2005; Buitink et al. 2006). A transcriptomic profiling of this re-induction of desiccation tolerance in *Medicago truncatula* demonstrated that a large number of genes was re-induced when hydrated radicles were submitted to a partial drying by an osmotic solution. Many of these genes are related to protection against a wide range of stresses. For instance, a number of genes encode regulatory genes that are typically expressed during abiotic/drought stresses as well as maturation. Furthermore, highly induced expression is found for genes encoding LEA proteins, detoxification enzymes and heat-shock proteins (Buitink et al. 2006). During this partial dehydration, a massive repression of genes occurred belonging to numerous classes, including cell cycle, biogenesis, primary and energy metabolism, suggesting that the re-establishment of DT in the germinated radicles goes together with an active regulation to prepare for the return to the quiescent state imposed by the incipient lack of water. Although in *M. truncatula*, the re-induction of desiccation tolerance does not re-induce dormancy in the seeds, it has been reported that rehydration–dehydration cycles can also re-induce dormancy. Batlla and Benech-Arnold (2006) demonstrated that seeds in weed seed banks under field conditions that were subjected to fluctuating soil water content regime generally showed an increase in their dormancy level after periods of storage under dry soil conditions, and a decrease in their dormancy level after periods of storage under moist soil conditions.

Dormant seeds that remain hydrated need also to be protected against sudden unfavourable environmental conditions. Indeed, protective mechanisms seem to be activated in imbibed seeds. For example, the seed coats of dormant barrel medic seeds remain devoid of any contaminating fungi and bacteria for months, whereas isolated seed coats are readily infected (personal observation W. Bolingue). Also, expression studies in dormant *Arabidopsis* seeds demonstrate that genes related to defence and protection are highly expressed (Cadman et al. 2006). We re-analysed the transcriptome data of imbibed dormant *Arabidopsis* seeds (D-dataset) from Cadman et al. (2006) to screen for genes encoding putative protective molecules. Several genes (6) encode LEA proteins, out of which two belong to group 1 (PF00477) and two to group 5 (seed maturation protein, PF04927). In addition, nine genes encode small heat shock proteins, HSP70 and chaperone proteins dnaJ. Furthermore, 14 genes involved in detoxification are highly expressed, such as metallothionein, aldo-reductase, glutathione reductase-S-transferase and peroxiredoxin. Interestingly, a similar set of genes are also highly expressed in relation to desiccation tolerance, indicating partially similar regulatory mechanisms underlying both dormancy and desiccation tolerance. Genes related to biotic stress are equally expressed (8), such as defensins or CC-NBS-LRR class disease resistance proteins. In barrel medic, a number of genes are up-regulated during imbibition in dormant seeds that are related to secondary metabolism and defence responses, whereas their expression remains low in imbibed seeds that are non-dormant and will readily germinate (W. Bolingue and J. Buitink, unpublished data). Apparently, regulation of gene expression related to protection and defence is constitutively activated in dormant seeds.

In conclusion, in order to survive long time in seed banks, dormant seeds need to be resistant in the dry as well as hydrated state against biotic and abiotic factors that they are likely to encounter. A number of these mechanisms are likely to overlap with those acquired during maturation, with the acquisition of desiccation tolerance and longevity. Armed with these mechanisms, seeds can overcome those times under which conditions are unfavourable for seedling establishment, and will as such assure the propagation of future generations.

4.4 Future Prospects

In the last decade, enormous progress has been made in the understanding of the mechanisms and regulation of seed dormancy and germination, with a strong focus on model systems, such as *A. thaliana* and *M. truncatula*. It has become clear that environmental cues modulate the levels and balance of the plant hormones ABA and GA in a complex way and thus determine the occurrence of dormancy and germination. Annual dormancy cycling is mostly driven by changes in seasonal temperatures, whereas the breaking of dormancy is influenced by environmental cues such as light, nitrate and temperature/time. The regulation of seed dormancy (cycling) and germination in *A. thaliana* at the molecular level is complex and involves at least several hundreds of genes (Finch-Savage et al. 2007). It is, therefore, likely that transcriptional networks and their associated transcription factors are operational in the control of dormancy and germination. In order to identify clusters within the network, co-expression analysis may be performed on the different gene sets associated with different dormant states. In addition, to identify potential transcription factors, sequence motifs in both the promoter and non-coding regions of co-expressed genes may be identified. In addition to this (transcriptional) network analysis, a system's biology approach of seed germination and dormancy appears timely. Such an approach would incorporate all levels of complexity, from molecules to cells, to tissues to the whole seed and to the environment. It would also include responses to biotic stresses that occur in parallel to the dormancy/germination response, indicating a tight association between stress and dormancy, as in many other organisms.

References

Achard P, Renou JP, Berthome R, Harberd NP, Genschik P (2008) Plant DELLAs restrain growth and promote survival of adversity by reducing the levels of reactive oxygen species. Curr Biol 18:656–660

Ali-Rachedi S, Bouinot D, Wagner MH, Bonnet M, Sotta B, Grappin P, Jullien M (2004) Changes in endogenous abscisic acid levels during dormancy release and maintenance of mature seeds: studies with the Cape Verde Islands ecotype, the dormant model of *Arabidopsis thaliana*. Planta 219:479–488

Alonso-Blanco C, Bentsink L, Hanhart CJ, Blankestijn de Vries H, Koornneef M (2003) Analysis of natural allelic variation at seed dormancy loci of *Arabidopsis thaliana*. Genetics 164:711–729

Amen RD (1968) A model of seed dormancy. Bot Rev 34:1–31

Bailly C (2004) Active oxygen species and antioxidants in seed biology. Seed Sci Res 14:93–107

Baskin CC, Baskin JM (1998) Seeds; ecology, biogeography and evolution of dormancy and germination. Academic, San Diego

Baskin JM, Baskin CC (2004) A classification system for seed dormancy. Seed Sci Res 14:1–16

Bassel GW, Zielinska E, Mullen RT, Bewley JD (2004) Down-regulation of DELLA genes is not essential for germination of tomato, soybean, and Arabidopsis seeds. Plant Physiol 136:2782–2789

Bassel GW, Mullen RT, Bewley JD (2006) ABI3 expression ceases following, but not during, germination of tomato and Arabidopsis seeds. J Exp Bot 57:1291–1297

Batlla D, Benech-Arnold RL (2006) The role of fluctuations in soil water content on the regulation of dormancy changes in buried seeds of Polygonum aviculare L. Seed Sci Res 16:47–59
Bentsink L, Koornneef M (2008) Seed dormancy and germination. In: Somerville CR, Meyerowitz EM (eds) The Arabidopsis book. American Society of Plant Biologists, Rockville, MD. doi:10.1199/tab.0119
Bentsink L, Jowett J, Hanhart CJ, Koornneef M (2006) Cloning of DOG1, a quantitative trait locus controlling seed dormancy in Arabidopsis. Proc Natl Acad Sci USA 103:17042–17047
Bethke PC, Libourel IGL, Aoyama N, Chung Y-Y, Still DW, Jones RL (2007) The Arabidopsis aleurone layer responds to nitric oxide, gibberellin, and abscisic acid and is sufficient and necessary for seed dormancy. Plant Physiol 143:1173–1188
Bewley JD (1997a) Breaking down the walls – a role for endo-β-mannanase in release from seed dormancy? Trends Plant Sci 2:464–469
Bewley JD (1997b) Seed germination and dormancy. Plant Cell 9:1055–1066
Bove J, Lucas P, Godin B, Ogé L, Jullien M, Grappin P (2005) Gene expression analysis by cDNA-AFLP highlights a set of new signaling networks and translational control during seed dormancy breaking in Nicotiana plumbaginifolia. Plant Mol Biol 57:93–612
Bradford KB, Nonogaki H (2007) Seed development, dormancy and germination. Annual plant reviews. Wiley-Blackwell, Oxford
Buitink J, Leprince O (2004) Glass formation in plant anhydrobiotes: survival in the dry state. Cryobiology 48:215–228
Buitink J, Leprince O, Hemminga MA, Hoekstra FA (2000) Molecular mobility in the cytoplasm: a new approach to describe and predict lifespan of dry germplasm. Proc Natl Acad Sci USA 97:2385–2390
Buitink J, Ly Vu B, Satour P, Leprince O (2003) A physiological model to study the re-establishment of desiccation tolerance in germinated radicles of *Medicago truncatula* Gaertn. seeds. Seed Sci Res 13:273–286
Buitink J, Leger JL, Guisle I, Ly-Vu B, Wuillème S, Lamirault G, Le Bars A, Le Meur N, Becker A, Küster K, Leprince O (2006) Transcriptome profiling uncovers metabolic and regulatory processes occurring during the transition from desiccation sensitive to -tolerant stages in Medicago truncatula seeds. Plant J 47:735–750
Cadman CSC, Toorop PE, Hilhorst HWM, Finch-Savage WE (2006) Gene expression profiles of Arabidopsis Cvi seed during cycling through dormant and non-dormant states indicate a common underlying dormancy control mechanism. Plant J 46:805–822
Cao D, Hussain A, Cheng H, Peng J (2005) Loss of function of four DELLA genes leads to light- and gibberellin-independent seed germination in Arabidopsis. Planta 223:105–113
Cao DN, Cheng H, Wu W, Soo HM, Peng JR (2006) Gibberellin mobilizes distinct DELLA-dependent transcriptomes to regulate seed germination and floral development. Plant Physiol 142:509–525
Carrera E, Holman T, Medhurst A, Peer W, Schmuths H, Footitt S, Theodoulou FL, Holdsworth MJ (2007) Gene expression profiling reveals defined functions of the ABC transporter COMATOSE late in phase II of germination. Plant Physiol 143:1669–1679
Carrera E, Holman T, Medhurst A, Dietrich D, Footitt S, Theodoulou FL, Holdsworth MJ (2008) Seed after-ripening is a discrete developmental pathway associated with specific gene networks in Arabidopsis. Plant J 53:214–224
Catusse J, Job C, Job D (2008a) Transcriptome- and proteome-wide analyses of seed germination. C R Biol 331:815–822
Catusse J, Strub J-M, Job C, Van Dorsselaer A, Job D (2008b) Proteome-wide characterization of sugarbeet seed vigour and its tissue specific expression. Proc Natl Acad Sci USA 105:10262–10267
Chiang GCK, Barua D, Kramer EM, Amasino RM, Donohue K (2009) Major flowering time gene, *FLOWERING LOCUS C*, regulates seed germination in *Arabidopsis thaliana*. Proc Natl Acad Sci USA 106:11661–11666
Chibani K, Ali-Rachedi S, Job C, Job D, Jullien M, Grappin P (2006) Proteomic analysis of seed dormancy in Arabidopsis. Plant Physiol 142:1493–1510

Chiwocha SDS, Cutler AJ, Abrams SR, Ambrose SJ, Yang J, Ross ARS, Kermode AR (2005) The etr1-2 mutation in Arabidopsis thaliana affects the abscisic acid, auxin, cytokinin and gibberellin metabolic pathways during maintenance of seed dormancy, moist-chilling and germination. Plant J 42:35–48

Clerkx EJM, El-Lithy ME, Vierling E, Ruys GJ, Blankestijn-De Vries H, Groot SP, Vreugdenhil D, Koornneef M (2004) Analysis of natural allelic variation of Arabidopsis seed germination and seed longevity traits between the accessions Landsberg erecta and Shakdara, using a new recombinant inbred line population. Plant Physiol 135:432–443

Corbineau F, Bianco J, Garello G, Come D (2002) Breakage of Pseudotsuga menziesii seed dormancy by cold treatment as related to changes in seed ABA sensitivity and ABA levels. Physiol Plant 114:313–319

da Silva EAA, Toorop PE, van Aelst AC, Hilhorst HWM (2004) Abscisic acid controls embryo growth potential and endosperm cap weakening during coffee (Coffea arabica cv. Rubi) seed germination. Planta 220:251–261

Debeaujon I, Koornneef M (2000) Gibberellin requirement for Arabidopsis seed germination is determined both by testa characteristics and embryonic abscisic acid. Plant Physiol 122:415–424

Debeaujon I, Léon-Kloosterziel KM, Koornneef M (2000) Influence of the testa on seed dormancy, germination, and longevity in Arabidopsis. Plant Physiol 122:403–413

Faria JMR, Buitink J, van Lammeren AAM, Hilhorst HWM (2005) Changes in DNA and microtubules during loss and re-establishment of desiccation tolerance in germinating Medicago truncatula seeds. J Exp Bot 56:2119–2130

Finch-Savage WE, Leubner-Metzger G (2006) Seed dormancy and the control of germination. New Phytol 171:501–523

Finch-Savage WE, Cadman CSC, Toorop PE, Lynn JR, Hilhorst HWM (2007) Seed dormancy release in Arabidopsis Cvi by dry after-ripening, low temperature, nitrate and light shows common quantitative patterns of gene expression directed by environmentally specific sensing. Plant J 51:60–78

Finkelstein R, Reeves W, Ariizumi T, Steber C (2008) Molecular aspects of seed dormancy. Annu Rev Plant Biol 59:387–415

Gallardo K, Job C, Groot SPC, Puype M, Demol H, Vandekerckhove J, Job D (2001) Proteomic analysis of Arabidopsis seed germination and priming. Plant Physiol 126:835–848

Grappin P, Bouinot D, Sotta B, Miginiac E, Jullien M (2000) Control of seed dormancy in Nicotiana plumbaginifolia: post-imbibition abscisic acid synthesis imposes dormancy maintenance. Planta 210:279–285

Hilhorst HWM (1995) A critical update on seed dormancy. I. Primary dormancy. Seed Sci Res 5:61–73

Hilhorst HWM (2007) Definition and hypotheses of seed dormancy. In: Bradford KJ, Nonogaki H (eds) Seed development, dormancy and germination. Annual plant reviews, vol 27. Blackwell, Sheffield, UK, pp 50–71

Holdsworth MJ, Bentsink L, Soppe WJJ (2008a) Molecular networks regulating Arabidopsis seed maturation, after-ripening, dormancy and germination. New Phytol 179:33–54

Holdsworth MJ, Finch-Savage WE, Grappin P, Job D (2008b) Post-genomics dissection of seed dormancy and germination. Trends Plant Sci 13:7–13

Ikuma H, Thimann KV (1963) The role of the seed-coats in germination of photosensitive lettuce seeds. Plant Cell Physiol 4:169–185

Iuchi S, Suzuki H, Kim Y-C, Iuchi A, Kuromori T, Ueguchi-Tanaka M, Asami T, Yamaguchi I, Matsuoka M, Kobayashi M, Nakajima M (2007) Multiple loss-of-function of Arabidopsis gibberellin receptor AtGID1s completely shuts down a gibberellin signal. Plant J 50:958–966

Job C, Rajjou L, Lovigny Y, Belghazi M, Job D (2005) Patterns of protein oxidation in Arabidopsis seeds and during germination. Plant Physiol 138:790–802

Karssen CM, Laçka E (1986) A revision of the hormone balance theory of seed dormancy: studies on gibberellin and/or abscisic acid-deficient mutants of Arabidopsis thaliana. In: Bopp M (ed) Plant growth substances 1985. Springer, Berlin, pp 315–323

Koornneef M, Bentsink L, Hilhorst H (2002) Seed dormancy and germination. Curr Opin Plant Biol 5:33–36

Koornneef M, Alonso-Blanco C, Vreugdenhil D (2004) Naturally occurring genetic variation in Arabidopsis thaliana. Ann Rev Plant Biol 55:141–172

Krock B, Schmidt S, Hertweck C, Baldwin IT (2002) Vegetation-derived abscisic acid and four terpenes enforce dormancy in seeds of the post-fire annual, Nicotiana attenuata. Seed Sci Res 12:239–252

Kucera B, Cohn MA, Leubner-Metzger G (2005) Plant hormone interactions during seed dormancy release and germination. Seed Sci Res 15:281–307

Kushiro T, Okamoto M, Nakabayashi K, Yamagishi K, Kitamura S, Asami T, Hirai N, Koshiba T, Kamiya Y, Nambara E (2004) The Arabidopsis cytochrome P450 CYP707A encodes ABA 8′-hydroxylases: key enzymes in ABA catabolism. EMBO J 23:1647–1656

Laserna MP, Sánchez RA, Botto JF (2008) Light-related loci controlling seed germination in *Ler* x Cvi and Bay-0 x Sha recombinant inbred-line populations of *Arabidopsis thaliana*. Ann Bot 102:631–642

Le Page-Degivry MT, Bianco J, Barthe P, Garello G (1996) Change in hormone sensitivity in relation to the onset and breaking of sunflower embryo dormancy. In: Lang GA (ed) Plant dormancy: physiology, biochemistry and molecular biology. CAB International, Wallingford, pp 221–231

Lee SC, Cheng H, King KE, Wang WF, He YW, Hussain A, Lo J, Harberd NP, Peng JR (2002) Gibberellin regulates Arabidopsis seed germination via RGL2, a GAI/RGA-like gene whose expression is up-regulated following imbibition. Genes Dev 16:646–658

Leprince O, Vertucci CW (1995) A calorimetric study of the glass-transition behaviors in axes of bean-seeds with relevance to storage stability. Plant Physiol 109:1471–1481

Leubner-Metzger G (2002) Seed after-ripening and over-expression of class I β-1, 3-glucanase confer maternal effects on tobacco testa rupture and dormancy release. Planta 215:959–968

Leubner-Metzger G (2003) Functions and regulation of β-1, 3-glucanase during seed germination, dormancy release and after-ripening. Seed Sci Res 13:17–34

Leubner-Metzger G (2005) β-1, 3-Glucanase gene expression in low-hydrated seeds as a mechanism for dormancy release during tobacco after-ripening. Plant J 41:133–145

Leubner-Metzger G, Meins F Jr (2000) Sense transformation reveals a novel role for class I β-1, 3-glucanase in tobacco seed germination. Plant J 23:215–221

Liu P-P, Koizuka N, Homrichhausen TM, Hewitt JR, Martin RC, Nonogaki H (2005a) Large-scale screening of Arabidopsis enhancer-trap lines for seed germination-associated genes. Plant J 41:936–944

Liu P-P, Koizuka N, Martin RC, Nonogaki H (2005b) The BME3 (Blue Micropylar End 3) GATA zinc finger transcription factor is a positive regulator of Arabidopsis seed germination. Plant J 44:960–971

Liu Y, Koornneef M, Soppe WJJ (2007) The absence of histone H2B monoubiquitination in the Arabidopsis hub1 (rod4) mutant reveals a role for chromatin remodeling in seed dormancy. Plant Cell 19:433–444

Manz B, Müller K, Kucera B, Volke F, Leubner-Metzger G (2005) Water uptake and distribution in germinating tobacco seeds investigated in vivo by nuclear magnetic resonance imaging. Plant Physiol 138:1538–1551

Miura K, Lin SY, Yano M, Nagamine T (2002) Mapping quantitative trait loci controlling seed longevity in rice (Oryza sativa L.). Theor Appl Gen 104:981–986

Müller K, Tintelnot S, Leubner-Metzger G (2006) Endosperm-limited Brassicaceae seed germination: abscisic acid inhibits embryo-induced endosperm weakening of *Lepidium sativum* (cress) and endosperm rupture of cress and *Arabidopsis thaliana*. Plant Cell Physiol 47:864–877

Nakabayashi K, Okamoto M, Koshiba T, Kamiya Y, Nambara E (2005) Genome-wide profiling of stored mRNA in Arabidopsis thaliana seed germination: epigenetic and genetic regulation of transcription in seed. Plant J 41:697–709
Nambara E, Marion-Poll A (2003) ABA action and interactions in seeds. Trends Plant Sci 8:213–217
Ogawa M, Hanada A, Yamauchi Y, Kuwahara A, Kamiya Y, Yamaguchi S (2003) Gibberellin biosynthesis and response during Arabidopsis seed germination. Plant Cell 15:1591–1604
Oh E, Kim J, Park E, Kim J-I, Kang C, Choi G (2004) PIL5, a phytochrome-interacting basic helix-loop-helix protein, is a key negative regulator of seed germination in Arabidopsis thaliana. Plant Cell 16:3045–3058
Ooms JJJ, Leon-Kloosterziel KM, Bartels D, Koornneef M, Karssen CM (1993) Acquisition of desiccation tolerance and longevity in seeds of Arabidopsis thaliana (a comparative study using abscisic acid-insensitive abi3 mutants). Plant Physiol 102:1185–1191
Oracz K, Bouteau HE-M, Farrant JM, Cooper K, Belghazi M, Job C, Job D, Corbineau F, Bailly C (2007) ROS production and protein oxidation as a novel mechanism for seed dormancy alleviation. Plant J 50:52–465
Peeters AJM, Blankestijn-DeVries H, Hanhart CJ, Leon-Kloosterziel KM, Zeevaart JAD, Koornneef M (2002) Characterization of mutants with reduced seed dormancy at two novel rdo loci and a further characterization of rdo1 and rdo2 in Arabidopsis. Physiol Plant 115:604–612
Penfield S, Josse E-M, Kannangara R, Gilday AD, Halliday KJ, Graham IA (2005) Cold and light control seed germination through the bHLH transcription factor SPATULA. Current Biol 15:1998–2006
Penfield S, Li Y, Gilday AD, Graham S, Graham IA (2006) Arabidopsis ABA INSENSITIVE4 regulates lipid mobilization in the embryo and reveals repression of seed germination by the endosperm. Plant Cell 18:1887–1899
Piskurewicz U, Jikumaru Y, Kinoshita N, Nambara E, Kamiya Y, Lopez-Molina L (2008) The gibberellic acid signaling repressor RGL2 inhibits Arabidopsis seed germination by stimulating abscisic acid synthesis and ABI5 activity. Plant Cell 20:2729–2745
Rajjou L, Gallardo K, Debeaujon I, Vandekerckhove J, Job C, Job D (2004) The effect of alpha-amanitin on the Arabidopsis seed proteome highlights the distinct roles of stored and neo-synthesized mRNAs during germination. Plant Physiol 134:1598–1613
Richards DE, King KE, Aitali T, Harberd NP (2001) How gibberellin regulates plant growth and development: a molecular genetic analysis of gibberellin signaling. Ann Rev Plant Physiol Plant Mol Biol 52:67–88
Rohde A, Kurup S, Holdsworth M (2000) ABI3 emerges from the seed. Trends Plant Sci 5:418–419
Schwachtje J, Baldwin IT (2004) Smoke exposure alters endogenous gibberellin and abscisic acid pools and gibberellin sensitivity while eliciting germination in the post-fire annual, Nicotiana attenuata. Seed Sci Res 14:51–60
Shen-Miller J, Mudgett MB, Schopf JW, Clarke S, Berger R (1995) Exceptional seed longevity and robust growth: ancient sacred lotus from China. Am J Bot 82:1367–1380
Sliwinska E, Jendrzejczak E (2002) Sugar-beet seed quality and DNA synthesis in the embryo in relation to hydration-dehydration cycles. Seed Sci Technol 30:597–608
Stupnikova I, Benamar A, Tolleter D, Grelet J, Borovskii G, Dorne A-J, Macherel D (2006) Pea seed mitochondria are endowed with a remarkable tolerance to extreme physiological temperatures. Plant Physiol 140:326–335
Tamura N, Yoshida T, Tanaka A, Sasaki R, Bando A, Toh S, Lepiniec L, Kawakami N (2006) Isolation and characterization of high temperature-resistant germination mutants of Arabidopsis thaliana. Plant Cell Physiol 47:1081–1094
Teng S, Rognoni S, Bentsink L, Smeekens S (2008) The Arabidopsis GSQ5/DOG1 Cvi allele is induced by the ABA-mediated sugar signalling pathway, and enhances sugar sensitivity by stimulating ABI4 expression. Plant J 55:372–381

Toorop PE, van Aelst AC, Hilhorst HWM (2000) The second step of the biphasic endosperm cap weakening that mediates tomato (Lycopersicon esculentum) seed germination is under control of ABA. J Exp Bot 51:1371–1379
Tyler L, Thomas SG, Hu J, Dill A, Alonso JM, Ecker JR, Sun TP (2004) DELLA proteins and gibberellin-regulated seed germination and floral development in Arabidopsis. Plant Physiol 135:1008–1019
Walters C (2004) Temperature dependency of molecular mobility in preserved seeds. Biophys J 86:1253–1258
Wen CK, Chang C (2002) Arabidopsis RGL1 encodes a negative regulator of gibberellin responses. Plant Cell 14:87–100
Yamaguchi S, Kamiya Y (2002) Gibberellins and light-stimulated seed germination. J Plant Growth Regul 20:369–376
Yamaguchi S, Smith MW, Brown RG, Kamiya Y, Sun T (1998) Phytochrome regulation and differential expression of gibberellin 3beta-hydroxylase genes in germinating Arabidopsis seeds. Plant Cell 10:2115–2126
Yamaguchi S, Kamiya Y, Sun TP (2001) Distinct cell-specific expression patterns of early and late gibberellin biosynthetic genes during Arabidopsis seed germination. Plant J 28:443–453
Yamauchi Y, Ogawa M, Kuwahara A, Hanada A, Kamiya Y, Yamaguchi S (2004) Activation of gibberellin biosynthesis and response pathways by low temperature during imbibition of Arabidopsis thaliana seeds. Plant Cell 16:367–378

Chapter 5
Bud Dormancy in Perennial Plants: A Mechanism for Survival

James V. Anderson, David P. Horvath, Wun S. Chao, and Michael E. Foley

Abstract Dormancy in vegetative buds of perennial plants plays an important role for surviving harsh environmental conditions. Identifying the genetic and physiological mechanisms regulating dormancy in these vegetative structures will allow manipulation of plant growth and development in both crops and weeds. Model plants have been used to study the physiological effects that photoperiod and temperature impart on dormancy regulation in perennial buds. At the molecular level, models derived through analysis of the transcriptome have shed new light on multiple cellular pathways and physiological processes associated with dormancy transitions and, in some cases, have revealed overlap with pathways regulating flowering and cold acclimation. In this chapter, we discuss proposed models based on advances to our understanding of physiological and molecular factors affecting dormancy regulation in vegetative buds of perennials.

5.1 Introduction

Vegetative buds contain meristems in various stages of growth and development, which posses the capacity to serve as reservoirs for potential vegetative and/or floral development following extremes in seasonal weather conditions. In perennials, these buds are formed on vegetative propagules such as bulbs, corms, roots, rhizomes, stolons, and tubers or, in the case of trees they also exist as apical or axillary buds during the growing and nongrowing season (Anderson et al. 2001). During the perennial life cycle, axillary and apical buds of most woody tree species and shrubs or underground adventitious buds of herbaceous species transition through the

J.V. Anderson (✉), D.P. Horvath, W.S. Chao, and M.E. Foley
Weed Biology Research Unit, USDA-Agricultural Research Service, 1605 Albrecht Blvd, Fargo, ND 58105-5674, USA
e-mails: james.anderson@ars.usda.gov; david.horvath@ars.usda.gov; wun.chao@ars.usda.gov; michael.foley@ars.usda.gov

E. Lubzens et al. (eds.), *Dormancy and Resistance in Harsh Environments*, Topics in Current Genetics 21,
DOI 10.1007/978-3-642-12422-8_5,

various stages of dormancy (para-, endo-, and eco-dormancy) as defined by Lang et al. (1987). During the growing season, meristematic development in paradormant axillary or adventitious buds is under the control of physiological signals generated external to the buds (this process is also referred to as correlative inhibition or apical dominance) (Fig. 5.1). In autumn, shortening photoperiod and/or cool temperature

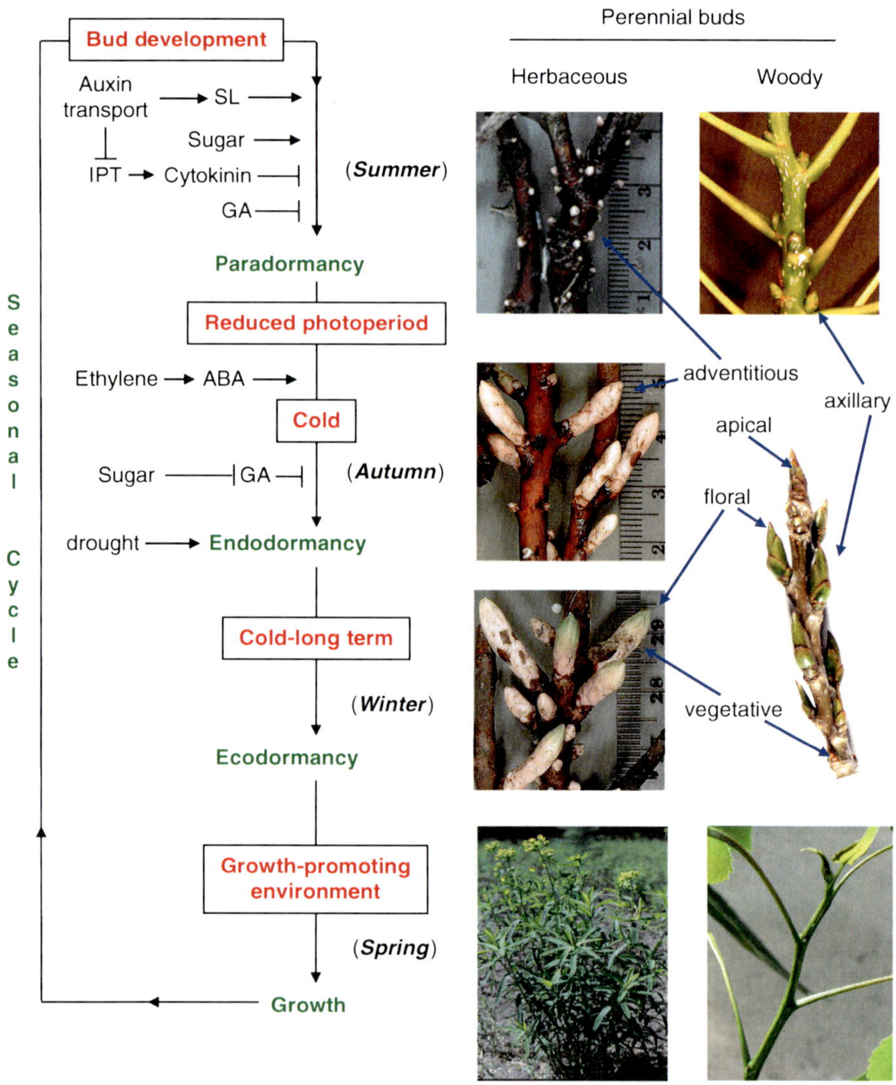

Fig. 5.1 Overview of dormancy status during seasonal bud development in herbaceous and woody perennials. Diagram shows developmental comparison of underground adventitious buds of herbaceous (leafy spurge) and apical or axillary buds of woody (poplar) plants during seasonal transitions in well-defined phases of para-, endo-, and eco-dormancy. Abbreviations: *ABA* abscisic acid, *GA* gibberellic acid, *IPT* isopentenyl transferase, *SL* strigolactone

induces endodormancy (also referred to as innate dormancy) through internal signals and processes within the bud that prevent growth even if external physiological signals are removed and the plants are returned to growth-promoting conditions. Establishment of endodormancy prevents initiation of new shoot growth from buds during autumn when the environment can rapidly fluctuate between growth-promoting and nonpromoting conditions. As a result, endodormancy prevents buds from establishing new shoot growth that would be susceptible to environmental stress (e.g., frost and dehydration) and would thus deplete plants of buds needed to generate vegetative growth the following growing season (Chao et al. 2007; Rohde and Bhalerao 2007; Volaire and Norton 2006). Extended periods of extreme temperatures are generally required to release buds from endodormancy. However, these environmental extremes also prevent further growth, a process known as ecodormancy. In this chapter, we will discuss the environmental factors that affect physiological and molecular signaling mechanisms associated with seasonal transitions in well-defined phases of plant bud dormancy; particularly the induction and release from endodormancy by environmental signals such as photoperiod and/or temperature, and its overlap with flowering pathways in woody and herbaceous perennials.

5.1.1 Bud Phenology in Model Perennials

Determination of vegetative or floral bud development is dependent on specific environmental signals such as exposure to extended cold temperatures or specific photoperiods (Chouard 1960; Rohde and Bhalerao 2007). However, across plant species, the developmental context of buds from annuals, woody perennials, and herbaceous perennials is different. For example, in most woody perennials, such as poplar (*Populus* spp.), the mature growing apices transition to overwintering buds, while in herbaceous perennials such as leafy spurge (*Euphorbia esula*), the mature growing apices senesce and die, which is similar to the model annual plant *Arabidopsis thaliana*. However, unlike Arabidopsis, where the whole plant dies, adventitious buds located on the persisting underground crown and root system of leafy spurge overwinter for renewed seasonal shoot growth. Since buds of model herbaceous and woody perennials exhibit differences in their phenotypic development (Fig. 5.1), each will be discussed separately. Although this chapter focuses on dormancy in perennials, where appropriate, information obtained from studies with Arabidopsis will be included to bridge gaps in proposed models.

5.1.1.1 Woody Perennials

Several model dicot systems have been used to study bud dormancy and flowering in woody perennials. In the spring, apical buds of poplar initiate growth (often referred to as bud break) to become the new growing shoot meristem (Yuceer et al.

2003). In poplar, preformed early leaves in the apical buds rapidly expand, while leaf primordia on the shoot continue to differentiate to become what is referred to as later leaves. Axillary buds formed at the base of early preformed leaves remain vegetative, while axillary buds that form at the base of the later leaves are programmed to become floral buds in adult trees (see Fig. 5.1). The early preformed leaves may provide the florigenic signal required for the axillary buds of the late-formed leaves to differentiate into floral organs (Yuceer et al. 2003). However, the floral buds that form at the base of later leaves will not flower until they have overwintered. After exposure to 1–2 weeks of short photoperiod, apical meristems cease growth and the leaf primordia differentiate into scales rather than leaves, a process known as bud set. If these buds are placed back under long photoperiod conditions, they still resume growth, indicating that the limited short photoperiods induce ecodormancy. However, subjecting apical buds to 4–6 weeks of short photoperiod prevents growth upon return to long photoperiod conditions, indicating that exposure of buds to extended short photoperiods induces endodormancy. These buds also become partially cold acclimated after 4–6 weeks under short photoperiod conditions; although additional cold temperatures enhance cold hardiness (Welling et al. 1997). Once buds are endodormant, they remain so until extended exposure to cold temperatures reestablish growth competence.

Other woody perennials used to study bud dormancy include chestnut (*Castanea sativa*) (Ramos et al. 2005), grape (*Vitis* spp.) (Fennell and Hoover 1991; Mathiason et al. 2008; Schnabel and Wample 1987; Wake and Fennell 2000), various other fruit crops such as apple (*Malus* spp.) (Foster et al. 2003; Heide and Prestrud 2005) and members of the *Prunus* family such as peach (Bielenberg et al. 2004; Li et al. 2009) and apricot (Yamane et al. 2008), and to lesser extents other woody perennials such as red osier dogwood (*Cornus sericea*) (Smithberg and Weiser 1968; Svendsen et al. 2007). Like poplar, these species follow similar bud developmental phenology, with the formation of an apical meristem (sometimes referred to as a terminal vegetative bud) and axillary buds forming either vegetative or floral buds depending on their maturity requirements, relative position on the stem, and/or light/temperature conditions during development. In grape, bud meristems that overwinter develop 6–9 nodes over the growing season; each node develops a leaf with a vegetative axillary bud along with a tendril or floral bud opposite the leaf. At nodes 4–6, the opposing bud generally forms a flower, while the other bud forms a tendril (Pratt 1971). In apple, the apical bud and those closest to the apical bud become committed to flowering, while more distal buds remain committed to vegetative production (Foster et al. 2003). In both grape and apple, floral bud development occurs in the growing season that precedes flowering, similar to poplar (Foster et al. 2003; Pratt 1971).

Endodormancy induction in some grape varieties require short photoperiod conditions, while other varieties require both short photoperiod and cold temperatures (Fennell and Hoover 1991; Schnabel and Wample 1987). A similar phenomenon has been noted in northern vs. southern ecotypes of red osier dogwood (Smithberg and Weiser 1968; Svendsen et al. 2007). In apple and pear (*Pyrus* spp.), temperature alone appears to regulate endodormancy induction and release

(Heide and Prestrud 2005). In all these systems, once endodormancy is released, buds develop into either vegetative meristems or flowers depending on their pre-determined state.

5.1.1.2 Herbaceous Perennials

Unlike most woody perennials, which develop new vegetative growth from apical or axillary buds located on above ground stems and branches, the aerial portions of herbaceous perennials seasonally die back to ground level, and development of new stems and shoots occurs from adventitious buds that are underground or at the soil surface. Compared to woody perennials, substantially fewer herbaceous perennial models are used to study well-defined phases of dormancy. Leafy spurge has emerged as a prominent model system to study dormancy transitions in herbaceous perennials (Chao et al. 2005).

Leafy spurge is a wild flower common to road sides and pasture lands in central and eastern Europe, but has become an invasive weed in the Northern Great Plains of the US and Canada. Leafy spurge reproduces sexually by seeds, or asexually by vegetative reproduction from an abundance of underground adventitious buds. Dormancy-imposed inhibition of new shoot growth from underground adventitious buds has long been considered a key characteristic leading to the persistence and invasiveness of herbaceous perennial weeds such as leafy spurge, Canada thistle (*Cirsium arvense*), field bindweed (*Convolvulus arvensis*), etc. As a result, the nonuniform emergence of vegetative shoots, resulting from dormancy-imposed inhibition of growth, is one of the key characteristics allowing many weedy plants to escape conventional control measures. Understanding the pathways and networks that regulate dormancy in weedy perennials could identify new targets for manipulating plant growth and reduce economic costs to land managers worldwide.

In herbaceous perennials like leafy spurge, bud and flower development, as well as dormancy transition, deviate from those described for woody perennials. In field settings, leafy spurge forms new adventitious buds on the underground portion of its stem (often referred to as the crown; see Fig. 5.1) and on roots. During late spring to early summer, usually after flowering has occurred, new crown buds form (Anderson et al. 2005). However, visible crown buds will also form on stems of greenhouse-grown, nonflowering plants within 2–3 months after propagation from shoot cuttings. Like axillary buds of other models, crown and root buds will not develop into new shoots during the growing season unless the above ground portion of the plant dies or is removed. During late summer and early autumn, these paradormant buds transition to a state of endodormancy which is often marked by bud expansion. Following extended cold treatment, endodormant buds transition to an ecodormant state at which point they simultaneously become both growth and floral competent (Anderson et al. 2005). Leafy spurge plants that have transitioned through both endo- and eco-dormancy generate new shoot growth followed by flowering after several weeks when returned to growth-conducive environments (Doğramaci et al. 2010; Foley et al. 2009).

Potato (*Solanum tuberosum*) is another model system used to study bud dormancy. In potato, short photoperiod exposure perceived by the leaves induces underground stolons to develop tubers with axillary buds (Rodríguez-Falcón et al. 2006). Once formed, these buds are endodormant (Sonnewald 2001). In addition to the substantial body of work on potato bud endodormancy at the physiological level (Suttle 2004, 2008), there is evidence that seasonal regulation of floral and tuber-inducing signaling pathways might also overlap at the molecular level (Rodríguez-Falcón et al. 2006) and that endodormancy maintenance may involve chromatin remodeling (Law and Suttle 2004).

5.2 Environmental Regulation

Light and/or temperature are important environmental signals affecting transitions between well-defined phases of dormancy and the underlying mechanisms regulating these transitions are a major focus in this review (Figs. 5.2 and 5.3). These environmental signals influence the physiology of buds and alter their ability to initiate new vegetative growth under growth-conducive conditions (Allona et al. 2008; Anderson et al. 2001; Cao et al. 2008; Franklin 2009; Horvath et al. 2003; Rohde and Bhalerao 2007). Interestingly, in leafy spurge, cold temperatures can induce endodormancy, but an extended period of cold temperature (referred to as vernalization in relation to flowering) signals a dual response that initiates growth- and floral-competence (Foley et al. 2009). In poplar, proteins normally associated with flowering in Arabidopsis, such as FLOWERING LOCUS T (FT), TERMINAL FLOWER 1 (TFL1), PHYTOCHROME A (PHYA), and CONSTANS (CO), have also been associated with regulating growth cessation and endodormancy (Böhlenius et al. 2006; Eriksson 2000; Ruonala et al. 2008). In Arabidopsis, genes encoding these proteins are impacted by both light and temperature (Franklin 2009; Kobayashi and Weigel 2007; Michaels 2009; Nozue and Maloof 2006; Penfield 2008). However, although it has been suggested that signal transduction pathways regulating flowering and endodormancy converge (Horvath 2009), there are still many unresolved questions on how these pathways interact.

5.2.1 Light

Photoperiod is generally considered the primary signal regulating endodormancy induction in many perennials (Howe et al. 1996; Jeknic and Chen 1999; Li et al. 2003; Smithberg and Weiser 1968; Wake and Fennell 2000), with the notable exceptions in several Rosaceae family members (Heide and Prestrud 2005). In Arabidopsis and poplar, photoperiod is perceived by photoreceptors such as phytochrome, which perceives the ratio of red to far-red light (Eriksson 2000; Franklin 2009), and in Arabidopsis cryptochrome (CRY) is also known to perceive and

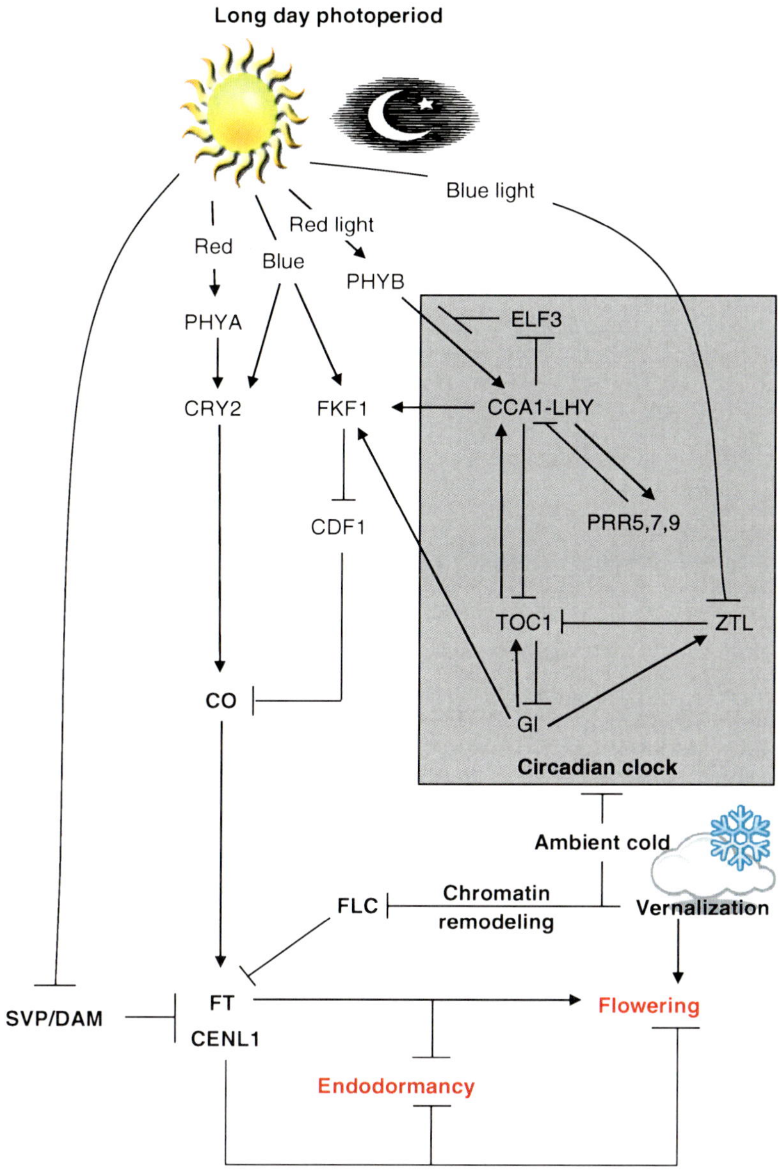

Fig. 5.2 Based on information gathered from both annuals and perennials, this proposed model represents a simplified review of the multiple pathways linking environmental input signals to dormancy regulation and flowering. Both phytochromes and cryptochromes couple input signals from photoperiod and temperature to circadian clock associated proteins. Circadian control involves feedback loop interactions among a host of proteins such as CCA1, ELF3, GI, LHY, PRR9, PRR7 and PRR5, TOC1, and ZTL. Output from entrainment of the circadian clock genes modify the expression of an additional signal transduction cascade that originates from PHYA and CRY2. This portion of the leaf photoperiod measuring mechanism acts through FKF1, GI, CDF1,

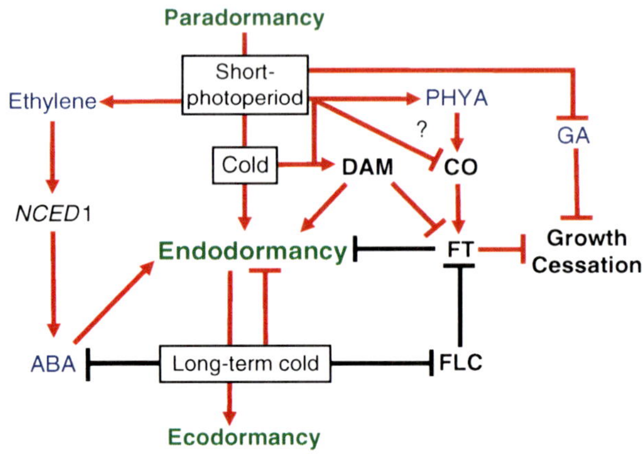

Fig. 5.3 Proposed model of environmental, physiological, and genetic factors impacting dormancy transitions in perennial buds. *Red arrows/bars* indicate genes, hormones, metabolites, or environmental condition that have proven, in at least one perennial system, to induce/inhibit effects on targets, while *black arrows/bars* indicate suspected induction/inhibition on targets. Short daylight affects PHYA and GA which impact growth cessation and endodormancy induction. Short daylight also induces ethylene production which may induce ABA (through induction of *NCED1*) to enhance and perhaps maintain endodormancy. Short daylight and or short-term cold are suspected of inducing *DAM* genes, the product of which likely act to inhibit *FT* expression to regulate endodormancy induction. Short-term cold also induced ABA accumulation in the buds which may enhance endodormancy. Long-term cold inhibits the expression of some (but not all) *DAM* genes, reduces ABA content, and releases buds from endodormancy. "?" represents contradictory data where short day clearly induces *PHYA* expression yet overexpression of transgenic *PHYA* inhibits endodormancy

transduce blue light signals (Cashmore et al. 1999; McClung 2006). These and other molecular input signals interact with and respond to components of the circadian clock (see Fig. 5.2), which controls diurnal rhythms in annuals and perennials (Harmer 2009; McClung 2006; Robertson et al. 2008; Rodríguez-Falcón et al. 2006).

In temperate ecosystems, photoperiod is the most reliable indicator of seasonal transitions. Thus, it is not surprising that light regulates dormancy and flowering via genes and signaling mechanisms involved in regulating the circadian clock (Smith 2000). For example, PHYA and PHYB, proteins involved in entrainment of the

---

Fig. 5.2. (continued) and the floral promoter CO, which partially controls expression of floral integrators like *FT*. FT is generally transported through the phloem to the apical meristem. Vernalization mediates the epigenetic repression of *FLC* or *FLC-LIKE*, which encodes a MADS-box transcription factor that also represses the expression of floral pathway integrators such as *FT*. Long photoperiods also inhibit expression of *SVP/DAM*, which encodes other MADS-box transcription factors that inhibits *FT* and *CENL1*. Although not shown in this diagram, there are multiple points of crosstalk between circadian regulation and cold responses that also modify the activation of floral and endodormancy responses

circadian clock (McClung 2006; Rodríguez-Falcón et al. 2006), have a significant impact on flowering in Arabidopsis (Fujiwara et al. 2008; Ishikawa et al. 2006; Mas and Yanovsky 2009; McClung 2006). PHYA also has an indisputable impact on endodormancy in poplar (Olsen et al. 1997; Ruonala et al. 2008) and PHYB plays a critical role in perception of short photoperiod and induction of tuberization in potato (Rodríguez-Falcón et al. 2006). Activation of Arabidopsis PHYB initiates a series of regulatory feedback loops formed by numerous circadian clock genes (McClung 2006). Additional research with Arabidopsis highlights the affect of photoperiod on other circadian regulating genes such as *CIRCADIAN CLOCK ASSOCIATED1* (*CCA1*), *EARLY FLOWERING3-4* (*ELF3* and *ELF4*), *GIGANTIA* (*GI*), *LATE ELONGATED HYPOCOTYL* (*LHY*), *PHYTOCHROME-INTERACTING FACTOR3* (*PIF3*), *PSEUDO-RESPONSE REGULATORS* (*PRR9*, *PRR7* and *PRR5*), *TIMING OF CAB1* (*TOC1*), *LUX ARRHYTHMO* (*LUX*), and *ZEITLUPE* (*ZTL*), (Harmer 2009; McClung 2006). These circadian clock genes interact with and modify the expression of an additional signal transduction cascade that originates from PHYA and CRY1/2. This portion of the photoperiod measuring mechanism acts through FLAVIN-BINDING, KELCH REPEAT, F-BOX1 (FKF1), CYCLING DOF FACTOR1 (CDF1), and CO (Fujiwara et al. 2008; Kobayashi and Weigel 2007; Michaels 2009; Sawa et al. 2007), and partially controls *FT* and *TFL1* in Arabidopsis (Kobayashi et al. 1999).

In poplar, FT and CENTRORADIALIS-LIKE 1 (CENL1) (orthologue of TFL1) regulate seasonal growth cessation (Böhlenius et al. 2006), and dormancy (Ruonala et al. 2008). In Arabidopsis and poplar, FT and CENL1 also regulate flowering with FT promoting flowering and TFL1/CENL1 inhibiting flowering (Böhlenius et al. 2006; Kobayashi et al. 1999; Mohamed 2006). Interestingly, whereas wild-type *Populus* spp. ceased growth and set buds under short photoperiod (Eriksson 2000; Olsen 1997), short photoperiod-induced growth cessation and *CO* repression did not occur in *PHYA* overexpressing lines (Böhlenius et al. 2006); even when placed in 6 h of daylight (Olsen et al. 1997). Thus, overexpression of *PHYA* in *Populus* spp. may prevent growth cessation and endodormancy induction by up-regulating *FT* and *CENL1* (Böhlenius et al. 2006; Ruonala et al. 2008).

Because photoperiod impacts circadian regulation and endodormancy induction, it is likely that circadian clock genes also affect endodormancy and *vice versa*. For example, during the transition from endodormancy to ecodormancy, numerous genes involved in circadian regulation were up-regulated in crown buds of leafy spurge (Horvath et al. 2008). A similar induction of genes involved in circadian regulation was also observed in grape, chestnut, and poplar during the transition from endodormancy to ecodormancy (Mathiason et al. 2008; Ramos et al. 2005; Ruttink et al. 2007). Interplay between several components of the circadian clock and floral-regulating MADS-box transcription factors FLOWERING LOCUS C (FLC) and SHORT VEGETATIVE PHASE (SVP) are known (Fujiwara et al. 2008; Michaels 2009; Salathia et al. 2006), and both of these factors have been shown to regulate *FT* in Arabidopsis (Helliwell et al. 2006; Lee et al. 2007; Searle et al. 2006).

5.2.2 Temperature

Temperature, like photoperiod, can also regulate endodormancy and flowering induction in some perennials (Foley et al. 2009; Heide and Prestrud 2005; Mouhu et al. 2009; Schnabel and Wample 1987; Smithberg and Weiser 1968; Svendsen et al. 2007). However, compared to photoperiod, considerably less is known about the role that cold plays in endodormancy induction. Cold temperatures are sufficient for inducing endodormancy in some dogwood ecotypes (Svendsen et al. 2007), Rosaceae spp. (Heide and Prestrud 2005), citrus (Moss 1969), and leafy spurge (Anderson et al. 2005; Foley et al. 2009). Interestingly, Foley et al. (2009) further discovered that crown buds must transition through a state of endodormancy to achieve fulfillment of flowering by vernalization in leafy spurge.

The circadian clock is also known to integrate low temperature responses in both annuals and perennials (Harmer 2009; Penfield 2008; Ramos et al. 2005; Samach and Wigge 2005). Thus, it is not surprising that temperature affects endodormancy in perennials (Heide 2008; Kwolek and Woolhouse 1982; Svendsen et al. 2007). In addition, there is evidence for cross talk between low temperature and phytochrome signaling (Allona et al. 2008; Benedict et al. 2006; Heschel et al. 2007; Kim et al. 2002; Olsen et al. 1997; Penfield 2008; Smith 2000), which may link temperature responses to *FT* and *CENL1* expression and endodormancy induction.

5.2.2.1 Cold-Hardening/Cold-Acclimation

Cool autumn temperatures not only initiate endodormancy in buds of perennials but also induce a phenomenon important for survival referred to as cold hardening or cold acclimation. Cold hardening protects the cellular constituents from damage as a result of dehydration and freezing (Guy 1990; Thomashow 2001), and in Arabidopsis cold-regulated gene expression likely occurs through a process involving circadian evening elements (Mikkelsen and Thomashow 2009). Several transcription factors induced by cold temperatures bring about global changes in gene expression that play an important role in the hardening/acclimation process. Among the best studied of these transcription factors is a family of AP2 DNA-binding transcription factors known as C-REPEAT BINDING FACTOR (CBF) 1-4, also known as DEHYDRATION RESPONSIVE ELEMENT BINDING FACTOR (DREB). CBFs are themselves believed to be regulated by another transcription factor known as INDUCER of CBF EXPRESSION (ICE) 1. *ICE1* is present at normal growing temperatures but is either activated or interacts with a protein that is activated by cold (Thomashow 2001). PHYA has been implicated as a transduction pathway mediator; either by PHYA-mediated phosphorylation, or by modulation of proteolysis of ICE1 (Benedict et al. 2006; also see Penfield 2008). Interestingly, overexpression of *PHYA* appears to inhibit short photoperiod-induced cold hardening in poplar (Olsen et al. 1997). In trees, PHYA may prevent short photoperiod induced cold hardening and growth cessation through maintenance of gibberellic acid (GA) levels (see Allona et al. 2008).

Transition from para- to endo-dormancy in leafy spurge crown buds, which coincided with decreasing night temperatures, are paralleled by up-regulation of cold-hardening transcripts with homology to Arabidopsis *DREB A4* (At2g35700) and *ICE1-LIKE* (Doğramaci et al. 2010; Horvath et al. 2008). Since cold-induced expressions of some DREB/CBF-family members are gated by the circadian clock in Arabidopsis (Fowler et al. 2005), these particular regulators of transcription might also have some impact or overlap with pathways affecting dormancy transitions in response to low temperature, particularly in plants such as apple or leafy spurge that rely primarily on cold-induced endodormancy.

5.2.2.2 Vernalization

Although cold hardening generally occurs concomitantly with endodormancy induction, extended cold temperatures both break endodormancy and induce floral competency in numerous perennials (Anderson et al. 2005; Chouard 1960; Foley et al. 2009; Nishikawa et al. 2007). Chouard (1960) originally hypothesized that there might be a connection between vernalization and release from endodormancy. This idea was later expanded to suggest that mechanisms such as chromatin modification might play an underling role in endodormancy in perennials such as leafy spurge, because buds need to "remember" that they are dormant, particularly during brief warm spells common during late autumn and early winter (Horvath et al. 2003). This mechanism was later shown to be true in the case of vernalization in Arabidopsis (Amasino 2004; Henderson and Dean 2004; Sung and Amasino 2004). Further support of this hypothesis in perennials comes from identification of differentially expressed chromatin-modifying genes during endodormancy induction and release in leafy spurge, poplar (in both apical and cambial meristems), grape, and potato (Campbell et al. 2008; Doğramaci et al. 2010; Druart et al. 2007; Horvath et al. 2008; Or et al. 2000; Ruttink et al. 2007).

In contrast to perennials, the mechanisms by which extended cold temperature induces floral competence in winter annuals, such as Arabidopsis, is well understood and information is accumulating on vernalization processes in other species (Alexandre and Hennig 2008). During vernalization, the chromatin structure in the promoter and 5′ coding regions of the floral repressor *FLC* is altered to prevent expression of *FLC*, even after multiple rounds of cell division (He and Amasino 2005). The key players in regulating chromatin modifications are VERNALIZATION INSENSITIVE3 (VIN3), VERNALIZATION1 (VRN1), and 2 (VRN2) (Amasino 2004; Sung and Amasino 2004). In Arabidopsis, extended cold temperatures up-regulate *VIN3* by unknown mechanisms and once induced, VIN3 along with VRN1, VRN2, and LIKE HETEROCHROMATIN PROTEIN1 (LHP1) specifically alter the methylation and acetylation of histones on the *FLC* promoter to block transcription (Henderson and Dean 2004; Sung and Amasino 2004; Sung et al. 2006).

Chromatin remodeling regulates other flowering genes besides *FLC*. Chromatin alterations also regulate *FT* to some extent and *TFL2*, another protein similar to

LHP1 (Takada and Goto 2003). Likewise, the gene *EARLY BOLTING IN SHORT DAYS* (*EBS*), that is similar to a group of bromo/homeodomain containing zinc finger proteins associated with chromatin-modifying complexes, impacts *FT* expression (Pineiro et al. 2003). Mutations in other chromatin-modifying proteins such as various members of the SWITCH/SUCROSE NONFERMENTABLE (SWI/SNF) protein complexes, POLYCOMB group proteins, and other known chromatin-modifying proteins also impact floral timing and development (Farrona et al. 2004; Henderson and Dean 2004; Noh and Noh 2006), and may also impact endodormancy maintenance in perennials (Doğramaci et al. 2010; Horvath et al. 2008). However, the role that epigenetic factors play in regulating endodormancy induction and release in perennial buds requires additional support to confirm the function of chromatin-modifying genes differentially expressed during these transitions.

5.3 Genetic/Physiological Model(s) for Regulation of Dormancy Transitions

Data presented so far highlights that signal transduction pathways affected by both cold and light signaling converge on the expression of *FT* and *CENL1* to affect endodormancy and flowering (Fig. 5.2 and Table 5.1). Both of these genes are also regulated by SVP and FLC in Arabidopsis, two specific floral-regulating MADS-box transcription factors (Fujiwara et al. 2008; Lee et al. 2007; Michaels 2009). In perennial species, *DORMANCY ASSOCIATED MADS-box* (*DAM*) genes, which are closely related to *SVP*, have been identified as playing a role in endodormancy induction. *DAM* genes were first identified as having a role in dormancy through map-based cloning of the *EVERGROWING* locus in peach (Bielenberg et al. 2004). The *evergrowing* (earlier called *evergreen*) peach varieties contain a mutation that prevents induction of dormancy under short daylight conditions (Li et al. 2009), in contrast to that observed for wild-type varieties (Diaz 1974). Sequencing of the *EVERGROWING* locus of wild type and mutant lines revealed a deletion of a series of *MADS-box* genes in the mutant line (Bielenberg et al. 2008).

It has been hypothesized that induction of *DAM* genes may be required for down-regulating *FT* and *CENL1* during the initiation of growth cessation and/or endodormancy in some perennial species (Horvath et al. 2008; Horvath 2009), since SVP down-regulates *FT* in Arabidopsis (Fujiwara et al. 2008; Lee et al. 2007) and *FT* and *CENL1* influences growth cessation and dormancy in trees (Böhlenius et al. 2006; Ruonala et al. 2008). This hypothesis is supported by potential *FT*- and *CENL1-LIKE* genes from leafy spurge being down-regulated concomitantly with *DAM* induction during endodormancy (Horvath et al. 2008). Additionally, Horvath et al. (2010) found that *FT* expression was down-regulated in transgenic Arabidopsis lines overexpressing leafy spurge *DAM1*, and these transgenic lines also had reduced bolt height and delayed flowering compared to wild type. Although the functionality of the putative *FT-LIKE* gene from leafy spurge has not been verified, the results were consistent with the hypothesis that DAM regulates *FT*/*CENL1*.

Table 5.1 Floral regulatory genes and their affect on flowering in the model annual Arabidopsis and the impact of suspected orthologs on dormancy in one or more perennial systems

Gene	Function	Flowering				Dormancy			
		RR[a]	I[b]	R[c]	AE[d]	RR[a]	I[b]	R[c]	AE[d]
APETALA1 (AP1)	FMI[e]	X	–	–	X	–	–	–	–
CONSTANS (CO)	PP[f]	X	X	–	X	X	–	–	X
FLOWERING LOCUS T (FT)	FI[g]	X	X	–	X	X	–	X	X
FLOWERING LOCUS C (FLC)	FR[h]	X	–	X	X	–	–	–	–
LEAFY (LFY)	FMI	X	–	–	X	–	–	–	–
PHYTOCHROME A (PHYA)	PP	X	X	–	X	X	–	X	X
SHORT VEGETATIVE PHASE/DORMANCY ASSOCIATED MADS BOX (SVP/DAM)	FI	X	–	X	X	X	X	–	X
SUPPRESSOR OF CONSTANS 1 (SOC1)	FI	X	–	–	X	–	–	–	–
TERMINAL FLOWERING1 (TFL1/CENL1)	FI	X	X	X	X	X	–	X	X
VERNALIZATION INSENSITIVE 3 (VIN3)	VRN[i]	X	X	–	X	–	–	–	–
VERNALIZATION INSENSITIVE 1/2 (VIN1/2)	VRN	X	X	–	–	–	–	–	–
Other chromatin modifiers	VRN	X	–	–	X	X	–	–	X
Circadian regulators	PP	X	X	X	X	X	–	X	X

[a]RR, denotes proven regulatory roles
[b]I, indicates if the gene induces flowering and/or dormancy
[c]R, if gene represses flowering and/or dormancy
[d]AE, indicates if environmental signals that regulate flowering and/or dormancy cause alter expression of the genes
[e]Floral meristem identity gene
[f]Photoperiod pathway gene
[g]Floral integrator gene
[h]Floral repressor
[i]Vernalization pathway gene

Research is continuing in multiple perennials to determine if *DAM* gene expression has any direct impact on dormancy, flowering, or *FT* expression.

The induction of *DAM* genes may also play a role in regulating transitional phases of dormancy in other perennials. For example, *DAM* genes from raspberry (*Rubus idaeus*) (Mazzitelli et al. 2007), potato (Campbell et al. 2008), apricot (Yamane et al. 2008), peach (Bielenberg et al. 2008; Li et al. 2009), and likely poplar (see Horvath et al. 2008) have been implicated in bud dormancy regulation. Differential expression of *DAM* genes has been observed during dormancy-inducing short photoperiod conditions in *Populus* spp. and *Prunus* spp. (Ruttink et al. 2007; Yamane et al. 2008) and during cold-induced endodormancy induction of leafy spurge (Horvath et al. 2010). Additionally, putative *cis*-acting elements similar to circadian-regulating evening elements, affecting diurnal regulation, are conserved in the promoters of *DAM* genes of poplar and leafy spurge (Horvath et al. 2008), but CBF-binding sites (Fowler et al. 2005) are only observed in the leafy spurge *DAM1* promoter (Horvath et al. 2008). Thus among perennial species, the presence of CBF sites in the promoters of *DAM* genes could explain why cold is the primary signal-inducing endodormancy in some perennial species, as opposed to other perennial species where photoperiod primarily regulates endodormancy induction.

Evidence presented in this review suggests that *PHYA* may play a role in seasonal growth cessation and endodormancy induction through regulation of *FT* and *CENL1*. However, release from endodormancy by extended cold temperatures likely does not involve these genes, at least in perennials such as leafy spurge, where *FT*- and *CENL1-LIKE* genes are not up-regulated following endodormancy release (Horvath et al. 2008). Likewise, *PHYA* signaling is unlikely in endodormant buds from most temperate perennials since leaves (the primary photoreceptor organs) are not present during the transition from endodormancy to ecodormancy. Observing the impact of *FT*- and *CENL1-LIKE* gene expression in buds of perennials to determine if *FT* and *CENL1* expression is sufficient for endodormancy release across multiple species should help to clarify their functional role in endodormancy maintenance. However, since some *DAM* genes are preferentially expressed only during endodormancy (Horvath et al. 2008; Li et al. 2009; Yamane et al. 2008), DAM could be part of a mechanism involved in controlling both induction and release of endodormancy in perennials.

5.3.1 Hormones

Hormones such as abscisic acid (ABA), GA, cytokinin, auxin, and ethylene are implicated in various aspects of growth cessation, bud set, and endodormancy in both woody and herbaceous perennials (Allona et al. 2008; Chao et al. 2006, 2007; Horvath et al. 2003, 2008; Horvath 2009; Olsen 2006; Rodríguez-Falcón et al. 2006; Rohde and Bhalerao 2007). Since photoperiod impacts endodormancy induction and affects circadian responses, it is likely that circadian responses could impact hormone levels affecting dormancy and vice versa (Alabadi and Blázquez 2009; Covington and Harmer 2007; Nozue and Maloof 2006; Robertson et al. 2008). In addition, cold night temperatures combined with inhibition of GA accumulation are sufficient for inducing ecodormancy and bud set in *PHYA* overexpressing poplar lines (Mølmann et al. 2005), and cold temperature also blocks growth through the action of hormones, such as ABA and GA (see Figs. 5.1 and 5.3). There is evidence that ABA levels as well as oxidative stress may play some role in endodormancy release in perennials (Arora et al. 2003; Destefano-Beltrán et al. 2006; Or 2009). Additionally, auxin responses are impacted by conditions that induce endodormancy (Anderson et al. 2005; Horvath et al. 2008; Schrader et al. 2004), and a node that links the clock and auxin networks has recently been identified (Rawat et al. 2009). For many years, auxin was considered the primary hormone regulating the transition from paradormancy to growth (Beveridge 2006). However, the lack of apically derived auxin transport into buds, and the contradictory role of auxin requirements for bud growth pose unanswered questions as to the mechanisms by which auxin controlled bud outgrowth. Cytokinins have been noted for promotion of cell division and organ formation that are invoked as key regulatory signals promoting axillary bud outgrowth when the apical meristem is removed, and auxin is known to negatively regulate cytokinin biosynthesis in

Arabidopsis (Nordstrom et al. 2004). A strigolactone, whose production appears to be dependent on polar auxin transport in annuals, has also been implicated in inhibiting shoot branching (Gomez-Roldan et al. 2008) and appears to directly impact bud growth (see Fig. 5.1). It will be interesting to determine if this new hormone plays any role in transitions between paradormancy and endodormancy in perennials.

Data obtained from poplar, leafy spurge, and potato provides evidence that a transient spike in ethylene (or ethylene perception) precedes, and is necessary for, the initiation of endodormancy (Horvath et al. 2008; Ruttink et al. 2007; Suttle 1998). Research done on ABA accumulation in *Citrus* suggests that ethylene may directly induce the key ABA biosynthetic gene *9-CIS-EPOXYCAROTENOID DIOXYGENASE (NCED1)* (Rodrigo and Alquezar 2006). In turn, catabolism of ABA during dormancy release in potato may, in part, be correlated to decreased levels of *NCED* (Destefano-Beltrán et al. 2006). In leafy spurge, ABA levels were elevated during endodormancy but dropped following the transition to ecodormancy (Horvath et al. 2008). At least ten other genes associated with ethylene production or ethylene responses were highly expressed during paradormancy, but were repressed later during endo- and eco-dormancy (Horvath et al. 2008).

5.3.2 Sugar

Sugar levels have been correlated with the transition of vegetative buds from paradormancy to endodormancy (Anderson et al. 2005; Arora et al. 2003; Chao et al. 2006). The effect of sugars on regulating well-defined phases of vegetative bud dormancy has been investigated using the model perennial weed, leafy spurge. Sugars from leaves of leafy spurge have been linked to suppression of underground adventitious bud growth during paradormancy (Horvath 1999), and further research has shown that glucose or sucrose can inhibit root bud growth in a mechanism reversed by GA (Chao et al. 2006; Horvath et al. 2002). In addition, sugar signaling is suspected to play an important role in maintaining paradormancy by affecting cell cycle progression at the G1/S phase (Horvath et al. 2002). Shoot removal causes a rapid degradation of starch and a decline in sucrose, concurrent with the transition from paradormancy to active shoot growth (Chao et al. 2006; Horvath et al. 2002). In contrast, during autumn senescence, conversion of starch to sucrose occurs in leafy spurge underground adventitious buds, which is also paralleled by a transition from paradormancy to endodormancy (Anderson et al. 2005). In parallel with these transitions, transcript levels of a leafy spurge starch degrading enzyme, β-amylase, were 1,000-fold higher in endodormant buds and 16,000-fold higher in ecodormant buds than that of paradormant buds (Chao and Serpe 2009). Genes involved in starch degradation were also up-regulated in the cambial meristems of poplar during endodormancy (Schrader et al. 2004). Based on these and other findings, it is proposed that both sugars and their metabolism may play a role in regulating vegetative bud dormancy through cross talk with hormones (Fig. 5.1) (Anderson et al. 2001, 2005; Horvath et al. 2003). Specifically, it has been suggested that sugars are antagonistic to GA perception and likely play a role in signaling pathways required for inducing endo- and eco-dormancy (Chao et al. 2007).

5.4 Conclusions

Vegetative buds have an essential role in the perennial life cycle, and also serve as a mechanism for plant survival after periods of harsh environmental stress. Evolution of this process allowed plants to adapt to seasonal changes in environment and to expand across vast areas of land in temperate regions of the world; an adaptation that could be adversely challenged by global warming. The ability of meristematic cells within these buds to enter and exit well-defined phases of dormancy in response to environmental signals ensures appropriate timing of both vegetative growth and flowering. Consequently, meristematic cells of perennial buds are governed by specific internal signal transduction pathways that respond in concert to external environmental cues. Advances in our understanding of these processes now suggests that photoperiod and temperature signal transduction pathways affecting dormancy likely converge and/or share components with signaling pathways regulating flowering, including similar transcription factors, chromatin remodeling genes, and biochemical signals and receptors.

It is not surprising that many of the genes involved in circadian response pathways have been linked to dormancy transitions, since buds of several woody and herbaceous perennial species use seasonal changes in photoperiod as a reliable signal for regulating endodormancy induction. Additional studies are still needed to identify what, if any, role circadian regulated genes play in dormancy transitions, and to understand why some circadian-regulating genes are up-regulated or constitutively expressed during cold periods associated with endo- and eco-dormancy (Horvath et al. 2008; Ramos et al. 2005). Clock regulators must undergo temporal synchronization to impact responses that allow plant buds to adjust to seasonal changes. As a consequence, understanding how photoperiod and temperature modify expression of specific clock genes could provide insight into the mechanisms through which these environmental factors regulate dormancy and survival.

Duplication of *DAM* genes may have led to specialization of these genes for unique roles in the regulation of dormancy and flowering. Nearly all of the genes mentioned in this review are members of gene families. Understanding the evolution of these gene families and the specific factors controlling both their regulation and function is needed to determine what role, if any, individual members have in dormancy regulation. Comparative transcriptomics indicates that numerous genes involved in multiple physiological, developmental, and biochemical responses show conserved patterns of expression during endodormancy transitions and during floral transitions. However, additional research is needed to determine the specific functions and interactions that these multiple signaling and molecular pathways have on transitions in well-defined phases of bud dormancy in woody and herbaceous perennial plants.

Acknowledgments Special thanks to Cetin Yuceer for providing the poplar pictures used in Fig. 5.1. The first and second authors contributed equally.

References

Alabadi D, Blázquez MA (2009) Molecular interactions between light and hormone signaling to control plant growth. Plant Mol Biol 69:409–417

Alexandre CM, Hennig L (2008) FLC or not FLC: the other side of vernalization. J Exp Bot 59(6):1127–1135

Allona I, Ramos A, Ibáñez C, Contreras A, Casado R, Aragoncillo C (2008) Review. Molecular control of winter dormancy establishment in trees. Span J Agric Res 6:201–210 (Special issue)

Amasino R (2004) Vernalization, competence, and the epigenetic memory of winter. Plant Cell 16(10):2553–2559

Anderson JV, Chao WS, Horvath DP (2001) A current review on the regulation of dormancy in vegetative buds. Weed Sci 49:581–589

Anderson JV, Gesch RW, Jia Y, Chao WS, Horvath DP (2005) Seasonal shifts in dormancy status, carbohydrate metabolism, and related gene expression in crown buds of leafy spurge. Plant Cell Environ 28(12):1567–1578

Arora R, Rowland LJ, Tanino K (2003) Induction and release of bud dormancy in woody perennials: a science comes of age. HortScience 38(5):911–921

Benedict C, Geisler M, Trygg J, Hunter N, Hurry V (2006) Consensus by democracy. Using meta-analyses of microarray and genomic data to model the cold acclimation signaling pathway in arabidopsis. Plant Physiol 141:1219–1232

Beveridge CA (2006) Axillary bud outgrowth: sending a message. Curr Opin Plant Biol 9:35–40

Bielenberg DG, Wang Y, Fan S, Reighard GL, Scorza R, Abbott AG (2004) A deletion affecting several gene candidates is present in the evergrowing peach mutant. J Hered 95(5):436–444

Bielenberg DG, Wang Y, Li Z, Zhebentyayeva T, Fan S, Reighard GL, Scorza R, Abbott AG (2008) Sequencing and annotation of the evergrowing locus in peach [*Prunus persica* (L.) Batsch] reveals a cluster of six MADS-box transcription factors as candidate genes for regulation of terminal bud formation. Tree Genet Genomes 4:495–507

Böhlenius H, Huang T, Charbonnel-Campaa L, Brunner AM, Jansson S, Strauss SH, Nilsson O (2006) CO/FT regulatory module controls timing of flowering and seasonal growth cessation in trees. Science 312:1040–1043

Campbell MA, Segear E, Beers L, Knauber D, Suttle J (2008) Dormancy in potato tuber meristems: chemically induced cessation in dormancy matches the natural process based on transcript profiles. Funct Integr Genomics 8:317–328

Cao Y, Dai Y, Cui S, Ma L (2008) Histone H2B monoubiquitination of the chromatin of FLOWERING LOCUS C regulates flowering time in *Arabidopsis*. Plant Cell 20:2586–2602

Cashmore AR, Jarillo JA, Wu J-Y, Liu D (1999) Cryptochromes: blue light receptors for plants and animals. Science 284:760–765

Chao WS, Serpe MD (2009) Changes in the expression of carbohydrate metabolism genes during three phases of bud dormancy in leafy spurge. Plant Mol Biol 73:227–239

Chao WS, Horvath DP, Anderson JV, Foley ME (2005) Potential model weeds to study genomics, ecology, and physiology in the 21st century. Weed Sci 53(6):929–937

Chao WS, Serpe MD, Anderson JV, Gesch RW, Horvath DP (2006) Sugars, hormones, and environment affect the dormancy status in underground adventitious buds of leafy spurge (*Euphorbia esula*). Weed Sci 54:59–68

Chao WS, Foley ME, Horvath DP, Anderson JV, Chao WS, Foley ME, Horvath DP, Anderson JV (2007) Signals regulating dormancy in vegetative buds. Int J Plant Dev Biol 1(1):49–56, ISSN 1749-4753

Chouard P (1960) Vernalization and its relations to dormancy. Annu Rev Plant Physiol 11: 191–238

Covington MF, Harmer SL (2007) The circadian clock regulates auxin signaling and responses in *Arabidopsis*. PLoS Biol 5(8):e222. doi:10.1371/journal.pbio.0050222

Destefano-Beltrán L, Knauber D, Huckle L, Suttle JC (2006) Effects of postharvest storage and dormancy status on ABA content, metabolism, and expression of genes involved in ABA biosynthesis and metabolism in potato tuber tissues. Plant Mol Biol 61:687–697

Diaz MD (1974) Vegetative and reproductive growth habits of evergreen peach trees in Mexico. Proceedings from the 19th International Horticulture Congress, vol 18, Warsaw, p 525

Doğramaci M, Horvath DP, Chao WS, Foley ME, Christoffers MJ, Anderson JV (2010) Extended low temperature impacts dormancy status, flowering competence, and transcript profiles in crown 571 buds of leafy spurge. Plant Mol Bio 73:207–226

Druart N, Johansson A, Baba K, Schrader J, Sjodin A, Bhalerao RR, Resman L, Trygg J, Moritz T, Bhalerao RP (2007) Environmental and hormonal regulation of the activity-dormancy cycle in the cambial meristem involves stage-specific modulation of transcriptional and metabolic networks. Plant J 50:557–573

Eriksson ME (2000) The role of phytochrome A and gibberellins in growth under long and short day conditions. Studies in hybrid aspen. Swedish University of Agricultural Sciences, Umea, ISSN 1401-6230

Farrona S, Hurtado L, Bowman JL, Reyes JC (2004) The Arabidopsis thaliana SNF2 homolog AtBRM controls shoot development and flowering. Development 131(20):4965–4975

Fennell A, Hoover E (1991) Photoperiod influences growth, bud dormancy, and cold-acclimation in Vitis-Labruscana and V-Riparia. J Am Soc Hortic Sci 116(2):270–273

Foley ME, Anderson JV, Horvath DP (2009) The effects of temperature, photoperiod, and vernalization on regrowth and flowering competence in *Euphorbia esula* (Euphorbiaceae) crown buds. Botany 87:986–992

Foster T, Johnston R, Seleznyova A (2003) A morphological and quantitative characterization of early floral development in apple (*Malus* x *domestica* Borkh.). Ann Bot 92:199–206

Fowler SG, Cook D, Thomashow MF (2005) Low temperature induction of Arabidopsis CBF1, 2, and 3 is gated by the circadian clock. Plant Physiol 137:961–968

Franklin KA (2009) Light and temperature signal crosstalk in plant development. Curr Opin Plant Biol 12:63–68

Fujiwara S, Oda A, Yoshida R, Niinuma K, Miyata K, Tomozoe Y, Tajima T, Nakagawa M, Hayashi K, Coupland G, Mizoguchi T (2008) Circadian clock proteins LHY and CCA1 regulate SVP protein accumulation to control flowering in *Arabidopsis*. Plant Cell 20:2960–2971

Gomez-Roldan V, Fermas S, Brewer PB, Puech-Pages V, Dun EA, Pillot JP, Letisse F, Matusova R, Danoun S, Portais JC, Bouwmeester H, Becard G, Beveridge CA, Rameau C, Rochange SF (2008) Strigolactone inhibition of shoot branching. Nature 455:189–195

Guy CL (1990) Cold acclimation and freezing tolerance: role of protein metabolism. Annu Rev Plant Phys Plant Mol Biol 41:187–223

Harmer SL (2009) The circadian system in higher plants. Ann Rev Plant Biol 60:357–377

He Y, Amasino RM (2005) Role of chromatin modification in flowering-time control. Trends Plant Sci 10(1):30–35

Heide OM (2008) Interaction of photoperiod and temperature in the control of growth and dormancy of Prunus species. Sci Hort 115:309–314

Heide OM, Prestrud AK (2005) Low temperature, but not photoperiod, controls growth cessation and dormancy induction and release in apple and pear. Tree Physiol 25(1):109–114

Helliwell CA, Wood CC, Robertson M, Peacock WJ, Dennis ES (2006) The Arabidopsis FLC protein interacts directly in vivo with SOC1 and FT chromatin and is part of a high-molecular-weight protein complex. Plant J 46:183–192

Henderson IR, Dean C (2004) Control of *Arabidopsis* flowering: the chill before the bloom. Development 131:3829–3838
Heschel MS, Selby J, Butler C, Whitelam GC, Sharrock RA, Donohue K (2007) A new role for phytochromes in temperature-dependent germination. New Phytol 174:735–741
Horvath DP, Sung S, Kim D, Anderson JV (2010) Characterization, expression and function of DORMANCY ASSOCIATED MADS-BOX genes from leafy spurge. Plant Mol Biol 73:169–179
Horvath DP (1999) Role of mature leaves in inhibition of root bud growth in Euphorbia esula L. Weed Sci 47:544–550
Horvath DP (2009) Common mechanisms regulate flowering and dormancy. Plant Sci 177: 523–531
Horvath DP, Chao WS, Anderson JV (2002) Molecular analysis of signals controlling dormancy and growth in underground adventitious buds of leafy spurge. Plant Physiol 128(4): 1439–1446
Horvath DP, Anderson JV, Chao WS, Foley ME (2003) Knowing when the grow signals regulating bud dormancy. Trends Plant Sci 8(11):534–539
Horvath DP, Chao WS, Suttle JC, Thimmapuram J, Anderson JV (2008) Transcriptome analysis identifies novel responses and potential regulatory genes involved in seasonal dormancy transitions of leafy spurge (*Euphorbia esula* L.). BMC Genomics 9:536–552
Howe GT, Gardner G, Hackett WP, Furnier GR (1996) Phytochrome control of short-day-induced bud set in black cottonwood. Physiol Plant 97:95–103
Ishikawa M, Kiba T, Chua NH (2006) The Arabidopsis SPA1 gene is required for circadian clock function and photoperiodic flowering. Plant J 46:736–746
Jeknic Z, Chen THH (1999) Changes in protein profiles of poplar tissues during the induction of bud dormancy by short-day photoperiods. Plant Cell Physiol 40(1):25–35
Kim H-J, Kim Y-K, Park J-Y, Kim J (2002) Light signaling mediated by phytochrome plays an important role in cold-induced gene expression through the C-repeat/dehydration responsive element (C/DRE) in *Arabidopsis thaliana*. Plant J 29(6):693–704
Kobayashi Y, Weigel D (2007) Move on up, it's time for change-mobile signals controlling photoperiod-dependent flowering. Genes Dev 21:2371–2384
Kobayashi Y, Kaya H, Goto K, Iwabuchi M, Araki T (1999) A pair of related genes with antagonistic roles in mediating flowering signals. Science 286:1960–1962
Kwolek AVA, Woolhouse HW (1982) Studies on the dormancy of *Calluna-Vulgaris* (L) Hull, during winter – the effect of photoperiod and temperature on the induction of dormancy and the annual cycle of development. Ann Bot 49(3):367–376
Lang GA, Early JD, Martin GC, Darnell RL (1987) Endo-, para-, and ecodormancy: physiological terminology and classification for dormancy research. HortScience 22:371–377
Law RD, Suttle JC (2004) Changes in histone H3 and H4 multi-acetylation during natural and forced dormancy break in potato tubers. Physiol Plant 120(4):642–649
Lee JH, Yoo SJ, Park SH, Hwang I, Lee JS, Ahn JH (2007) Role of SVP in the control of flowering time by ambient temperature in Arabidopsis. Genes Dev 21:397–402
Li C, Junttila O, Ernstsen A, Heino P, Palva ET (2003) Photoperiodic control of growth, cold acclimation and dormancy development in silver birch (*Betula pendula*) ecotypes. Physiol Plant 117:206–212
Li Z, Reighard GL, Abbott AG, Bielenberg DG (2009) Dormancy-associated MADS genes from the EVG locus of [Purnus persica (L.) Batsch] have distinct seasonal and photoperiodic expression patterns. J Exp Bot 60:3521–3530
Mas P, Yanovsky MJ (2009) Time for circadian rhythms: plants get synchronized. Curr Opin Plant Biol 12:574–579
Mathiason K, He D, Grimplet J, Venkateswari J, Galbraith DW, Or E, Fennell A (2008) Transcript profiling in *Vitis riparia* during chilling requirement fulfillment reveals coordination of gene expression patterns with optimized bud break. Funct Integr Genomics. doi:10.1007/s10142-008-0090-y

Mazzitelli L, Hancock RD, Haupt S, Walker PG, Pont SDA, McNicol J, Cardle L, Morris J, Viola R, Brennan R, Hedley PE, Taylor MA (2007) Co-ordinated gene expression during phases of dormancy release in raspberry (*Rubus idaeus* L.) buds. J Exp Bot 58:1035–1045
McClung CR (2006) Plant circadian rhythms. Plant Cell 18:792–803
Michaels SD (2009) Flowering time regulation produces much fruit. Curr Opin Plant Biol 12: 75–80
Mikkelsen MD, Thomashow MF (2009) A role for circadian evening elements in cold-regulated gene expression in Arabidopsis. Plant J 60:328–339
Mohamed R (2006). Expression and function of *Populus* homologs to *TERMINAL FLOWER1* genes: roles in onset of flowering and shoot phenology. PhD dissertation, Oregon State University, Corvallis, OR
Mølmann JA, Asante DKA, Jensen JB, Krane MN, Ernstsen A, Junttila O, Olsen JE (2005) Low night temperature and inhibition of gibberellin biosynthesis override phytochrome action and induce bud set and cold acclimation, but not dormancy in *PHYA* overexpressors and wild-type of hydrid aspen. Plant Cell Environ 28:1579–1588
Moss GI (1969) Influence of temperature and photoperiod on flower induction and inflorescence development in sweet orange – (*Citrus Sinensis* L Osbeck). J Hort Sci Biotech 44(4):311–320
Mouhu K, Hytönen T, Folta K, Rantanen M, Paulin L, Auvinen P, Elomaa P (2009) Identification of flowering genes in strawberry, a perennial SD plant. BMC Plant Biol 9:122. doi:10.1186/1471-2229-9-122
Nishikawa F, Endo T, Shimada T, Fujii H, Shimizu T, Omura M, Ikoma Y (2007) Increased CiFT abundance in the stem correlates with floral induction by low temperature in Satsuma mandarin (*Citrus unshiu* Marc.). J Exp Bot 58(14):3915–3927
Noh B, Noh YS (2006) Chromatin-mediated regulation of flowering time in Arabidopsis. Physiol Plant 126(4):484–493
Nordstrom A, Tarkowski P, Tarkowska D, Norbaek R, Astot C, Dolezal K, Sandberg G (2004) Auxin regulation of cytokinin biosynthesis in *Arabidopsis thaliana*: a factor of potential importance for auxin-cytokinin-regulated development. Proc Natl Acad Sci 101:8039–8044
Nozue K, Maloof J (2006) Diurnal regulation of plant growth. Plant Cell Environ 29:396–408
Olsen JE (2006) Mechanisms of dormancy regulation. Acta Hortic 727:157–165
Olsen JE, Junttila O, Nilsen J, Eriksson ME, Martinussen I, Olsson O, Sandberg G, Moritz T (1997) Ectopic expression of oat phytochrome A in hybrid aspen changes critical daylength for growth and prevents cold acclimatization. Plant J 12(6):1339–1350
Or E (2009) Grape bud dormancy release-the molecular aspect. In: Roubelakis-Angelakis KA (ed) Grapevine molecular physiology and biotechnology, 2nd edn. Springer Science and Business Media B.V. doi 10.1007/978-90-481-2305-6_1.
Or E, Vilozny I, Eyal Y, Ogrodovitch A (2000) The transduction of the signal for grape bud dormancy breaking induced by hydrogen cyanamide may involve the SNF-like protein kinase GDBRPK. Plant Mol Biol 43(4):483–494
Penfield S (2008) Temperature perception and signal transduction in plants. New Phytol 179: 615–628
Pineiro M, Gomez-Mena C, Schaffer R, Martinez-Zapater JM, Coupland G (2003) EARLY BOLTING IN SHORT DAYS is related to chromatin remodeling factors and regulates flowering in Arabidopsis by repressing FT. Plant Cell 15(7):1552–1562
Pratt C (1971) Reproductive anatomy in cultivated grapes – A Review. Am J Enol Vitic 22(2): 92–109
Ramos A, Perez-Solis E, Ibanez C, Casado R, Collada C, Gomez L, Aragoncillo C, Allona I (2005) Winter disruption of the circadian clock in chestnut. Proc Natl Acad Sci USA 102(19):7037–7042
Rawat R, Schwartz J, Jones MA, Sairanen I, Cheng Y, Andersson CR, Zhao Y, Ljung K, Harmer SL (2009) REVEILLE1, a Myb-like transcription factor, integrates the circadian clock and auxin pathways. Proc Natl Acad Sci USA 106:16883–16888
Robertson FC, Skeffington AW, Gardner MJ, Webb AAR (2008) Interactions between circadian and hormonal signaling in plants. Plant Mol Biol 69:419–427

Rodrigo MJ, Alquezar B (2006) Cloning and characterization of two 9-cis-epoxycarotenoid dioxygenase genes, differentially regulated during fruit maturation and under stress conditions, from orange (*Citrus sinensis* L. Osbeck). J Exp Bot 57(3):633–643

Rodríguez-Falcón M, Bou J, Prat S (2006) Seasonal control of tuberization in potato: conserved elements with the flowering response. Annu Rev Plant Biol 57:151–180

Rohde A, Bhalerao RP (2007) Plant dormancy in the perennial context. Trends Plant Sci 12: 217–223

Ruonala R, Rinne PLH, Kangasjarvi J, van der Schoot C (2008) *CENL1* expression in the rib meristem affects stem elongation and the transition to dormancy in *Populus*. Plant Cell 20:59–74

Ruttink T, Arend M, Morreel K, Storme V, Rombauts S, Fromm J, Bhalerao RP, Boerjan W, Rohde A (2007) A molecular timetable for apical bud formation and dormancy induction in poplar. Plant Cell 19(8):2370–2390

Salathia N, Davis SJ, Lynn JR, Michaels SD, Amasino RM, Millar AJ (2006) FLOWERING LOCUS C-dependent and -independent regulation of the circadian clock by the autonomous and vernalization pathways. BMC Plant Biol 6:10. doi:10.1186/1471-2229-6-10

Samach A, Wigge PA (2005) Ambient temperature perception in plants. J Plant Biol 8:483–486

Sawa M, Nusinow DA, Kay SA, Imaizumi T (2007) FKF1 and GIGANTEA complex formation is required for day-length measurement in *Arabidopsis*. Science 318:261–265

Schnabel BJ, Wample RL (1987) Dormancy and cold hardiness in Vitis-Vinifera L Cv White Riesling as influenced by photoperiod and temperature. Am J Enol Vitic 38(4):265–272

Schrader J, Moyle R, Bhalerao R, Hertzberg M, Lundeberg J, Nilsson P, Bhalerao RP (2004) Cambial meristem dormancy in trees involves extensive remodeling of the transcriptome. Plant J 40:173–187

Searle I, He Y, Turck F, Vincent C, Fornara F, Kröber S, Amasino RA, Coupland G (2006) The transcription factor FLC confers a flowering response to vernalization by repressing meristem competence and systemic signaling in Arabidopsis. Gene Dev 20:898–912

Smith H (2000) Phytochromes and light signal perception by plants – an emerging synthesis. Nature 407:585–591

Smithberg MH, Weiser CJ (1968) Patterns of variation among climatic races of red-osier dogwood. Ecology 49(3):495–505

Sonnewald U (2001) Control of potato tuber sprouting. Trends Plant Sci 6(8):333–335

Sung S, Amasino RM (2004) Vernalization and epigenetics: how plants remember winter. Curr Opin Plant Biol 7:102–108

Sung S, He Y, Eshoo TW, Tamada Y, Johnson L, Nakahigashi K, Goto K, Jacobsen SE, Amasino RM (2006) Epigenetic maintenance of the vernalized state in *Arabidopsis thaliana* requires LIKE HETEROCHROMATIN PROTEIN 1. Nat Genet 38(6):706–710

Suttle JC (1998) Involvement of ethylene in potato microtuber dormancy. Plant Physiol 118(3): 843–848

Suttle JC (2004) Physiological regulation of potato tuber dormancy. Am J Potato Res 81:253–262

Suttle JC (2008) Effects of synthetic phenylurea and nitroguanidine cytokinins on dormancy break and sprout growth in russet burbank minitubers. Am J Potato Res 85:121–128

Svendsen E, Wilen R, Stevenson R, Liu R, Tanino KK (2007) A molecular marker associated with low-temperature induction of dormancy in red osier dogwood (*Cornus sericea*). Tree Physiol 27(3):385–397

Takada S, Goto K (2003) TERMINAL FLOWER2, an Arabidopsis homolog of HETEROCHROMATIN PROTEIN1, counteracts the activation of FLOWERING LOCUS T by CONSTANS in the vascular tissues of leaves to regulate flowering time. Plant Cell 15(12):2856–2865

Thomashow MF (2001) So what's new in the field of plant cold acclimation? Lots! Plant Physiol 125:89–93

Volaire F, Norton M (2006) Summer dormancy in perennial temperate grasses. Ann Bot 98: 927–933

Wake CMF, Fennell A (2000) Morphological, physiological and dormancy responses of three Vitis genotypes to short photoperiod. Physiol Plant 109:203–210

Welling A, Kaikuranta P, Rinne P (1997) Photoperiodic induction of dormancy and freezing tolerance in *Betula pubescens*. Involvement of ABA and dehydrins. Physiol Plant 100:119–125

Yamane H, Kashiwa Y, Ooka T, Tao R, Yonemori K (2008) Suppression subtractive hybridization and differential screening reveals endodormancy-associated expression of an SVP/AGL24-type MADS-box gene in lateral vegetative buds of Japanese apricot. J Am Soc Hortic Sci 133(5):708–716

Yuceer C, Land SB Jr, Kubiske ME, Harkess RL (2003) Shoot morphogenesis associated with flowering in *Populus deltoides* (Salicaceae). Am J Bot 90:196–206

Chapter 6
LEA Proteins: Versatility of Form and Function

Alan Tunnacliffe, Dirk K. Hincha, Olivier Leprince, and David Macherel

Abstract LEA proteins represent one of the functional elements thought to be important in maintaining viability of organisms and biological structures in the ametabolic dry state. They are found in plant tissues, seeds and pollen, anhydrobiotic invertebrates and some desiccation-tolerant micro-organisms. Recent findings suggest that LEA proteins play various, possibly multiple, roles in the drying cell: they are implicated in the homeostasis of proteins and nucleic acids, in stabilizing cell membranes, in redox balancing and in the formation and stability of the glassy state. This striking versatility might derive from the largely unstructured nature of LEA proteins in solution and the associated structural and functional plasticity.

6.1 Introduction

6.1.1 Anhydrobiosis and Desiccation Tolerance

Certain plants, animals and micro-organisms are able to dry out completely and yet remain viable, a phenomenon known as anhydrobiosis ("life without water") or desiccation tolerance (Fig. 6.1). Examples of anhydrobiotic organisms include microorganisms such as bakers' yeast; tardigrades, bdelloid rotifers and some nematodes among the invertebrates; and the resurrection plants – the latter so-called because they are apparently dead in the dry state, where water contents can

A. Tunnacliffe (✉)
Department of Chemical Engineering and Biotechnology, University of Cambridge, Cambridge, UK
e-mail: at10004@biotech.cam.ac.uk

D.K. Hincha
Max-Planck-Institute of Molecular Plant Physiology, Potsdam, Germany

O. Leprince and D. Macherel
UMR 1191 Physiologie Moléculaire des Semences, Université d'Angers/Agrocampus-Ouest/INRA, 49045 Angers, France

E. Lubzens et al. (eds.), *Dormancy and Resistance in Harsh Environments*, Topics in Current Genetics 21,
DOI 10.1007/978-3-642-12422-8_6, © Springer-Verlag Berlin Heidelberg 2010

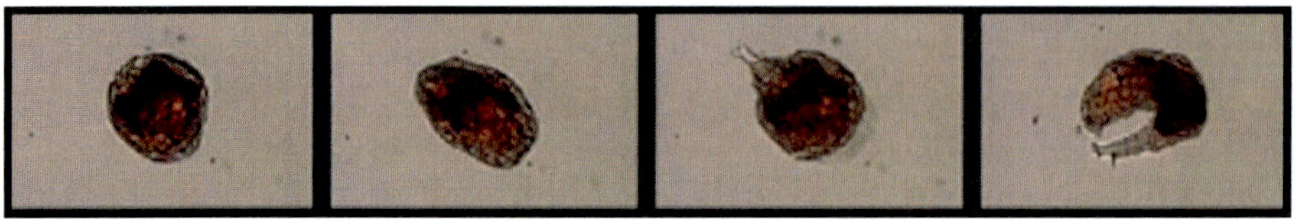

Fig. 6.1 The bdelloid rotifer, *Philodina roseola*, recovering from desiccation after immersion in water. Images at 5–10 min intervals. The foot appears in the third frame, and the animal uncurls fully in the final frame with a head-to-toe length of ~250 μm

be <10% of normal, but resume life processes on rehydration. Indeed, most plant seeds are desiccation tolerant, as are other propagules such as pollen (Hoekstra et al. 2001; Clegg 2001; Alpert 2006).

Given that water is essential for life, it is remarkable that biological mechanisms have evolved allowing essentially complete desiccation without loss of viability: although dried anhydrobiotic organisms are in a state of suspended animation, they are fully metabolically active after rehydration. Even more striking, in the dry state, desiccation tolerant organisms are highly resistant to environmental stress, surviving extremes of temperature and pressure, for example. Moreover, these organisms are able to remain in the anhydrobiotic state for long periods: invertebrates have been revivified successfully after decades of inactivity (Guidetti and Jönsson 2002), and lotus and date palm seeds have germinated after 1,300 and 2,000 years, respectively (Shen-Miller et al. 1995; Sallon et al. 2008). This contrasts dramatically with desiccation-sensitive cell types and organisms which suffer damage to all main cell components on drying: proteins lose their correct conformation and can form toxic aggregates; membranes undergo phase transitions causing leakiness and fusion; RNA is likely to experience structural rearrangements, and RNA, DNA and chromatin stability will be compromised (Leprince et al. 1995; Kanias and Acker 2006).

6.1.2 Mechanisms of Desiccation Tolerance

Survival of desiccation and the dry state is one of the most intriguing phenomena in nature and is becoming increasingly well understood. The non-reducing disaccharides, trehalose (animals, fungi) and sucrose (plants), which accumulate in diverse anhydrobiotic organisms, are thought to be involved and have been shown to function as both water replacement molecules and vitrification agents (Crowe et al. 1998; Cacela and Hincha 2006). Intracellular glasses, which confer long-term stability in the dry state, are more complex systems than those of simple sugar glasses, however, and have quite different properties (Buitink and Leprince 2004). Nevertheless, non-reducing disaccharides are not an absolute requirement for anhydrobiosis, being absent from bdelloid rotifers (Lapinski and Tunnacliffe 2003; Caprioli et al. 2004; McGee 2006) and some tardigrades (Hengherr et al. 2008), and not essential in bakers' yeast (Ratnakumar and Tunnacliffe 2006). Other molecules are likely to contribute significantly to desiccation tolerance, therefore, and gene discovery programmes to determine key adaptations in desiccation tolerance are well advanced in a number of organisms, where the "anhydrobiotic gene set" is becoming more clearly defined.

Perhaps the main feature distinguishing naturally anhydrobiotic organisms is that they produce many examples of highly hydrophilic proteins (a subset of which have been called "hydrophilins"; Garay-Arroyo et al. 2000; Battaglia et al. 2008) in preparation for severe dehydration. This has been observed in resurrection plants (Ramanjulu and Bartels 2002), in seeds (e.g. Boudet et al. 2006), in anhydrobiotic nematodes (Solomon et al. 2000; Browne et al. 2002; Goyal et al. 2003; Tyson et al. 2007), in bdelloid rotifers (Pouchkina-Stantcheva et al. 2007), in brine shrimp cysts (Hand et al. 2006; Menze et al. 2009; see Chap. 10) and in micro-organisms (Garay-Arroyo et al. 2000; Battista et al. 2001; Dure 2001). The best characterized of these are the LEA (late embryogenesis abundant) proteins, originally described in plant seeds, but now known to occur in invertebrates and micro-organisms also (see several recent reviews: Rorat 2006; Tunnacliffe and Wise 2007; Battaglia et al. 2008; Shih et al. 2008). For many years it has been known that LEA proteins accumulate in maturing plant seeds as they acquire desiccation tolerance; their discovery in invertebrates suggests that similar mechanisms govern anhydrobiosis in both animals and plants.

6.1.3 Unstructured, Highly Hydrophilic Proteins

Several nomenclature systems for LEA proteins have been suggested since they were originally described in cotton seeds (Dure et al. 1981; Table 6.1). Most LEA proteins are highly hydrophilic and rather small, but there are exceptions in both properties. In the model plant *Arabidopsis thaliana*, nine different groups have recently been proposed based on amino acid sequence analysis (Hundertmark and Hincha 2008; Bies-Ethève et al. 2008). [See also the reviews cited above for a discussion of nomenclature.] The largest of these groups is group 3 (Pfam LEA_4), which is also the main group of LEA proteins found in non-plant organisms, suggesting that the other groups have arisen later in evolution: for example, group 1 proteins are prevalent in plants, but rare in metazoa (Sharon et al. 2009). Another group of LEA proteins that has been studied extensively in recent years is group 2; these LEA proteins are also frequently referred to as dehydrins (Pfam dehydrin) and might function as phosphorylated forms (Rohrig et al. 2006, 2008). LEA proteins seem to be distributed in several cell compartments in plants (see Table 4 in Tunnacliffe and Wise 2007) and probably also in animals (Menze et al. 2009). Of the many different LEA proteins identified in plants (51 in *Arabidopsis*; Hundertmark and Hincha 2008; Bies-Ethève et al. 2008), however, only a small minority has been functionally and structurally characterized to date.

Probably because of their highly hydrophilic nature, most LEA proteins and other hydrophilins are predicted to be largely unstructured in the hydrated state and

Table 6.1 PFAM families (http://pfam.sanger.ac.uk/ (April 09)), sequence number and species distribution

PFAM (seq. nr)	Historical names	Species distribution	A. t. nr	Main reported features
LEA_5 (81)	Group 1[a] D-19[b]	78 Plant 2 Metazoa 1 Bacteria	2	Hydrophilic 20-residue motif – Mainly IDPs – Stabilization of macromolecules
Dehydrin (481)	Group 2 D-11	114 Plant	10	Mainly IDPs – Several distinctive sequence motifs (K, Y, S segments) – Stabilization of macromolecules – Binding to membranes – Chaperone activity – Metal binding – Antioxidant – DNA/RNA binding
LEA_4 (176)	Group 3 Group 5 D-7 D-29	124 Plant 20 Metazoa 3 Fungi 28 Bacteria 1 Archaea	18	The most widespread group – Repeat motives – Mainly IDPs – Transition to amphipathic α-helix during desiccation – Stabilization of macromolecules and membranes – Ion binding
LEA_1 (106)	Group 4 D-113	22 Plant	3	Highly hydrophilic proteins – Predicted to be IDPs – Largely uncharacterized
LEA_2 (128)	Lea14 D-95	82 Plant 20 Bacteria	3	Predicted not to be IDPs – 3D structure from NMR available for At1g01470 – No functional information available
LEA_3 (69)	Lea5 D-73	27 Plant	4	Mostly predicted not to be IDPs – Structural information available for one Lea5 protein suggests unstructured, however[c]
SMP (103)	Group 6 D-34	102 Plant 1 Fungi	6	Least hydrophilic group of LEA proteins – Mostly predicted not to be IDPs – Largely uncharacterized

A. t. nr refers to the number of corresponding sequences in the genome of *Arabidopsis thaliana* (Hundertmark and Hincha 2008), *IDP* intrinsically disordered protein
[a]According to Dure et al. (1989) and Bray (1993)
[b]According to Dure (1993a)
[c] Haaning et al. 2008

are thus part of the large set of natively unfolded proteins which form a substantial proportion of proteomes: 11% of proteins in the SwissProt database and 6–17% of various proteomes are fully disordered; 35–51% of eukaryotic proteins have at least one long (>50 residues) unstructured region (citations in Uversky et al. 2000; Tompa 2002; Tompa et al. 2005). Most unfolded proteins become more structured on binding partner molecules, but such molecular partners have not been described for those involved in desiccation tolerance, so at least some might instead function in the unstructured state.

Alternatively, LEA proteins might adopt a defined structure under those conditions where they are supposed to function, i.e. during partial or complete dehydration brought about by freezing and/or drying. There are several examples in the literature now of both animal and plant LEA proteins that are unstructured in dilute solution but fold into mainly alpha-helical structures in the dry state (see Hundertmark and Hincha 2008 for references). While this could be of functional relevance for LEA proteins in anhydrobiotic organisms, the relevance during partial water loss caused by freezing or mild dehydration is doubtful. Many LEA proteins that are accumulated in the vegetative organs of higher plants under mild stress conditions must function in partially dehydrated systems if they are behaving as effective protectants under physiologically relevant conditions. For three such proteins, the cold induced *Arabidopsis* dehydrins COR47, LTI29 and LTI30, it has recently been shown that simulated osmotic stress or molecular crowding did not induce appreciable folding in vitro (Mouillon et al. 2008). The authors suggested that therefore dehydrins probably function as unstructured proteins. However, a combination of the effects of dehydration and the presence of as yet unidentified target molecules on LEA protein structure cannot be excluded on the basis of our present knowledge. In addition, the low temperatures associated with freezing stress could influence the structure of LEA proteins involved in plant freezing tolerance. An increase in alpha-helicity with decreasing temperature has, for example, been shown for type I fish anti-freeze proteins (Haymet et al. 1999, 2001).

Expression of LEA proteins has been associated with tolerance to water stress caused by desiccation, salt or low temperature in many studies (Cuming 1999; Tunnacliffe and Wise 2007). Furthermore, reduction of expression of LEA protein genes by RNA interference in nematodes (Gal et al. 2004), or by mutation in bacteria (Battista et al. 2001), has been shown to decrease desiccation tolerance. Therefore, it seems likely that hydrophilic proteins play an important role in anhydrobiosis and other forms of water stress.

6.2 LEA Protein Function

6.2.1 Protein Protection

Until recently, the precise molecular function of LEA proteins and other hydrophilins has been unclear, although they have been suggested to act as protein stabilizers, hydration buffers, membrane protectants, antioxidants, organic glass formers and ion sinks/chelators (Tunnacliffe and Wise 2007). Evidence is accumulating to support some of these roles, particularly that of protein stabilization, since LEA proteins and other hydrophilins can preserve enzyme activity in vitro after desiccation or freezing (Sanchez-Ballesta et al. 2004; Goyal et al. 2005; Grelet et al. 2005; Reyes et al., 2005, 2008; Haaning et al. 2008; Nakayama et al. 2008). One mechanism for the protection observed is the prevention of water stress-induced aggregation of sensitive proteins (Goyal et al. 2005; Chakrabortee et al. 2007; Kovacs et al. 2008; Nakayama et al. 2008).

Many proteins, including the enzymes citrate synthase (CS) and lactate dehydrogenase (LDH), form insoluble aggregates when dried or frozen, but aggregation is markedly reduced in the presence of LEA proteins from any of the three major groups (Goyal et al. 2005; Pouchkina-Stantcheva et al. 2007; Kovacs et al. 2008; Nakayama et al. 2008). Group 2 proteins, at least, also prevent protein aggregation on heat stress (Kovacs et al. 2008). This protein anti-aggregation activity can also extend to the protection of complex mixtures of proteins, such as the water-soluble proteomes of human and nematode cells (Chakrabortee et al. 2007). Probably due to their hydrophilic, unstructured nature, LEA proteins themselves are not susceptible to aggregation on desiccation, freezing or even boiling.

If LEA proteins have a physiological role in protecting a wide range of proteins against aggregation, this should be demonstrable in living cells. Accordingly, a group 3 LEA protein from the nematode *A. avenae*, when introduced into mammalian cells expressing aggregation-prone proteins containing long polyglutamine (polyQ) or polyalanine (polyA) sequences, is able to reduce aggregate formation in vivo (Fig. 6.2; Chakrabortee et al. 2007). The polyQ or polyA expansion proteins are associated with a number of human neurodegenerative diseases including Huntington's disease and oculopharyngeal muscular dystrophy. Such an experimental system makes it possible to circumvent the technical difficulties involved in establishing an aggregation assay in dried cells. It also demonstrates anti-aggregation activity in the hydrated state, suggesting that extreme water loss is not required before LEA proteins exhibit their protein protection function. In turn, this suggests that LEA proteins can carry out this function in a disordered configuration, unless association with "client" proteins induces some folding.

The anti-aggregation activity of LEA proteins is reminiscent of that of molecular chaperones, whose function in the cell is to facilitate the correct folding and assembly of proteins and protein complexes. The "classical" molecular chaperones, such as the Hsp70 family and the Hsp60 chaperonin complexes, are commonly perceived as "heat shock proteins" (Hsps), being upregulated by heat stress, although they also play a vital role in folding nascent proteins under non-stress conditions (Ellis and Hartl 2003). Their function as stress combatants is to potentiate refolding of polypeptide structures which have become partially denatured. Besides such "folding" chaperones, both the eukaryotic and prokaryotic cytoplasm also contain "holding" and "disaggregating" chaperones. Holding chaperones include small Hsps, some of which form large multimeric "windowed" complexes capable of passively stabilizing protein species in a partially unfolded state, preventing aggregation until stress has abated and refolding, by Hsp70 and Hsp60 teams or spontaneously, becomes possible (e.g. Haslbeck et al. 1999). Disaggregating chaperones include the Hsp100 family which are thought to disentangle protein aggregates, which then serve as clients for folding chaperones (Ben-Zvi and Goloubinoff 2001).

LEA proteins show properties similar to those of the holding chaperones, including their ability to function in the absence of ATP hydrolysis, unlike the folding chaperones which generally require ATP. However, LEA proteins have been distinguished from holding chaperones due both to their lack of structure and their relatively poor ability to prevent heat-induced protein aggregation (Goyal et al. 2005; see, however, Kovacs et al. 2008 for a recent example of plant LEA proteins that function during heat stress). Another point of difference is that molecular

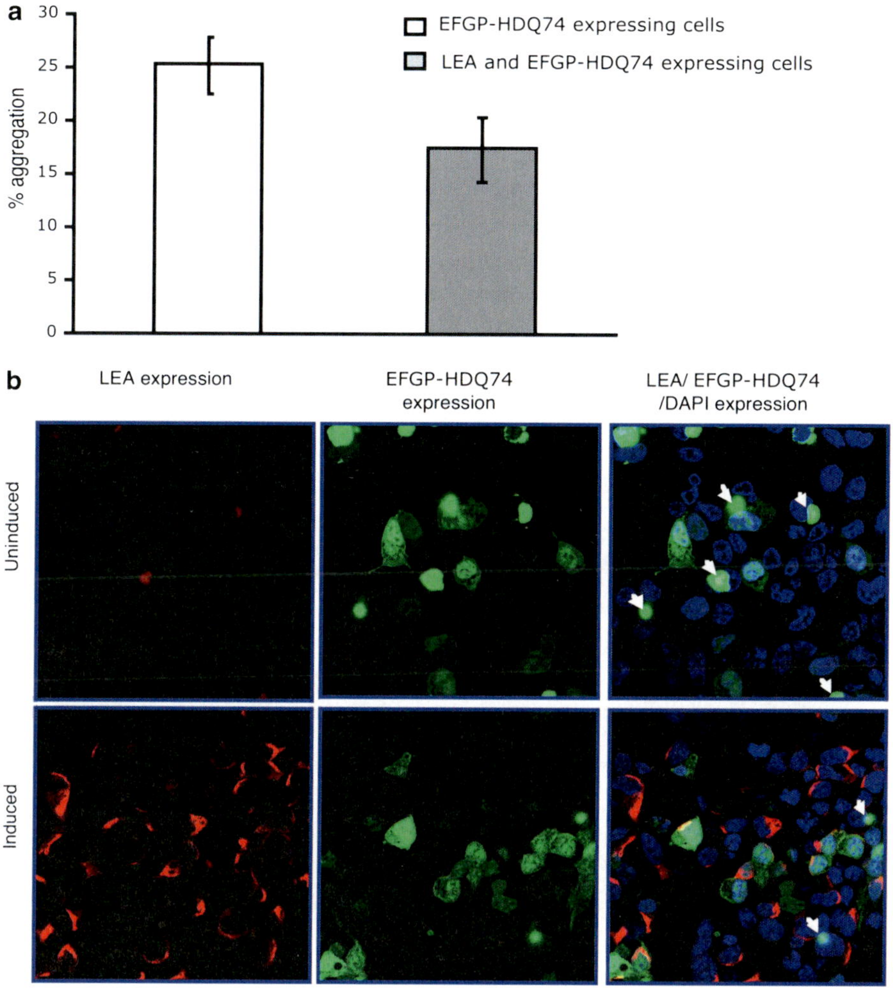

Fig. 6.2 Nematode LEA protein AavLEA1 reduces aggregation of EGFP-HDQ74, a GFP-tagged polyQ expansion protein derived from huntingtin, in human cells stably transfected with an inducible form of the LEA protein gene. (**a**) Percentage of cells showing aggregation of EFGP-HDQ74 where expression of LEA protein is either induced or not: significantly different at $p < 0.005$ by odds ratio analysis; (**b**) confocal images of the same human cells showing LEA protein (immunostained *red*), EFGP-HDQ74 (*green*) and nuclear DNA (DAPI-stained *blue*). LEA protein expression was induced 24 h after transient transfection of EFGP-HDQ74; only cells clearly expressing both EFGP-HDQ74 and LEA protein were counted. EFGP-HDQ74 aggregates are indicated by *arrows* in the *right-hand* composite panels

chaperones form transient complexes with their client proteins, often through interaction of hydrophobic surfaces; such specific interactions are thought to be unlikely with unstructured LEA proteins, particularly since they are highly hydrophilic in nature, but no unequivocal experimental evidence for or against such interactions has been published to date. Finally, LEA protein genes are not usually upregulated by heat stress (Browne et al. 2004; Hundertmark and Hincha 2008), unlike many of the classical molecular chaperones.

The reality of LEA protein activity might well be simpler than that of molecular chaperones: as entropic chains, LEA proteins will exert an excluded volume effect which, in the increasingly crowded environment of the dehydrating cytoplasm, could serve to decrease interaction between partially denatured polypeptides with the potential to aggregate. This has been called "molecular shield" activity (Wise and Tunnacliffe 2004; Goyal et al. 2005). An analogous effect is well known to physical chemists interested in polymer stabilization of colloidal dispersions (i.e. "depletion stabilization"; Napper 1983). Molecular shield function is also similar to that of the entropic bristles of MAP2, tau and neurofilament side arms (Mukhopadhyay et al. 2004), which act as spacers of cytoskeletal filaments, except that shield proteins are not necessarily tethered to a surface. Association of molecular shields with the surface of other proteins is also a possibility, however, and would lead to a steric or electrosteric stabilization effect, again familiar to colloid scientists. Finally, shield proteins might have a broader space-filling role and help to prevent collapse of the cell as its water is lost.

Interestingly, such ideas hark back to early work on molecular chaperone function, before current models became prevalent, when it was proposed that chaperones reduced the rate of interaction of aggregation-prone protein species (Minton et al. 1982). Indeed, the distinction between molecular shield and molecular chaperone is becoming somewhat blurred, particularly after recent contributions from Peter Tompa's laboratory. Thus, like molecular chaperones, two group 2 LEA proteins (dehydrins) are able to reduce heat-induced aggregation of CS and luciferase (Kovacs et al. 2008). In addition, it seems that some protein and RNA chaperones require unstructured domains for optimum function (Tompa and Csermely 2004). It is argued by these authors that at least a component of chaperone function could be driven by entropy transfer mechanisms: unstructured chaperone regions might interact with inappropriately folded domains of client proteins leading to increased folding of the chaperone together with partial unfolding of the client, allowing the client to overcome the energy barrier to correct folding. LEA proteins could also behave in this way, and this might be their predominant mode of action, since they lack defined secondary structure. However, an entropic space filling function is also an important component of the Tompa/Csermely model, as it is in the molecular shield model. An attractive hypothesis which is supported by the current experimental evidence is that protein anti-aggregation activity comprises several sub-activities – neutralization of exposed ("sticky") hydrophobic surfaces by specific interaction, potential refolding of client proteins driven by entropy transfer or ATP hydrolysis (or both), electrosteric or depletion stabilization of crowded colloidal suspensions, and volume exclusion – which are manifested to different degrees in different protectant proteins. The latter two functions might be particularly important under conditions of water stress where water activity in the cell is markedly reduced, with concomitant danger of collapse of cell structure. In the context of desiccation or freeze tolerance, these activities are expected to serve as damage avoidance or limitation mechanisms and it is anticipated that conventional folding and disaggregating chaperones, together with the various protease and autophagy systems, will be involved in repairing or removing damaged proteins once the water stress has receded.

6.2.2 Membrane Protection

Membrane protection is essential to preserve cellular and organellar integrity during desiccation in anhydrobiotes. Because of their hydrophilic character, LEA proteins are highly soluble polypeptides, which are therefore not expected a priori to interact with cellular membranes. However, the molecular landscape within cells is dramatically affected by dehydration, and rules which prevail under hydrated conditions are challenged in the drying environment. This is nicely illustrated by the example of protective sugars, which in spite of their hydrophilic character contribute to membrane protection by preserving the fluidity of membranes in the dry state (Hoekstra et al. 2001). The sugar molecules provide a network of H-bonds with the polar head-groups of phospholipids that maintain spacing and thus molecular order of the membrane, preventing transition to a deleterious gel phase. In spite of being macromolecules with associated structural constraints, it is conceivable that some LEA proteins might play a similar role, or perhaps contribute with sugars to H-bond networking. Although this remains to be demonstrated, it has previously been suggested that an LEA protein from pollen contributes to a tight H-bonding network in the dehydrated cytoplasm (Wolkers et al. 2001).

In a dehydrating cell, LEA proteins may also be driven to interact more deeply with membranes and possibly provide protection in the dry state. Since LEA proteins are generally devoid of sequences with transmembrane segment properties, it is unlikely a typical LEA polypeptide could integrate into a membrane as an intrinsic protein. LEA proteins were recognized early on for their propensity to form amphipathic alpha-helices (Dure 1993b). Therefore, it cannot be excluded that some LEA proteins form inter- or intra-molecular helix bundles within membranes. Nevertheless, such an insertion, which would have to occur at low hydration, is a challenging idea since such a process is highly regulated during normal membrane protein biogenesis (von Heijne 2006).

Among LEA proteins, plant-specific dehydrins have been localized in the vicinity of membranes (Danyluk et al. 1998; Egerton-Warburton et al. 1997), and it has been suggested that the K-segment, a 15-mer Lys-rich sequence which is a signature of dehydrins, might be involved in membrane binding (Close 1996). The K-segment is predicted to form an amphipathic class A helix, a motif which is involved in apolipoprotein binding to low density lipoprotein vesicles (Segrest et al. 1994). Indeed, DHN1, a maize dehydrin, was shown to bind liposomes containing anionic phospholipids, and the binding stimulated an increase in helicity, which led to the hypothesis of a membrane-stabilizing function (Koag et al. 2003). Since all dehydrins share one or more K segments, this interesting hypothesis could have wide significance. Two other *Arabidopsis* dehydrins (ERD10 and ERD14) were also shown to bind anionic phospholipid vesicles, without inducing modifications in the fluidity of membranes in the fully hydrated state, suggesting binding through peripheral electrostatic interactions (Kovacs et al. 2008).

Functional studies of a group 3 LEA protein (LEAM) found in pea seed mitochondria revealed another mechanism by which an LEA protein could interact with a membrane and afford protection in the dry state (Tolleter et al. 2007). LEAM is an intrinsically unstructured polypeptide located in the matrix space of mitochondria but is able to fold into amphipathic alpha-helices during drying. Protein structure modelling revealed a peculiar arrangement of the residues of the class A helix motif. Thus, at low hydration, the LEAM is proposed to insert into the inner leaflet of the mitochondrial membrane, parallel to the plane, exposing the hydrophobic face of the helix to the core of the membrane and being stabilized by electrostatic interactions between the charged residues of the protein and of the polar moiety of phospholipids (Fig. 6.3).

The occurrence of interactions between LEAM and phospholipids in the dry state was confirmed by differential scanning calorimetry, and the protective effect of LEAM was demonstrated using a liposome desiccation assay (Tolleter et al. 2007). With such a fully reversible membrane-association mechanism, LEAM could provide protection to the membrane in the dry state, without hampering its primary energy-transducing function in the hydrated state. Since class A motifs are prominent in several LEA proteins from different origins, such a membrane protection role might be common in exemplars other than LEAM. For example, interactions of invertebrate LEA proteins with phospholipid membranes in the dry state were also demonstrated using Fourier-transform infrared (FTIR) spectroscopy

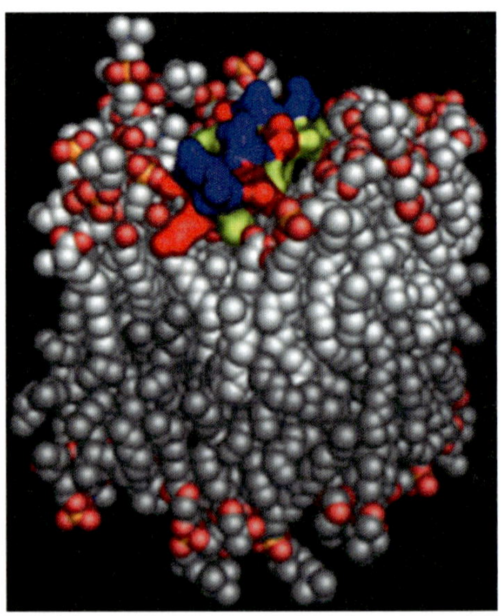

Fig. 6.3 Model of a group 3 LEA protein peptide embedded in a phospholipid membrane. An alpha-helical 29-residue peptide (DAKERTKEAANRAA-ENADSAGVKSRDYAY) from pea seed mitochondrial protein LEAM embedded in a bilayer of DOPC (di-oleoyl phosphatidylcholine). The negatively charged Asp and Glu residues form a crest (*blue*) on the top of the helix, while positively charged residues Arg and Lys (*red*) stabilize the structure by interacting with the phosphate in the head-group of DOPC

(Pouchkina-Stantcheva et al. 2007). In addition, a cold-responsive chloroplast LEA-like protein (COR15am) from *Arabidopsis* was shown to increase freezing tolerance by preventing the appearance of interbilayer hexagonal II phase at low temperature (Artus et al. 1996; Steponkus et al. 1998). This effect was likely mediated by modification of the intrinsic curvature of the chloroplast inner envelope membrane by the amphipathic LEA-like protein.

In conclusion, although their biophysical features, namely their lack of structure and hydrophilicity, suggest LEA proteins should act in the soluble compartments of cells, membranes might be very important targets for these unusual proteins. It is anticipated that on folding during desiccation, or perhaps freezing, specific LEA proteins could indeed be targeted to membranes and contributed to their protection. Maintaining the integrity of cellular membranes during acute stress is certainly a major issue, for which LEA proteins might be essential actors.

6.2.3 The Glassy State

In anhydrobiotes, drying induces a supersaturation of cytosolic components and reduces molecular mobility by over five orders of magnitude. Eventually, at around 10% moisture (on a dry weight basis), the cytoplasm vitrifies and enters into the so-called glassy state (references in Buitink and Leprince 2004). Although glass formation is not a mechanism that initially confers anhydrobiosis, the formation of intracellular glasses is indispensable for survival of the dry state, at least in plant propagules. There is now mounting evidence that LEA proteins which accumulate to high levels (typically 2–4% of the water soluble proteome; Roberts et al. 1993) increase the density of the sugar glasses. Table 6.2 shows data for three distinct LEA proteins: D7, a group 3 (Pfam LEA_4) protein from *Typha latifolia* pollen (Wolkers et al. 2001); PM16, a group 4 (Pfam LEA_1) protein from *Glycine max* seeds (Shih et al. 2004); and PM25, a group 5 (Pfam SMP) protein found in *Medicago truncatula* seeds (Boudet et al. 2006). The presence of LEA proteins increases the glass transition temperature (Tg) of a glassy matrix made of sucrose and strengthens the H-bonding network, assessed by the wavenumber–temperature

Table 6.2 Glass transition temperature (Tg) and Wavenumber–Temperature Coefficient (WTC) of glassy matrices formed with sucrose and LEA proteins D7, PM16 or PM25 (mass ratio 1:1)

Glass characteristics	Sucrose			
	Alone	+ D7	+ PM25	+ PM16
Tg (°C)	65.5	79.0	80.2	100.8
WTC ([cm^{-1}]/°C)	0.235	0.135	0.094	0.07

The sugar/protein glasses were formed by rapidly drying 10 μl droplets under an airflow of 3% RH. Values were determined by Fourier-transform infrared (FTIR) spectroscopy from the relationship between the wavenumber of the OH stretching vibration and the heating temperature. This vibration reveals information about the strength and length of hydrogen bonding within the glass. Data for D7 were obtained from Wolkers et al. (2001) and for PM16 from Shih et al. (2004)

coefficient (WTC) determined in FTIR experiments (Table 6.2), thereby increasing the density of the sugar glass. Similar observations were made with a range of plant seeds and pollens: biological glasses in those anhydrobiotes also show a denser H-bonding network than sugar glasses (Buitink and Leprince 2004). The advantage of a denser glass could be to improve stability in the dry state and to increase capacity to preserve macromolecules by direct interactions between molecules. Proteins themselves also vitrify during desiccation; thus, PM25 was found to form a glass at ambient temperature upon drying (Tg at 3% relative humidity (RH) is around 60°C; Boudet, Hoekstra, Leprince, unpublished data). Thus, the presence of LEA proteins might aid the formation of an amorphous solid-like matrix for those anhydrobiotes that do not accumulate non-reducing sugars. More research is needed to confirm the importance of LEA proteins in biological glasses. For example, the question as to whether LEA proteins have special properties with regard to glass formation, and whether different LEA proteins have different properties, as implied by the large difference in WTC found between the three LEA proteins (Table 6.2), remains to be answered.

6.3 The Versatility of LEA Proteins

As more LEA proteins are studied in detail, a number of unusual examples are coming to light. Thus, two group 3 (Pfam LEA_4) LEA proteins were described in the bdelloid rotifer *Adineta ricciae* which were rather less hydrophilic than most, having a grand average hydropathy (GRAVY; Consensus scale, Eisenberg 1984) score of −0.46 (the average score for group 3 LEA proteins is −0.97; Wise 2003), which is comparable to that of many "normal" globular proteins such as serum albumin (−0.43). The first of these proteins, ArLEA1A, behaves like other group 3 LEA proteins: it is largely unstructured in solution, but gains structure on drying; and it reduces aggregation of CS in in vitro desiccation assays. In marked contrast, however, the second example, ArLEA1B, despite having a very similar sequence to ArLEA1A, is structured in solution, being largely folded as alpha-helix, and has no anti-aggregation activity; indeed, it itself aggregates on drying. However, ArLEA1B shows a higher propensity to associate with phospholipid bilayers than ArLEA1A (Pouchkina-Stantcheva et al. 2007). Therefore, the two bdelloid LEA proteins display complementary characteristics of both structure and function. Closely related pairs of LEA proteins are, in fact, a frequent occurrence. For *Arabidopsis* it has been shown that 33% of all genes encoding LEA proteins form tandem repeats in the genome, while 43% are part of homeologous pairs derived from ancient genome duplications (Hundertmark and Hincha 2008). Many of these gene pairs show diversification in their expression patterns, but studies investigating their functional diversification have so far not been published.

While some very similar LEA proteins show different properties, a recurring theme in the field is that of a single LEA protein with more than one function. For example, the chloroplast LEA-like protein, COR15am, protects both membranes and proteins (Lin and Thomashow 1992; Steponkus et al. 1998; Nakayama et al. 2008); the mitochondrial group 3 protein, LEAM, also exhibits the same two functions (Grelet et al. 2005; Tolleter et al. 2007); and a group 2 LEA protein from *Citrus* shows both ion binding (Hara et al. 2005) and antioxidant (Hara et al. 2004) properties, and most recently nucleic acid binding activity (Hara et al. 2009). This versatility might be a general feature of LEA proteins. Performance of more than one function, known as "moonlighting", is not uncommon among proteins, and it has been argued that moonlighting is more likely to evolve in unfolded, rather than in conventional, folded proteins (Tompa et al. 2005).

If the functional properties of LEA proteins relate to their unstructured, hydrophilic nature, we might expect proteins with related characteristics to perform similarly. Indeed, examples are already known in the literature, such as those formally characterized as hydrophilins (Reyes et al. 2005). Perhaps the best characterized are the abscisic acid stress ripening (ASR) proteins: these are small plant-specific proteins with a high degree of hydrophilicity and low degree of secondary structure whose expression is induced by water stress (Carrari et al. 2004). A chaperone-like activity reduces freeze- or heat-inactivation of several enzymes, and aggregation due to heat stress is alleviated (Konrad and Bar-Zvi 2008). Intriguingly, both desiccation and zinc ion binding can lead to increased structural complexity, and this is associated with DNA binding (Kalifa et al. 2004; Goldgur et al. 2007). Furthermore, overexpression studies in transgenic plants suggest a role for ASR proteins in gene regulation (Frankel et al. 2007). Although ASR proteins are reported only in plants, animals seem to have proteins which share some properties with both LEA proteins and ASR proteins. An example is the nematode protein, anhydrin, whose gene is induced by desiccation and osmotic stress, like LEA protein genes (Browne et al. 2004). Anhydrin is small (10 kDa), highly basic, with a pI of 10.6, and highly hydrophilic, with a GRAVY score of -1.56. It shows a potent anti-aggregation activity in in vitro desiccation experiments with CS (McGee 2006). When expressed in mammalian cells, it accumulates primarily in the nucleus and can reduce formation of aggregates of a polyA protein with a nuclear localisation (Chakrabortee et al. unpublished).

The new data from Hara et al. (2009), showing that a *Citrus* dehydrin binds to both DNA and RNA in vitro in a sequence non-specific manner, lends support to the idea of a role for LEA proteins in stabilizing nucleic acids and points out the fact that it is unclear to what extent nucleic acids are specifically protected in desiccation tolerant organisms. Cell DNA certainly becomes damaged after prolonged storage in the dry state, probably as a result of exposure to reactive oxygen species (Cheah and Osborne 1978; Potts 1994; Mattimore and Battista 1996). For example, in lettuce seeds, loss of viability during storage is associated with an increased frequency of surviving cells showing chromosome damage upon imbibition (Rao et al. 1987). With regard to RNA, strand breaks are also to be expected, but the low water activity associated with desiccation is likely to affect secondary structure, as does temperature. Nevertheless, RNA chaperone function in the context of anhydrobiosis has not received attention previously. RNA chaperones are ubiquitous in nature (Cristofari and Darlix 2002) and frequently contain disordered regions, more so than any other protein class (Tompa and Csermely 2004). Moreover, they are also often small and highly basic, particularly where they interact with nucleic acids (e.g. Ivanyi-Nagy et al. 2008). Some LEA proteins and other hydrophilic proteins, including the ASR proteins and anhydrin, have similar characteristics and could play a role in protection of the genome and transcriptome.

It therefore seems likely that a range of proteins will be uncovered in both plants and non-plant organisms which complement the activities of LEA proteins and share many of their characteristics. It should be emphasized in this context that the term "late embryogenesis abundant", i.e. LEA, is of a historical rather than systematic nature. A more systematic, structure- and function-based nomenclature will require further progress in the characterization of those proteins now considered as LEA proteins and their counterparts among the hydrophilins.

References

Alpert P (2006) Constraints of tolerance: why are desiccation-tolerant organisms so small or rare? J Exp Biol 209:1575–1584

Artus NN, Uemura M, Steponkus PL, Gilmour SJ, Lin C, Thomashow MF (1996) Constitutive expression of the cold-regulated *Arabidopsis thaliana* COR15a gene affects both chloroplast and protoplast freezing tolerance. Proc Natl Acad Sci USA 93:13404–13409

Battaglia M, Olvera-Carrillo Y, Garciarrubio A, Campos F, Covarrubias AA (2008) The enigmatic LEA proteins and other hydrophilins. Plant Physiol 148:6–24

Battista JR, Park MJ, McLemore AE (2001) Inactivation of two homologues of proteins presumed to be involved in the desiccation tolerance of plants sensitizes *Deinococcus radiodurans* R1 to desiccation. Cryobiology 43:133–139

Ben-Zvi AP, Goloubinoff P (2001) Review: mechanisms of disaggregation and refolding of stable protein aggregates by molecular chaperones. J Struct Biol 135:84–93

Bies-Ethève N, Gaubier-Comella P, Debures A, Lasserre E, Jobet E, Raynal M, Cooke R, Delseny M (2008) Inventory, evolution and expression profiling diversity of the LEA (late embryogenesis abundant) protein gene family in *Arabidopsis thaliana*. Plant Mol Biol 67:107–124

Boudet J, Buitink J, Hoekstra FA, Rogniaux H, Larré C, Satour P, Leprince O (2006) Comparative analysis of the heat stable proteome of radicles of *Medicago truncatula* seeds during germination identifies late embryogenesis abundant proteins associated with desiccation tolerance. Plant Physiol 140:1418–1436

Bray EA (1993) Molecular responses to water deficit. Plant Physiol 103:1035–1040

Browne JA, Tunnacliffe A, Burnell AM (2002) Plant desiccation gene found in a nematode. Nature 416:38

Browne JA, Dolan KM, Tyson T, Goyal K, Tunnacliffe A, Burnell AM (2004) Dehydration-specific induction of hydrophilic protein genes in the anhydrobiotic nematode *Aphelenchus avenae*. Eukaryot Cell 3:966–975

Buitink J, Leprince O (2004) Glass formation in plant anhydrobiotes: survival in the dry state. Cryobiology 48:215–220

Cacela C, Hincha DK (2006) Low amounts of sucrose are sufficient to depress the phase transition temperature of dry phosphatidylcholine, but not for lyoprotection of liposomes. Biophys J 90:2831–2842

Caprioli M, Katholm AK, Melone G, Ramløv H, Ricci C, Santo N (2004) Trehalose in desiccated rotifers: a comparison between a bdelloid and a monogonont species. Comp Biochem Physiol 139:527–532

Carrari F, Fernie AR, Iusem ND (2004) Heard it through the grapevine? ABA and sugar cross-talk: the ASR story. Trends Plant Sci 9:57–59

Chakrabortee S, Boschetti C, Walton LJ, Sarkar S, Rubinsztein DC, Tunnacliffe A (2007) Hydrophilic protein associated with desiccation tolerance exhibits broad protein stabilization function. Proc Natl Acad Sci USA 104:18073–18078

Cheah KSE, Osborne DJ (1978) DNA lesions occur with loss of viability in embryos of ageing rye seed. Nature 272:593–599

Clegg JS (2001) Cryptobiosis – a peculiar state of biological organization. Comp Biochem Physiol B 128:613–624

Close TJ (1996) Dehydrins: emergence of a biochemical role of a family of plant dehydration proteins. Physiol Plant 97:795–803

Cristofari G, Darlix JL (2002) The ubiquitous nature of RNA chaperone proteins. Prog Nucleic Acid Res Mol Biol 72:223–268

Crowe JH, Carpenter JF, Crowe LM (1998) The role of vitrification in anhydrobiosis. Annu Rev Physiol 60:73–103

Cuming AC (1999) LEA proteins. In: Shewry PR, Casey R (eds) Seed proteins. Kluwer, Dordrecht, NL, pp 753–780

Danyluk J, Perron A, Houde M, Limin A, Fowler B, Benhamou N, Sarhan F (1998) Accumulation of an acidic dehydrin in the vicinity of the plasma membrane during cold acclimation of wheat. Plant Cell 10:623–638

Dure L III (1993a) Structural motifs in Lea proteins. In: Close TJ, Bray EA (eds) Plant responses to cellular dehydration during environmental stress. The American Society of Plant Physiologists, Rockville, MD, pp 91–103

Dure L III (1993b) A repeating 11-mer amino acid motif and plant desiccation. Plant J 3:363–369

Dure L III (2001) Occurrence of a repeating 11-mer amino acid sequence motif in diverse organisms. Protein Pept Lett 8:115–122

Dure L III, Greenway SC, Galau GA (1981) Developmental biochemistry of cottonseed embryogenesis and germination: changing messenger ribonucleic acid populations as shown by in vitro and in vivo protein synthesis. Biochemistry 20:4162–4168

Dure L III, Crouch M, Harada J, Ho T-HD, Mundy J, Quatrano R, Thomas T, Sung ZR (1989) Common amino acid sequence domains among the LEA proteins of higher plants. Plant Mol Biol 12:475–486

Egerton-Warburton LM, Balsamo RA, Close TJ (1997) Temporal accumulation and ultrastructural localization of dehydrins in *Zea mays*. Physiol Plant 101:545–555

Eisenberg D (1984) Three-dimensional structure of membrane and surface proteins. Ann Rev Biochem 53:595–623

Ellis RJ, Hartl F-U (2003) Protein folding and chaperones. In: Cooper DN (ed) Nature encyclopedia of the human genome. Nature, London, pp 806–810

Frankel N, Nunes-Nesi A, Balbo I, Mazuch J, Centeno D, Iusem ND, Fernie AR, Carrari F (2007) ci21A/Asr1 expression influences glucose accumulation in potato tubers. Plant Mol Biol 63:719–730

Gal TZ, Glazer I, Koltai H (2004) An LEA group 3 family member is involved in survival of *C. elegans* during exposure to stress. FEBS Lett 577:21–26

Garay-Arroyo A, Colmenero-Flores JM, Garciarrubio A, Covarrubias AA (2000) Highly hydrophilic proteins in prokaryotes and eukaryotes are common during conditions of water deficit. J Biol Chem 275:5668–5674

Goldgur Y, Rom S, Ghirlando R, Shkolnik D, Shadrin N, Konrad Z, Bar-Zvi D (2007) Desiccation and zinc binding induce transition of tomato abscisic acid stress ripening 1, a water stress- and salt stress-regulated plant-specific protein, from unfolded to folded state. Plant Physiol 143:617–628

Goyal K, Tisi L, Basran A, Browne J, Burnell A, Zurdo J, Tunnacliffe A (2003) Transition from natively unfolded to folded state induced by desiccation in an anhydrobiotic nematode protein. J Biol Chem 278:12977–12984

Goyal K, Walton LJ, Tunnacliffe A (2005) LEA proteins prevent protein aggregation due to water stress. Biochem J 388:151–157

Grelet J, Benamar A, Teyssier E, Avelange-Macherel M-H, Grunwald D, Macherel D (2005) Identification in pea seed mitochondria of a late-embryogenesis abundant protein able to protect enzymes from drying. Plant Physiol 137:157–167

Guidetti R, Jönsson KI (2002) Long-term anhydrobiotic survival in semi-terrestrial micrometazoans. J Zool 257:181–187

Haaning S, Radutoiu S, Hoffmann SV, Dittmer J, Giehm L, Otzen DE, Stougaard J (2008) An unusual intrinsically disordered protein from the model legume *Lotus japonicus* stabilizes proteins in vitro. J Biol Chem 283:31142–31152

Hand SC, Jones D, Menze MW, Witt TL (2006) Life without water: expression of plant LEA genes by an anhydrobiotic arthropod. J Exp Zool 305A:1–5

Hara M, Fujinaga M, Kuboi T (2004) Radical scavenging activity and oxidative modification of citrus dehydrin. Plant Physiol Biochem 42:657–662

Hara M, Fujinaga M, Kuboi T (2005) Metal binding by citrus dehydrin with histidine-rich domains. J Exp Bot 56:2695–2703

Hara M, Shinoda Y, Tanaka Y, Kuboi T (2009) DNA binding of citrus dehydrin promoted by zinc ion. Plant Cell Environ 32:532–541

Haslbeck M, Walke S, Stromer T, Ehrnsperger M, White HE, Chen S, Saibil HR, Buchner J (1999) Hsp26: a temperature-regulated chaperone. EMBO J 18:6744–6751

Haymet ADJ, Ward LG, Harding MM (1999) Winter flounder "antifreeze" proteins: synthesis and ice growth inhibition of analogues that probe the relative importance of hydrophobic and hydrogen-bonding interactions. J Am Chem Soc 121:941–948

Haymet ADJ, Ward LG, Harding MM (2001) Hydrophobic analogues of the winter flounder "antifreeze" protein. FEBS Lett 491:285–288

Hengherr S, Heyer AG, Koehler HR, Schill RO (2008) Trehalose and anhydrobiosis in tardigrades – evidence for divergence in responses to dehydration. FEBS J 275:281–288

Hoekstra FA, Golovina EA, Buitink J (2001) Mechanisms of plant desiccation tolerance. Trends Plant Sci 6:431–438

Hundertmark M, Hincha DK (2008) LEA (late embryogenesis abundant) proteins and their encoding genes in *Arabidopsis thaliana*. BMC Genomics 9:118

Ivanyi-Nagy R, Lavergne J-P, Gabus C, Ficheux D, Darlix J-L (2008) RNA chaperoning and intrinsic disorder in the core proteins of Flaviviridae. Nucleic Acids Res 36:712–725

Kalifa Y, Gilad A, Konrad Z, Zaccai M, Scolnik PA, Bar-Zvi D (2004) The water- and salt-stress regulated Asr1 gene encodes a zinc-dependent DNA-binding protein. Biochem J 381:373–378

Kanias T, Acker JP (2006) Mammalian cell desiccation: facing the challenges. Cell Preserv Technol 4:253–277

Koag MC, Fenton RD, Wilkens S, Close TJ (2003) The binding of maize DHN1 to lipid vesicles. Gain of structure and lipid specificity. Plant Physiol 131:309–316

Konrad Z, Bar-Zvi D (2008) Synergism between the chaperone-like activity of the stress regulated ASR1 protein and the osmolyte glycine-betaine. Planta 227:1213–1219

Kovacs D, Kalmar E, Torok Z, Tompa P (2008) Chaperone activity of ERD10 and ERD14, two disordered stress-related plant proteins. Plant Physiol 147:381–390

Lapinski J, Tunnacliffe A (2003) Anhydrobiosis without trehalose in bdelloid rotifers. FEBS Lett 553:387–390

Leprince O, Colson P, Houssier C, Deltour R (1995) Changes in chromatin structure associated with germination of maize and their relation with desiccation tolerance. Plant Cell Environ 18:619–629

Lin C, Thomashow MF (1992) A cold-regulated Arabidopsis gene encodes a polypeptide having potent cryoprotective activity. Biochem Biophys Res Commun 183:1103–1108

Mattimore V, Battista JR (1996) Radioresistance of *Deinococcus radiodurans*: functions necessary to survive ionizing radiation are also necessary to survive prolonged desiccation. J Bacteriol 178:633–637

McGee BM (2006) Hydrophilic proteins in the anhydrobiosis of bdelloid rotifers. PhD thesis, University of Cambridge

Menze MA, Boswell L, Toner M, Hand SC (2009) Occurrence of mitochondria-targeted late embryogenesis abundant (LEA) gene in animals increases organelle resistance to water stress. J Biol Chem 284:10714–10719

Minton KW, Karmin P, Hahn GM, Minton AP (1982) Nonspecific stabilization of stress-susceptible proteins by stress-resistant proteins: a model for the biological role of heat shock proteins. Proc Natl Acad Sci USA 79:7107–7111

Mouillon J-M, Eriksson SK, Harryson P (2008) Mimicking the plant cell interior under water stress by macromolecular crowding: disordered dehydrin proteins are highly resistant to structural collapse. Plant Physiol 148:1925–1937

Mukhopadhyay R, Kumar S, Hoh JH (2004) Molecular mechanisms for organizing the neuronal cytoskeleton. Bioessays 26:1017–1025

Nakayama K, Okawa K, Kakizaki T, Inaba T (2008) Evaluation of the protective activities of a late embryogenesis abundant (LEA) related protein, Cor15am, during various stresses in vitro. Biosci Biotechnol Biochem 72:1642–1645

Napper DH (1983) Polymeric stabilization of colloidal dispersions. Academic, London

Potts M (1994) Desiccation tolerance of prokaryotes. Microbiol Rev 58:755–805

Pouchkina-Stantcheva NN, McGee BM, Boschetti C, Tolleter D, Chakrabortee S, Popova AV, Meersman F, Macherel D, Hincha DK, Tunnacliffe A (2007) Functional divergence of former alleles in an ancient asexual invertebrate. Science 318:268–271

Ramanjulu S, Bartels D (2002) Drought- and desiccation-induced modulation of gene expression in plants. Plant Cell Environ 25:141–151

Rao NK, Roberts EH, Ellis RH (1987) Loss of viability in lettuce seeds and the accumulation of chromosome damage under different conditions of storage. Ann Bot 60:85–96

Ratnakumar S, Tunnacliffe A (2006) Intracellular trehalose is neither necessary nor sufficient for desiccation tolerance in yeast. FEMS Yeast Res 6:902–913

Reyes JL, Rodrigo M-J, Colmenero-Flores JM, Gil J-V, Garay-Arroyo A, Campos F, Salamini F, Bartels D, Covarrubias AA (2005) Hydrophilins from distant organisms can protect enzymatic activities from water limitation effects in vitro. Plant Cell Environ 28:709–718

Reyes JL, Campos F, Wei H, Arora R, Yang Y, Karlson DT, Covarrubias AA (2008) Functional dissection of hydrophilins during in vitro freeze protection. Plant Cell Environ 31:1781–1790

Roberts JK, DeSimone NA, Lingle WL, Dure L III (1993) Cellular concentrations and uniformity of cell-type accumulation of two Lea proteins in cotton embryos. Plant Cell 5:769–780

Rohrig H, Schmidt J, Colby T, Brautigam A, Hufnagel P, Bartels D (2006) Desiccation of the resurrection plant *Craterostigma plantagineum* induces dynamic changes in protein phosphorylation. Plant Cell Environ 29:1606–1617

Rohrig H, Colby T, Schmidt J, Harzen A, Facchinelli F, Bartels D (2008) Analysis of desiccation-induced candidate phosphoproteins from *Craterostigma plantagineum* isolated with a modified metal oxide affinity chromatography procedure. Proteomics 8:3548–3560

Rorat T (2006) Plant dehydrins – tissue location, structure and function. Cell Mol Biol Lett 11:536–556

Sallon S, Solowey E, Cohen Y, Korchinsky R, Egli M, Woodhatch I, Simchoni O, Kislev M (2008) Germination, genetics, and growth of an ancient date seed. Science 320:1464

Sanchez-Ballesta MT, Rodrigo MJ, Lafuente MT, Granell A, Zacarias L (2004) Dehydrin from *Citrus*, which confers in vitro dehydration and freezing protection activity, is constitutive and highly expressed in the flavedo of fruit but responsive to cold and water stress in leaves. J Agric Food Chem 52:1950–1957

Segrest JP, Garber DW, Brouillette CG, Harvey SC, Anantharamaiah GM (1994) The amphipathic alpha helix: a multifunctional structural motif in plasma apolipoproteins. Adv Protein Chem 45:303–369

Sharon MA, Kozarova A, Clegg JS, Vacratsis PO, Warner AH (2009) Characterization of a group 1 late embryogenesis abundant protein in encysted embryos of the brine shrimp *Artemia franciscana*. Biochem Cell Biol 87(2):415–430

Shen-Miller J, Mudgett MB, Schopf W, Clarke S, Berger R (1995) Exceptional seed longevity and robust growth: ancient sacred lotus from China. Am J Bot 82:1367–1380

Shih M-D, Lin SC, Hsieh J-S, Tsou C-H, Chow T-Y, Lin T-P, Hsing Y-C (2004) Gene cloning and characterization of a soybean (*Glycine max* L.) LEA protein, GmPM16. Plant Mol Biol 56:689–703

Shih M, Hoekstra F, Hsing Y (2008) Late embryogenesis abundant proteins. Adv Bot Res 48:211–255

Solomon A, Solomon R, Paperna I, Glazer I (2000) Desiccation stress of entomopathogenic nematodes induces the accumulation of a novel heat-stable protein. Parasitology 121:409–416

Steponkus PL, Uemura M, Joseph RA, Gilmour SJ, Thomashow MF (1998) Mode of action of the COR15a gene on the freezing tolerance of *Arabidopsis thaliana*. Proc Natl Acad Sci USA 95:14570–14575

Tolleter D, Jaquinod M, Mangavel C, Passirani C, Payet N, Avelange-Macherel M-H, Macherel D (2007) Drying reveals structure and function of a plant mitochondrial protein. Plant Cell 19:1580–1589

Tompa P (2002) Intrinsically unstructured proteins. Trends Biochem Sci 27:527–533

Tompa P, Csermely P (2004) The role of structural disorder in the function of RNA and protein chaperones. FASEB J 18:1169–1175

Tompa P, Szász C, Buday L (2005) Structural disorder throws new light on moonlighting. Trends Biochem Sci 30:484–489

Tunnacliffe A, Wise MJ (2007) The continuing conundrum of the LEA proteins. Naturwissenschaften 94:791–812

Tyson T, Reardon W, Browne JA, Burnell AM (2007) Gene induction by desiccation stress in the entomopathogenic nematode *Steinernema carpocapsae* reveals parallels with drought tolerance mechanisms in plants. Int J Parasitol 37:763–776

Uversky VN, Gillespie JR, Fink AL (2000) Why are "natively unfolded" proteins unstructured under physiologic conditions? Proteins 41:415–427

von Heijne G (2006) Membrane-protein topology. Nat Rev Mol Cell Biol 7:909–918

Wise MJ (2003) LEAping to conclusions: a computational reanalysis of late embryogenesis abundant proteins and their possible roles. BMC Bioinformatics 4:52

Wise MJ, Tunnacliffe A (2004) POPP the question: what *do* LEA proteins do? Trends Plant Sci 9:13–17

Wolkers WF, McCready S, Brandt WF, Lindsey GG, Hoekstra FA (2001) Isolation and characterization of a D-7 LEA protein from pollen that stabilizes glasses in vitro. Biochim Biophys Acta 1544:196–206

Chapter 7
A Role for Molecular Studies in Unveiling the Pathways for Formation of Rotifer Resting Eggs and Their Survival During Dormancy

Nadav Y. Denekamp, Koushirou Suga, Atsushi Hagiwara, Richard Reinhardt, and Esther Lubzens

Abstract Rotifers are minute aquatic invertebrates inhabiting mainly continental waters with environmental conditions restricting their long-term survival. Formation of dormant forms, diapausing embryos (resting eggs), is a strategy that facilitates their genetic pool survival beyond unfavorable environmental conditions. Resting eggs are formed during the sexual reproductive cycle. Usually, rotifers display cyclic parthenogenesis, with asexually reproducing females (known as amictic females) forming diploid eggs (called amictic eggs) that are responsible for fast population growth. Specific environmental cues known as mixis induce sexual reproduction. Recent studies implicate crowding as the inducing signal for onset of sexual reproduction in brachionid monogonont rotifers. Sexual reproduction starts with the commencement of meiosis leading to the formation of haploid males and followed by the formation of diploid resting eggs. Rotifer resting eggs differ structurally from asexual or amictic eggs and can remain dormant for decades. Following hatching a new cycle of asexual reproduction ensues, leading to population growth and colonization of the specific environmental niche. In some cases, however, sexual reproduction may occur shortly after hatching. Very few molecular studies were performed on monogonont rotifers so far and most were limited to single gene studies. Large scale expressed sequence tags resources were formed recently for the rotifer *Brachionus plicatilis*. They are providing important platforms for molecular analyses on the events leading to the formation of males, resting eggs and on dormancy in this species. Genes known to be associated with

N.Y. Denekamp and E. Lubzens (✉)
Department of Marine Biology, Israel Oceanographic and Limnological Research, Haifa, Israel
e-mail: esther@ocean.org.il

K. Suga
Faculty of Fisheries, Nagasaki University, Nagasaki, Japan

A. Hagiwara
Graduate School of Science and Technology, Nagasaki University, Nagasaki, Japan

R. Reinhardt
Molecular Genetics, Max Planck Institute, Berlin-Dahlem, Germany

E. Lubzens et al. (eds.), *Dormancy and Resistance in Harsh Environments*, Topics in Current Genetics 21,
DOI 10.1007/978-3-642-12422-8_7,

desiccation tolerance during dormancy in other organisms (e.g., yeast spores, *Artemia* cysts, and seed plants) were also identified in resting eggs, even though they do not necessarily fully desiccate during the dormant period. Future studies should be directed to investigate the regulation of processes associated with the entrance and exit from dormancy, in this important metazoan model.

7.1 Introduction

Rotifers are minute metazoans (~50–2,000 μm in length; ~1,000 cells), and one of the major groups of zooplankton inhabiting streams, rivers, lakes, and estuaries, functioning as secondary producers in the aquatic ecosystem (Wallace 2002; Wallace et al. 2006). Over 2,000 rotifer species are reported and most of them inhabit freshwater. Taxonomically, phylum Rotifera is allied with other microinvertebrate groups in the superphylum of Gnathifera, a sister or basal group to the Lophotrochozoa (Giribert et al. 2000; Mark Welch 2001; Segerss 2007). Rotifers populate water bodies where environmental conditions restrict their long-term survival, including temporary ponds that may evaporate, ponds or lakes with unfavorable low or high temperatures, appearance and vanishing of suitable food, and advent of predators. Consequently, rotifers appear and die out rapidly, with large fluctuations in the live population, but the strategy of forming dormant resting eggs facilitates the genetic pool survival beyond these unfavorable conditions in the form of a genetic bank.

There are basically two main dormant forms displayed by rotifers: whole body desiccation displayed by bdelloid rotifers and formation of long-term surviving eggs resistant to adverse conditions. The term "dormant" or "dormancy" refers to a temporary suspension of visible metabolic activity or arrested development and encompasses the phenomena of diapause and quiescence, as suggested by Ricci (2001). Bdelloid rotifers enter dormancy (or cryptobiosis; another term used in connection with suspended metabolism; Clegg 2001) by desiccation, at any stage of their life cycle; the ova, the embryo, juveniles, or adults. They will resume activity after rehydration (Örstan 1998; Ricci 1998, 2001; Ricci and Caprioli 2005). This spectacular phenomenon was already observed by Leeuwenhoek at the beginning of the eighteenth century (reviewed in Tunnacliffe and Lapinski 2003). Monogonont rotifers, however, may display dormancy at the asexual parthenogenic egg stage or as an encysted diapausing embryo (resting egg), formed from fertilized ova. Only three cases so far were reported for asexually diapausing eggs, including the pseudosexual eggs of *Keratella hienalis*, *Notholca squaumla*, and *Synchaeta pectinata*. It was proposed that these eggs are produced in species and strains that do not reach sufficient population densities in their natural environment, as relatively high population densities are needed to ensure successful encounters between males and females during sexual reproduction and the formation of resting eggs. At least 15 monogonont species (including *Brachionus* species, *Epiphanes* species, *Rhinoglena frontalis*, *Asplachna* species, *Notommata* species, *Trichocerca rattus*, and one strain of *Hexartha*) are known to enter dormancy or diapause via the production of fertilized resting eggs (Gilbert 2007a) and the dormant stage is an encased embryo. Several organisms (Table 7.1) display dormancy at embryonic developmental stages, where the embryo discontinues its development during the dormant period. They resume development with the onset of external and/or internal cues and exit from the dormant stage. The ecological significance of dormancy in rotifers has been reviewed extensively in recent years (Ricci 2001; Gilbert and Schröder 2004; Schröder 2005; Gilbert 2007a, b). The present review aims at discussing how a rotifer resting egg is formed and highlights where molecular tools and information from studies in other organisms can assist in filling gaps in our knowledge by identifying pathways associated with sexual reproduction and the long survival of resting eggs during dormancy. An emphasis is made on resting-egg production of the *Brachionus plicatilis* complex species, as molecular data bases have become available only for these resting-egg producing species and they can serve for functional genomics studies in the near future.

Studies performed on natural populations from worldwide locations revealed that *B. plicatilis*, originally considered as one species, represents a complex of several cryptic species, as shown by molecular, behavioral, and physiological studies. Early morphotypes distinguished as large (L) and small (S; Fukusho and Iwamoto 1981) were later classified as *B. plicatilis* and *Brachionus rotundiformis*, respectively (Segers 1995), based on morphological, physiological, and genetic differences and additional SS morphotypes were added later (Fu et al. 1991a, b; Rumengan et al. 1991; Hagiwara et al. 1995a; Ciros-Perez et al. 2001; Kotani et al. 2001). Phylogenic analyses based on sequence information from two genes, an mtDNA gene cytochrome c oxidase I (COI) and a nuclear gene, ribosomal internal transcribed spacer (ITS1), revealed nine concordant genetically divergent lineages (Gómez 2005). Species identification of the *B. plicatilis* complex is still an ongoing effort (Suatoni et al. 2006). The various populations or species respond differently to external cues that lead to mixis and production of resting eggs. Morphological differences between SS, S, and L type rotifers carrying parthenogenic eggs are demonstrated in Fig. 7.1.

7.2 Switching from Asexual to Sexual Reproduction and the Onset of Meiosis

7.2.1 The Rotifer Life Cycle

The production of resting eggs requires a switch from the asexual to sexual reproduction (Fig. 7.2). Several monogonont rotifers display a heterogonic life cycle with two main types of females: (a) the first type is an asexually reproducing female known as amictic female and produces diploid amictic eggs. These eggs develop directly into diploid females by ameiotic parthenogenesis and (b) the

Table 7.1 A list of organisms displaying dormancy and the details on the cues facilitating the onset and exit from the dormant stage

Organism	Dormant form	Induction	Onset	Obligatory period	Exit from dormancy
A. ovalisporum (Cyanobacterium)	Akinets	K^+ depletion	Formation of akinets		Various environmental factors
S. cerevisiae (yeast)	Haploid spores	Absence of fermentable carbon	Meiosis	No	Fermentable carbon and nitrogen
A. thaliana (plant)	Seed (embryo)		Sexual reproduction	Primary dormancy	Light and temperature
C. elegans (nematode)	Dauer juvenile	Population density, lack of food	Larval dauer stage	No	Favorable condition
Bdelloid rotifers	Desiccated animals (all developmental stages)	Drought	Desiccation of the whole animal; ova	No	Rehydration
B. plicatilis (monogonont rotifer)	Resting eggs (embryo)	Mixis signal hormones	Sexual reproduction	Yes	Light
D. magna (Daphnia)	Resting eggs (embryo)	Stress: food availability, temp. Photoperiod	Sexual reproduction		Light, high temperature, high oxygen levels
A. franciscana (brine shrimp)	Cysts (embryo)	High salinity, desiccation, heat, food	Oviparous formation of cysts		Favorable conditions
Copepods	Resting eggs, copepodites, adults	Stress: food availability, temp. Photoperiod	Temperature, photoperiod (?)		Favorable conditions
O. arcticus (Insect)	Desiccated animals	Sub zero temperatures	Protective desiccation	No	Temperature

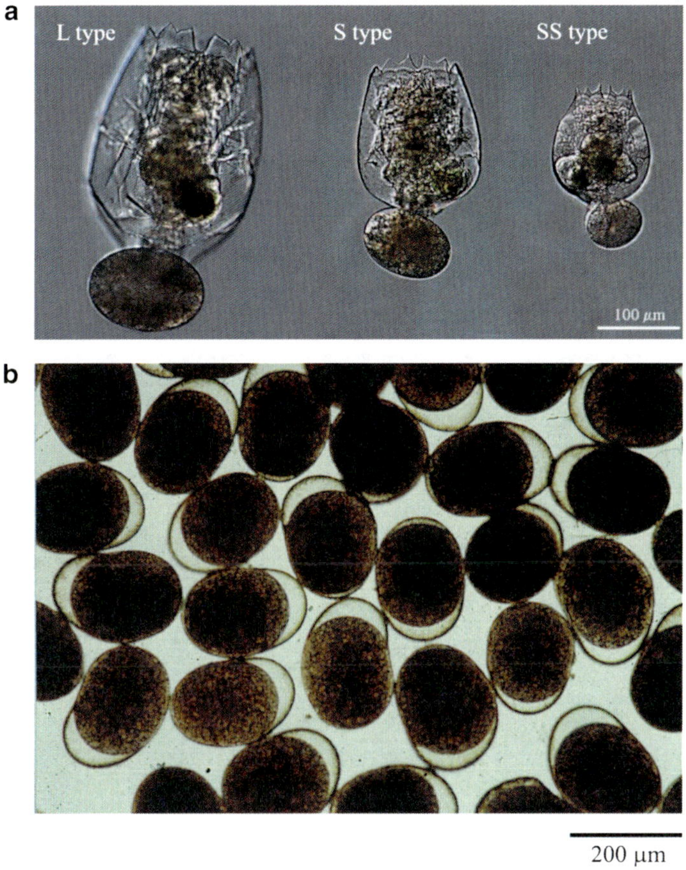

Fig. 7.1 (**a**) *Brachionus plicatilis* L, S, and SS type rotifers carrying amictic eggs. (**b**) Resting eggs from L-type rotifers

second type of female is known as a mictic female that is morphologically indistinguishable from amictic female, but with meiosis taking place within the vitellarium, leading to the production of haploid oocytes. If the haploid oocytes are not fertilized, they will be extruded as eggs from the female and develop into males. The male of *B. plicatilis* is haploid, its body size is smaller than the female, and its biological functions are limited to swimming and reproduction as the male lacks a functional digestive system (Ruttner-Kolisko 1974; Wallace and Snell 1991). If a haploid male encounters a juvenile mictic female carrying haploid oocytes in the vitellarium, internal fertilization can take place, resulting in diploid eggs that will develop into resting eggs (reviewed in Ruttner-Kolisko 1974). The age of the female is critical for successful fertilization and production of the resting eggs (Buchner et al. 1967; Snell and Childress 1987; Gómez and Serra 1996). The switch from asexual to sexual reproduction depends on an external cue or "mixis". The response to a "mictic signal" occurs in females during oogenesis (Gilbert 2007b)

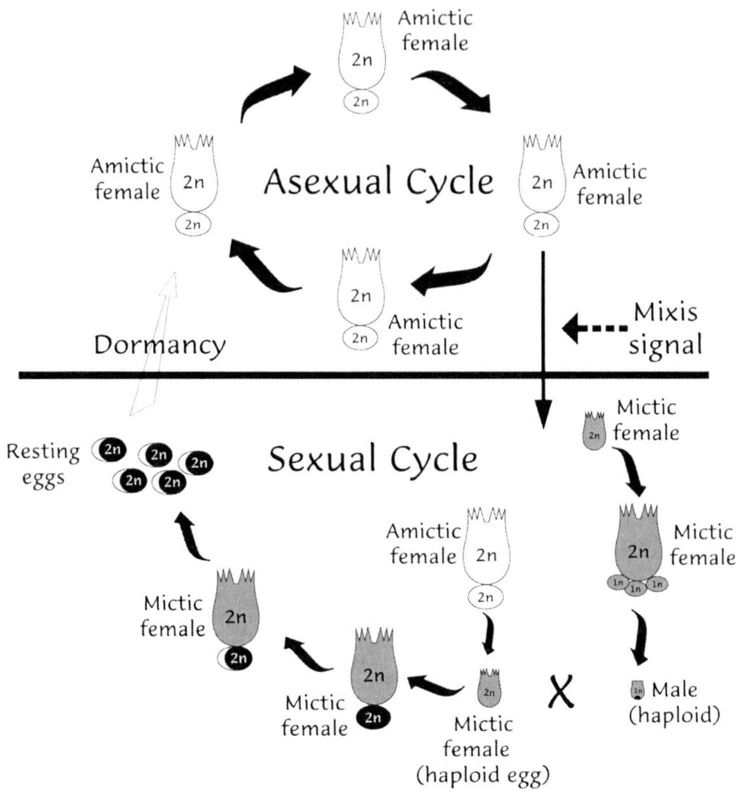

Fig. 7.2 The life cycle of *Brachionus plicatilis* showing asexual and sexual reproduction and formation of resting eggs. In the asexual life cycle, diploid amictic females produce parthenogenetic diploid amictic eggs. A mixis signal initiates the occurrence of a sexual cycle, whereby, diploid mictic females produce haploid eggs via meiosis. The haploid eggs develop into either haploid males or, if fertilized, they form diploid dormant (or diapausing) resting eggs. The internal insemination of diploid mictic females carrying haploid eggs is possible for only a few hours after birth. Mictic females are shaded in *gray* and include mictic females producing male eggs or mictic females that form diploid resting eggs. All females are diploid while males are haploid (adapted from Gilbert 2004b)

and is transmitted from the maternal female to the developing oocyte (Snell et al. 2006).

The transition in the life cycle seems more complex as a third category of females known as amphoteric females (producing amictic and haploid mictic eggs) was reported by several investigators, but their abundance seems to be usually extremely low (Gilbert 2007a). It should be highlighted that rotifers are eutelic and lack cell division as adults (like nematodes) and all the oocytes within female are present at time of birth of this female. There are some differences in the way rotifers from different populations or strains constituting the *B. plicatilis* species complex, carry their resting eggs. The eggs can be carried outside of the female's body (external to the lorica) in the same way amictic or male eggs are carried or they can be retained within the body of the female (Serra et al. 1998; Lubzens and Zmora 2003). If the eggs are carried outside the body, the female can produce several resting eggs and each one is released after it is fully formed. However, in case the eggs are retained within the body of the female, only one or two eggs are formed and they sink to the sediment after the death of the female.

7.2.2 Formation of Resting Eggs

Resting eggs appear several days after the onset of the extrinsic or intrinsic mictic cues (Gilbert 2007a) and may require two generations: first with the appearance of males and then the fertilization of juvenile females forming the diploid resting egg. Production of a resting egg is believed to involve higher energy investment than the parthenogenic amictic egg, with its additional egg shell layers, glycogen, and lipid droplets (Wurdak et al. 1977, 1978; Gilbert and Schröder 2004). Evidence to support this hypothesis comes from the fact that only a few resting eggs (1–5) can be produced by one female and their production takes longer (~42 h), in contrast to 17–24 diploid amictic eggs produced by an amictic female or 9–19 of haploid unfertilized mictic male eggs (Hagiwara et al. 1988b; Lubzens and Minkoff 1988; Lubzens and Zmora 2003). This means that successful production of resting eggs depends on the abundance of some food sources that would assure the production of an adequate number of rotifers and a longer time for producing the resting eggs. Because parthenogenesis continues side by side with the onset of resting-egg production, this can be viewed as a bet-hedging strategy which ensures the survival of the population's genes in case of habitat deterioration (Segers and Brockmann 1987; Philipi and Segers 1989; Schröder and Gilbert 2004). Or in other words, at the time of affluence, it is a good strategy to make investments in the bank (Carmona et al. 1995; Serra and King 1999; reviewed in Schröder 2005). In this case, it is the gene bank of the rotifer population. This strategy may change between species or populations adapted to specific environmental conditions as stem females hatching from dried sediment of temporary desert rock pools have a high probability of being mictic in species of *Hexathra* (Gilbert 2007a).

7.2.3 Factors Affecting the Formation of Resting Eggs

The detailed cascade of events leading to the onset of sexual reproduction in *B. plicatilis* has yet to be elucidated. Several environmental cues were found implicated in inducing mixis. Gilbert (2007a) mentions three main environmental species-specific signals that induce amictic rotifer females to produce mictic daughters and these are crowding, dietary tocopherol. and long photoperiods and in most cases only one of them is sufficient for any specific species. The formation of resting eggs by the rotifer *B. plicatilis* has been extensively investigated as this species has been commonly used, for decades, to feed marine fish larvae in commercial hatcheries (Ito 1960; Hirata 1980; Watanabe et al. 1983; Lubzens 1987; Yoshimura et al. 1996; Hagiwara et al. 2001, 2007; Lubzens and Zmora 2003 and several others).

7.2.3.1 External Factors

Most studies so far were aimed at identifying environmental changes eliciting the formation of mictic females and resting eggs. Variation in the response to temperature, salinity, nutrition (type and amount of food), and age of amictic females was reported extensively for various members of the *B. plicatilis* species complex and of *B. rotundiformis* (Lubzens 1981; Hino and Hirano 1984; Lubzens et al. 1985; Snell 1986; Hagiwara et al. 1988a, b; 1989, 2005; Hagiwara and Hino 1990; Hino and Hirano 1984, 1985, 1988; Lubzens and Minkoff 1988), indicating differences between members of this species complex. In addition, it was shown that specific bacteria promote resting-egg production (Hagiwara et al. 1994). Among the most discussed signals in recent years is the effect of crowding on the onset of mixis in *B. plicatilis* (Hino and Hirano 1977; Gilbert 2004a; Schröder and Gilbert 2004; Schröder 2005; Gilbert 2007a). To ensure successful encounters between males and females at the appropriate age or developmental stage, a certain population density is required (ranging from 6.6 to 70 l^{-1} in natural environments; reviewed in Snell et al. 2006). A water soluble fraction from rotifers cultured at high densities was reported to increase the occurrence of sexual reproduction and resting-egg formation (Hagiwara et al. 1994). A ~39 kDa protein, named as the mixis inducing protein (MIP), was isolated from *B. plicatilis* cultures and 17 amino acids sequenced from the N-terminus show strong similarity to a steroidogenesis-inducing protein (SIP), that was isolated from human ovarian follicular fluid (Snell et al. 2006). The full identity of this protein, its receptors, the exact location, and timing of synthesis within rotifers has yet to be determined. Crowding chemicals released into the culture media are not species specific within the closely related *B. plicatilis* species complex (Stelzer and Snell 2006) and their activity may function similarly to quorum sensing in bacteria (Kubanek and Snell 2008).

7.2.3.2 Intrinsic Factors

The predisposition of rotifers to respond to external signals or cues depends on some additional factors. Brachionid females hatching from resting eggs (stem females) have a lower response to the mictic signals (Serra et al. 2005; Schröder and Gilbert 2004; Schröder 2005; Gilbert 2007a). Starvation during initial period of life of stem females hatching from resting eggs resulted in a higher percentage of mixis in offspring while the same treatment of females hatching from amictic eggs did not affect mixis induction indicating some specific maternal factors in the cytoplasm of resting eggs that are absent in amictic eggs (Hagiwara et al. 2005).

Age-related changes in the propensity to produce mictic daughters were reported in some rotifer clones (Gilbert and Schröder 2007). Differential response of rotifers stemming from single resting eggs (that were produced in a clonal culture) suggests genetically based segregation in the response to the onset of sexual reproduction (Hino and Hirano 1977, 1985; Lubzens 1989). Intraclonal variation and age-related response for mixis suggest an interaction between several factors that is difficult to resolve experimentally (Gilbert and Schröder 2007).

7.2.3.3 Hormones

Several investigations were performed on the assumption that external signaling molecules elicit formation of hormones similar to those of other animals (of vertebrate and invertebrate origin), in evoking the response of inducing meiosis in females. These investigations included pig growth hormone, human chorionic gonadotropins, 17β-estradiol, triiodothyrionin, 20-hydroxuedcysone, 5-hydroxy-tryptamine, gamma amino-butiric acid, juvenile hormone (Gallardo et al. 1997, 1999, 2000), and more recently progesterone (Snell and DesRosiers 2008). While some effects were found on either increasing the incident of mictic female production or resting-egg production, the affecting compounds were not isolated or traced within rotifers, nor their putative receptors. In most cases, these compounds were applied at relatively high concentrations. Moreover, these compounds may have an indirect effect as they are supplied in the culture medium and could be chemically altered by bacteria or the food organisms or within the rotifers. Alternatively, they may elicit the production of other compounds that are biologically active. The direct effect of tocopherol on the onset of the mictic response in *Asplanchna* (Gilbert and Thompson 1968; Gilbert 1974) was demonstrated many years ago and is still the only study showing a direct involvement of a specific compound in eliciting meiosis in rotifers. Recent studies in mammalian species indicate that retinoic acid is the signaling molecule whereby germ cells in an ovary enter meiosis, during embryogenesis (Bowles and Koopman 2007). It remains to be shown whether a similar pathway occurs also in other organisms such as rotifers.

7.2.3.4 Future Directions

Extensive molecular studies were performed on the cascade of events leading to the formation of a dormant dauer form in the nematode *Caenorhabditis elegans* and these could be helpful in future studies on the molecular basis for the onset of the formation of dormant resting eggs, in rotifers. In response to high population density, *C. elegans* enters an alternative larval stage known as the dauer stage that can be considered a dormant or diapausing stage. Dauers are adapted for survival under harsh conditions, have a thickened cuticle, often remain motionless, do not feed and survive off fat storage. *C. elegans* senses its population density through small molecules consisting of several derivatives of the dideoxysugar acarylose, which are synthesized in the intestine and released or excreted into its surroundings (Butcher et al. 2009). The dauer pheromone is detected by sensory neurons in the head (Schackwitz et al. 1996) and signals through multiple pathways, including the TGFβ pathway and the insulin/insulin-like growth factor-1 (IGF-1) pathway (Shaw et al. 2007). Molecular genetic studies suggest that when hormones (serotonin, TGFβ, insulin, and steroids) are upregulated, reproductive growth is the selected pathway. However, when these hormones are downregulated, the animals enter dauer diapause (Motola et al. 2006; Gerisch et al. 2007; Fielenbach and Antebi 2009). While some recent studies on IGF-related pathways were performed in rotifers and showed their putative role in longevity (especially of genes functioning downstream of the IGF pathway; Yoshinaga et al. 2003, 2005; Kaneko et al. 2006), there is no information so far on the association of the TGFβ and IGF-1 pathways in formation of rotifer resting eggs.

The key event in the onset of sexual reproduction is the occurrence of meiosis in mictic females. While a "box tool" for identifying meiosis has been offered for genomic studies (Schurko and Logsdon 2008), universal functional molecular tools are not yet available, especially not for rotifers. Recent studies in *Daphnia pulex*, avian, and mammalian species (Bowles et al. 2006; Balthus et al. 2006; Bowles and Koopman 2007; Schurko et al. 2009; Smith et al. 2008) suggest a few directions for future studies on entry into meiosis. These studies revealed several meiosis-specific genes for which homologs in most model organisms are expressed only during meiosis and mutants containing null alleles are defective only in meiosis. These include sister chromatid cohesion genes, genes associated with meiotic interhomolog recombination and genes associated with regulation of crossover. It will be, however, necessary to show that the expression pattern of these genes is specific to meiosis in rotifers, as several may also be expressed during asexual reproduction (Schurko et al. 2009).

7.3 The Resting Egg and Its Morphology

Information on the morphological structure of resting eggs can provide some clues for future investigations regarding a search for genes associated with dormancy in these eggs, since the structure and formation of the resting egg differ from those of amictic egg. In the case of amictic reproduction, the diploid oocytes within the vitellarium of a female will accumulate yolk and the oocyte will develop into embryos within the pseudocoelom. Most members of the *B. plicatilis* species complex extrude the embryos encased within an oblate-spheroid egg shell and they are carried by the female until hatching. They are relatively transparent and the progress of development can be observed under relatively low magnification (Snell and Hoff 1988).

There is insufficient information on the early stages of formation of diapausing eggs within sexually reproducing mictic females. Externally, a mictic female producing a resting egg cannot be distinguished from an a mictic female, at early stages of resting-egg formation. The resting-egg forming females can be identified at later stages of development, when the egg content turns dark yellow to brown (unpublished observations). At the last stages of development an extraembryonic space is noticed (Fig. 7.2), between two shell layers (Wurdak et al. 1978), in contrast to the one shell cover of amictic eggs. The extraembryonic space is filled with fluid as long as the egg remains immersed. The studies reported by Wurdak et al. (1978) and the review by Gilbert (1989) on *Brachionus calyciflorus* provide some insight into embryonic development in rotifers. Externally, the egg in *B. calyciflorus* is delineated by an operculum, a lid like opening, through which the fully developed embryo hatches. Scanning and transmission electron microscope studies on *B. plicatilis* revealed a surface pattern of anastomizing smooth and wavy ridges (Munuswamy et al. 1996). In cross section under TEM, the alveolar and dense sublayers constitute the outer (S1) and inner (S2) egg membrane or shell membranes. S1 and S2 (the inner egg membrane) layers are derived from shell-secreting granules. These granules are synthesized in the vitellarium of the fertilized mictic female and transported to the growing oocyte. These membrane bound granules migrate to the periphery of the egg and release their contents to form the alveolar and inner sublayers in *B. calyciflorus* (Wurdak et al. 1978). The S1 and S2 membranes were suggested to protect the embryo from adverse conditions or mechanical injury during internal as well as during the dormant or diapause period. The embryo is surrounded by the S2 shell layer that separates it from the extraembryonic space and this layer was suggested to be homologous to the single shell layer of parthenogenic eggs since both contain chitin. A third shell layer probably forms during resumed embryogenesis, after dormancy is broken and this layer may protect the embryo during the hatching process from the egg. Changes occur within the resting eggs of *B. plicatilis* during early stages after ovulation and they include changes in the thickness of the first shell layer (from 1.2 to 4.0 μm) during the first 8 days (at 25°C) and the appearance of mucopolysaccharide that may have a protective role. Also, the total number of nuclei in an embryo increases from 22 on Day 2 after ovulation, to 39 on Day 6 (Hagiwara et al. 1995b).

There is also insufficient information on the early events of cleavage and gastrulation, following internal fertilization, to give us a complete understanding of the embryonic development in diapausing embryos and this information is important for determining the stage of the onset of dormancy (Gilbert 1989). Wurdak et al. (1978) examining the resting-egg embryos of *Asplanchna intermedia* and *B. calyciflorus* observed that these embryos were composed of an inner syncytial mass, surrounded by an outer syncytial layer. The outer layer is considered as ectoderm while the inner mass as entoderm, with some inner nuclei forming the precursors of the germovitellarium in the fully formed female. The two regions were delimited from each other, partially by membranes connection but are joined by structures resembling gap-junctions indicating close chemical communication between the layers. There is a massive breakdown of cell membranes in resting eggs that does not occur in the parthenogenically developing embryo, where cell boundaries are retained until after the cessation of nuclear divisions (Birky et al. 1967). This is rather surprising as resting eggs would have to undergo nuclear division cycle(s), after resumed embryonic development. Also, numerous lipid droplets in the inner layer are larger and denser than those observed in the ectoderm and alpha and beta-glycogen islets were also observed. This is in contrast to fewer lipid droplets and beta-glycogen storage in parthenogenic developing embryos indicating their importance for the resting egg. Lower protein synthesis activity and protein secretory activity in resting eggs were suggested by the lack of elongated cisternae in resting eggs and the reduced Golgi apparatus (Wurdak et al. 1978).

Gilbert (1989) raises interesting questions with regard to postcleavage development of the diapausing embryos. Specifically, these are related to gastrulation and early organogenesis stages, prior to the breakdown of the cell membranes and the functional significance of the inner and outer syncytial layers and their organization. These events in the embryo are of significance in attempting to understand the mechanism of entry to dormancy and in revealing the functional genes associated with them. While the morphological studies direct our attention to structural and functional differences between amictic and diapausing embryos, there is as yet, insufficient information on the functional differences at the molecular level.

7.4 Diapause and Hatching of Resting Eggs

Rotifer resting eggs remain dormant under extreme environmental condition such as desiccation and low temperature. Resting eggs of S-type strains could be hatched successfully from the sediment of Kai-ike pond in Koshiki island, Japan, 65 years ago after their production (Kotani et al. 2001). Resting eggs formed in the laboratory almost 30 years ago hatched after storage at 4°C (Denekamp et al. 2009 and unpublished data). Interestingly, the rotifer resting eggs maintain their dormancy in a nondesiccated form, as they contain ~70% water (Hagiwara 1996). They survive though, desiccation as demonstrated by Balompapueng et al. (1997). This raises a question on the long-term survival of these eggs in a nondesiccated form.

The hatching pattern rotifer resting eggs (of L-type) changes with exposure to different temperature (5–25°C) and light regimes (at 24L:0D and 0L:24D). When eggs were exposed to light just after formation, resting eggs hatched sporadically over months. No hatching was observed for 6 months when eggs were preserved under dark conditions regardless of the temperature (Hagiwara and Hino 1989). According to Minkoff et al. (1983), resting eggs from L-type rotifers hatched after an obligatory dormant period (when stored in the dark) after exposure to light (even for as little as 10 min was sufficient to induce hatching), but the extent of hatching depended on the salinity, temperature and was slightly modified by the presence of algae. The threshold of light intensity for hatching was estimated to be 4,400 lux for 30 min using a halogen lamp. Hatching of resting eggs depends on the wavelength of irradiation and is highest at 350–400 nm. Irradiation at more than 350 nm caused 1–25% hatching, but it reached 50–60% at a wavelength of 250–310 nm. Hatching level decreased with longer wavelength irradiation. It was also found that the addition of hydrogen peroxide or prostaglandins (E_1, E_2, or $F_{2\alpha}$) to the hatching media resulted in resting-egg hatching even in the dark. These results may suggest that production of peroxide in seawater caused by light as well as the oxidation of fatty acid to prostaglandins within the embryo are possible mechanisms involved in resting-egg hatching (Hagiwara et al. 1995b). Brine shrimp cysts hatch at 400–600 nm (Van der Linden et al. 1985) and may be associated with light absorption of hemopigment in the embryo (Van der Linden et al. 1986). There is no information, however, on the identity of the receptors associated with perceiving the signals for hatching in rotifers or the cascade of events leading to resumed development. Extensive studies performed on plant seed dormancy and germination may assist in paving similar directions for studies on genes associated with the onset, duration and resumed development of rotifer resting eggs (Raz et al. 2001; Bentsink and Koornneef 2008; Holdsworth et al. 2008a, b).

7.5 Molecular Aspects of Rotifer Functional Biology and Dormancy

Molecular data will contribute greatly to improve our knowledge on resting-egg formation, on the differences between the female and egg types and on dormancy. Studies on single gene function were made available in recent years (Wheelock et al. 1999; Kaneko et al. 2002, 2005), and increasing molecular information is now available on bdelloid rotifers (Mark Welch 2001, 2005; Mark Welch and Meselson 2001; Mark Welch and Mark Welch 2005; Mark Welch et al. 2004, 2008; Fontaneto et al. 2007; Pouchkina-Stantcheva et al. 2007; Suga et al. 2007b, 2008 and others), including studies related to the function of Late embryogenesis abundant proteins (LEAs) during desiccation (Pouchkina-Stantcheva et al. 2007). Larger transcriptome data in the form of expressed sequence tags (ESTs) were recently made available for the *B. plicatilis* complex species (Suga et al. 2007a; Denekamp et al. 2009). A partially sequenced cDNA library from all life stages of *B. plicatilis* generated a database of 2,300 ESTs corresponding to more than 450 transcripts. It revealed relatively abundant transcripts with no significant similarity to sequences in databases (Suga et al. 2007a). A similar conclusion was drawn from ~48,000 ESTs forming ~18,000 putative transcripts of *B. plicatilis* obtained from normalized and subtractive cDNA libraries. These were prepared from samples of all life stages, including resting eggs and resting eggs during hatching, but only 28–55% (depending on the library) produced significant matches to known data bases (Denekamp et al. 2009).

In an effort to reveal molecular aspects associated with onset of sexual reproduction, molecular data was generated recently on males as males are formed first after the onset of meiosis. A normalized cDNA library was constructed and about 37% of the total ESTs, constituting 1,521 sequences, had no significant similarity (score <50) to BLAST database (Table 7.2; Suga et al. in preparation). Only eight clones were found with significant similarity to male specific and/or testis-related genes (Table 7.3; Suga et al. in preparation). These include the spermidine synthase,

Table 7.2 Male EST database properties

Category	Number of clone
Total reads	5,568
ESTs after quality screening	5,254
Total number of isolated ESTs (Redundancy 1.36)	4,084
ESTs matched to database genes (Score >100)	1,552
ESTs did not matched to database genes (Score <50)	1,521

The cDNA library was constructed using the SMART cDNA library construction kit (Takara Bio, Japan) according to the manufacturer's instructions. The Duplex Specific Nuclease (DSN; Evrogen, Russia) was used to eliminate highly expressed genes. Plasmid DNA was sequenced for 5′ region of the clone with ABI Big Dye 3.1 chemistry using 5′ sequencing primer (Takara Bio) and eluted on ABI PRISM 310 or 3730xl Genetic Analyzers (Applied Biosystems, USA). The normalized cDNA library was constructed using about 7,000 males, which were harvested by filtration (plankton net mesh, 45 μm), and the contaminating small females and neonates were removed manually. From this cDNA library, 5,568 clones were randomly sequenced with an average read length of 622 ± 148 bp and 5,254 clones were obtained after quality screening. Further analysis eliminated overlapping genes and the sequenced clones were assembled into 4,084 clones that were compared with NCBI databases

Table 7.3 Male-specific genes by BLAST search

Estimated gene function	Score	E value
Sperm associated antigen 16	267	6×10^{-70}
Granulin	201	3×10^{-50}
Vertebrate male sterility domain	199	9×10^{-50}
Testis specific	190	6×10^{-47}
Vertebrate male sterility domain	171	5×10^{-55}
Testis-specific serine kinase 1	127	7×10^{-28}
Sperm associated antigen 17	114	5×10^{-24}
Spermatogenesis associated 17	100	4×10^{-20}

Clones showing significant similarity for male-specific genes using BLASTX nr database with a score >100 and E-value $<1.0 \times 10^{-20}$.

spermatogenesis, and sperm-specific transcripts and also testis-specific transcripts. A search in Illumina sequenced cDNA libraries from males, amictic and mictic females, amictic and resting eggs, revealed reads (Table 7.4; Denekamp et al. in preparation) matching some of these EST sequences, but others were not identified. It is possible that these differences between the EST library and Illumina sequencing are due to age or the genetic origin of the rotifer cultures. Surprisingly, resting eggs showed higher abundance of some transcripts (reads; e.g., CL3986Contig1) even more than males. Some contigs (CL389Contig1 and rotifera-CL4209 Contig1) showed higher expression levels in amictic females than in males. Since there is little molecular data for the Phylum Gnathifera and, no information on male rotifers, these unknown clones may represent genes species-, class-, phylum- or even superphylum-specific. Further analysis of these genes may shed new light on the molecular mechanism of rotifer male function, spermatogenesis, and male-specific hormones.

Table 7.4 The occurrence of male-specific transcripts Illumina reads of libraries prepared from amictic females (FA), mictic females carrying male eggs (FM), mictic females carrying resting eggs (FRE), males (M), amictic eggs (AE), and resting eggs (RE)

Function	EST id	Illumina profile (log2)						
		E value	FA	FM	FRE	M	AE	RE
Spermidine synthase	CL389Contig1	1E-100	0.48	0.86	−0.20	0.02	0.08	−0.34
Sperm associated antigen 16	sb104P0005E20_F.ab1	1E-16	0.73	0.66	−1.97	0.76	1.69	−5.50
Granulin	CL3786Contig1	1E-36	0.32	0.10	0.06	0.38	−1.16	0.72
Vertebrate male sterility domain	CL2507Contig1	1E-54	0.69	0.47	−0.31	0.84	0.70	−1.34
	CL2932Contig1	1E-98	0.80	0.42	−0.29	1.20	0.87	−0.95
	CL3917Contig1	2E-50	−0.08	−0.08	2.46	0.69	−0.16	−2.08
	CL3978Contig1	1.2E-62	0.76	0.79	−0.47	0.76	0.70	−1.39
	CL4366Contig1	1.1E-53	0.59	0.61	−0.11	1.26	0.68	−0.89
	CL6423Contig1	1.4E-48	0.69	0.58	−1.19	0.52	1.63	−3.29
Testis specific	rotifera-CL4209Contig1	1.1E-37	0.79	0.34	−0.83	0.07	0.43	−1.35
	sb103P0014L02_F.ab1	1.8E-63	−0.21	0.41	−2.21	0.30	1.61	−3.02
Testis-specific serine kinase 1	rotifera-CL485Contig1	1.7E-74	−1.62	−0.13	0.71	0.10	0.18	−0.78
	rotifera-CL485Contig2	1.3E-16	−2.47	−0.56	0.32	−0.28	−0.12	−0.69
Sperm associated antigen 17	Not found							
Spermatogenesis associated 17	Not found							

The number of Illumina reads in each library was normalized to the median and the log2 for each transcripts was calculated. Negative values indicate samples with number of Illumina reads that is lower than the median. Data from Denekamp et al. (in preparation)

There is almost no information on gene regulation during the brachionid rotifer life cycle and during dormancy and exit from it. Clues for biological processes associated with dormant stages can be inferred from spores, seed plants, and cysts where metabolic rate is extremely low (Keilin 1959; Clegg 2001) and dormancy is mostly associated with desiccation. Emerging evidence shows that there may be considerable commonality among mechanisms facilitating desiccation tolerance across the spectrum of organisms that display this trait, irrespective of their phylogeny (Berjak 2006). Some of the common features are protection against reactive oxygen species (ROS) and detoxification, the function of specific proteins assisting in maintaining the conformation of intracellular protein structures, implication of LEAs and sugars in the formation of the intracellular glassy state, and the removal and uptake of water associated with aquaporins and lipid metabolism (Clegg 2001; Storey 2003; Wang and Kim 2003; Kranner and Birtić 2005; Tunnacliffe and Wise 2007; Liberek et al. 2008). Transcripts associated with these functions were found in the ESTs generated from cDNA libraries of resting eggs and mictic females carrying resting eggs (Denekamp et al. 2009). Since these resting eggs do not necessarily desiccate, it is not clear whether the proteins associated with tolerance to desiccation contribute to survival of resting eggs during nondesiccated dormancy. Some of these transcripts were more abundant in resting eggs than in amictic eggs, such as those coding for LEA proteins, small heat-shock proteins (*shsp-3*), manganese superoxide dismutase (*mn-sod-2*), and glutathione

S-transferase (*gst-8*). Except for mn-sod-2, they also showed higher expression levels in females carrying resting eggs compared with amictic females. Other transcripts were more highly expressed in females carrying resting eggs such as those coding for copper or zinc superoxide dismutase (*cu/zn-sod-1*) and glutathione (*gst-2*). Surprisingly, the transcript coding for trehalose phosphate synthase (*tps*) associated with the trehalose metabolism did not change significantly in resting eggs relative to amictic eggs or in females carrying resting eggs relative to amictic females, suggesting that this gene may not be associated with resting-egg production. It cannot be discounted, however, that trehalose synthesis could be regulated at the translational level or enzyme activity, rather than the transcriptional level. This raises a question regarding stability during desiccation of resting eggs, as sugars participate together with LEAs in the formation of the glassy matrix, conferring stability in desiccated tolerant organisms in the dry state (Berjak 2006). This question is also relevant to bdelloid rotifers, since trehalose was not detected in desiccated forms of this rotifer (Caprioli et al. 2004; Tunnacliffe et al. 2005). The EST resources available now open the way for global expression profiling and could assist in identifying genes with functions in the onset, during and hatching of resting eggs.

7.6 Conclusions and Future Directions

There is a large gap in our knowledge on the cascade of events leading to the formation of resting eggs, their survival during dormancy and hatching. While extensive studies were published on genetic, environmental, and culture conditions promoting resting-egg formation, there is no information on the molecular and metabolic functions underlying these events. A partially characterized "mixis induction protein" is assumed to elicit the signal for entry into meiosis in the eggs developing within an amictic female, in a crowded culture of rotifers. Several studies revealed that the crucial stage for the onset of meiosis in the vitellarium should occur in oocytes that are still carried within the mother, before oviposition (Snell et al. 2006), but this needs to be confirmed. Moreover, there is no information on the identity of the internal signaling molecule and whether it involves tocopherol or retinoids or other compounds. After meiosis, haploid males are formed and molecular data indicate that they express male-specific genes. Surprisingly, the male-specific transcripts were highly abundant in resting eggs but not in the females forming resting eggs and they occur in amictic females and amictic eggs. Unfortunately, we do not have a good explanation for these observations. Following fertilization, transcripts with higher abundance in females carrying resting eggs and in resting eggs could be identified. Several of these transcripts, such as those putatively coding for glutathione-S-transferase, ferritin, small heat-shock proteins, LEA proteins, and transcripts of genes associate with lipid and fatty acid metabolism also appear in other forms displaying dormancy (Table 7.5). Yet, we are still far from the detailed picture available for *C. elegans*, where information on the

Table 7.5 Genes associated with dormancy in rotifer resting eggs and their appearance in *Saccharomyces cerevisiae* spores, *Artemia franciscana* cysts, and *Arabidopsis thaliana* seeds

Description	*B. plicatilis*		*S. Cerevisiae*		*A. franciscana*		*A. thaliana*	
Aldo-keto reductase	FM945201	+	NP_010159	+	–	–	NM_104751	+
Ferittin	FM945178	+	-		AY062896	+	[a]NM_115467	–
Small heat-shock protein	FM944934	+	NP_009628	+	AF031367	+	NM_104282	+
Glutathion-S transferase	FM944312	+	NP_011670	+	[b]ES503727	+	NM_105660	+
Late embryonic abundant protein	FM939178	+	[c]NP_116640	+	ES492663	+	NM_115164	+
Trehalose-phosphate synthase	[d]FM929156	–	NP_009684	–	–		NM_202376	–
Aquaporin	[d]FM916559	–	NP_015518	+	–		NM_101644	+
Superoxide dismutase	AB111351	+	NP_011872	+	ES524755	+	NM_111929	+
Cathepsin L	FM904985	+	[e]NP_014026	+	ES516910	+	NM_100251	–
	[d]FM940157	–						

Gene bank accession numbers are given for each gene

The "+" sign indicates overexpression in the dormant form. The "–" sign indicates that the gene is not overexpressed in the dormant form

The data for *S. cerevisiae* were taken from expression experiment for sporulation (Chu et al. 1998) and germination (Joseph-Strauss et al. 2007). The data for *A. franciscana* were taken from EST libraries for *A. franciscana* cysts (Qiu et al. 2007; Chen et al. 2009) and from single gene studies (Chen et al. 2003; Qiu et al. 2006). The data for *A. thaliana* were taken from expression experiments for seed germination (Cadman et al. 2006; Finch-Savage et al. 2007)

[a]Ferritin transcript is expressed in plant seeds, but a higher expression level is found after ripening of seeds

[b]Glutathione-S transferase transcript was found in a diapause destined cDNA library, however, its specificity to seeds needs confirmation

[c]This accession number refers to HSP12 which is similar to LEA proteins (Mtwisha et al. 1998)

[d]The expression of these genes is probably ubiquitous and not specific to resting eggs

[e]This accession number represents serine carboxypeptidase rather than Cathepsin L

molecular time course and metabolic basis of entry and exit from dormancy in the dauer stage is available (Wang and Kim 2003; Jeong et al. 2009). Among the main obstacles in deciphering the molecular events in rotifers is the lack of genomic and functional information. Only ~28% of the sequenced ESTs from a resting-egg subtractive library could be matched to available data bases (Denekamp et al. 2009). Moreover, functional genomic studies have not been performed to a large extent in rotifers, although siRNA technology has been developed (Shearer and Snell 2007). A fascinating question is the regulation of the onset of dormancy in the developing resting egg. From morphological data it is not fully clear when exactly the embryo stops its development. This is an intriguing question since several organisms enter dormancy during embryonic development (Table 7.1). Recent studies in *C. elegans* implicate microRNAs in regulating development of the dauer stage (Bethke et al. 2009). Moreover, it has been shown in zebrafish that microRNA430 (miR430) promotes the transition between maternal and zygotic gene expression repertoire (Giraldez et al. 2006), raising the question whether microRNAs regulate the onset and exit from dormancy in embryos in organisms displaying dormancy at this transition phase. This line of future research direction, associating miRNAs in regulation of dormancy, is supported by recent results showing differential regulation of miRNA levels during mammalian hibernation (Morin et al. 2008).

Acknowledgment The studies of NYD and EL in this review were financially supported by the European Commission (NEST #012674; Sleeping Beauty). We would like to thank anonymous reviewers for their helpful comments on the manuscript.

References

Balompapueng MD, Hagiwara A, Nozaki Y, Hirayama K (1997) Preservation of resting eggs of the euryhaline rotifer Brachionus plicatilis O.F. Müller by canning. Hydrobiologia 358: 163–166

Balthus AE, Menke DB, Hu YC, Goodheart ML, Carpenter AE, de Rooij DG, Page DC (2006) In germ cells of mouse embryonic ovaries, the decision to enter meiosis precedes premeiotic DNA replication. Nat Genet 38:1430–1434

Bentsink L, Koornneef M (2008) Seed dormancy and germination. In: The arabidopsis book. American Society of Plant Biologists, Rockville, MD, pp 1–18. doi:10.1199/tab.0119

Berjak B (2006) Unifying perspectives of some mechanisms basic to desiccation tolerance across life forms. Seed Sci Res 16:1–15

Bethke A, Fielenbach N, Wang Z, Mangelsdorf DJ, Antebi A (2009) Nuclear hormone receptor regulation of microRNAs control development progression. Science 324:95–98

Birky CW, Bignami RZ, Bentfield MJ (1967) Nuclear and cytoplasmic DNA synthesis in adult and embryonic rotifers. Biol Bull 133:502–509

Bowles J, Koopman P (2007) Retinoic acid, meiosis and germ cell fate in mammals. Development 134:3401–3411

Bowles J, Knight D, Smith C, Wilhelm D, Richman JM, Mamiya S, Yashiro K, Chawengsaksophak K, Wilson MJ, Rossant J, Hamada H, Koopman P (2006) Sex-specific regulation of retinoidc acid levels in developing mouse gonads determine germ cell fate. Science 312: 596–600

Buchner H, Mutschler C, Kiechle H (1967) Die Determination der Männchen-und Dauereiproduktion bei Asplnchna sieboldi. Biologische Zentral 86:599–621

Butcher RA, Ragains JR, Li W, Ruvkun G, Clardy J, Mak HY (2009) Biosynthesis of the Caenorhabditis elegans dauer pheromone. Proc Natl Acad Sci USA 106:1875–1879

Cadman CS, Torop PE, Hilhorst HW, Finch-Savage WE (2006) Gene expression profiles of Arabidopsis Cvi seeds during dormancy cycling indicate a common underlying dormancy control mechanism. Plant J 46:805–822

Caprioli M, Katholm AK, Melone G, Ramlov H, Ricci C, Santo N (2004) Trehalose in desiccated rotifers: a comparison between a bdelloid and a monogonont species. Comp Biochem Physiol A Mol Integr Physiol 139:527–532

Carmona MJ, Gomez A, Serra M (1995) Mictic patterns of the rotifer Brachionus plicatilis Müller in small ponds. Hydrobiologia 313(314):365–371

Chen T, Amons R, Clegg JS, Warner AH, MacRae TH (2003) Molecular characterization of artemin and ferritin from Artemia franciscana. Eur J Biochem 270:856–867

Chen WH, Ge X, Wang W, Yu J, Hu S (2009) A gene catalogue for post-diapause development of an anhydrobiotic arthropod Artemia franciscana. BMC Genomics 10:52

Chu S, DeRisi J, Eisen M, Mulholland J, Botstein D, Brown PO, Herskowitz I (1998) The transcriptional program of sporulation in budding yeast. Science 282:699–705

Ciros-Perez J, Gomez A, Serra M (2001) On the taxonomy of three sympatric sibling species of the Brachionus plicatilis (Rotifera) complex from Spain, with the description of B. ibericus n. sp. J Plankton Res 23:311–1328

Clegg JS (2001) Cryptobiosis – a peculiar state of biological organization. Comp Biochem Physiol B Biochem Mol Biol 128:613–624

Denekamp NY, Thorne MAS, Clark MS, Kube M, Reinhardt R, Lubzens E (2009) Discovering genes associated with dormancy in the monogonont rotifer Brachionus plicatilis. BMC Genomics 10:108

Fielenbach N, Antebi A (2009) C. elegans dauer formation and the molecular basis of plascticity. Genes Dev 22:2149–2165

Finch-Savage WE, Cadman CS, Toorop PE, Lynn JR, Hilhorst HW (2007) Seed dormancy release in Arabidopsis Cvi by dry after-ripening, low temperature, nitrate and light shows common quantitative patterns of gene expression directed by environmentally specific sensing. Plant J 51:60–78

Fontaneto D, Hernlou AE, Boschetti C, Caprioli M, Melone G, Ricci C, Barraclough TG (2007) Independently evolving species in asexual bdelloid rotifers. PLoS Biol 4:e87

Fu Y, Hirayama K, Natsukari Y (1991a) Morphological differences between two types of the rotifer Brachionus plicatilis O.F. Muller. J Exp Mar Biol Ecol 151:29–41

Fu Y, Hirayama K, Natsukari Y (1991b) Genetic divergence between S and L type strains of the rotifer Brachionus plicatilis O.F. Muller. J Exp Mar Biol Ecol 151:43–56

Fukusho K, Iwamoto H (1981) Polymorphosis in size of rotifer, Brachionus plicatilis, cultured with various feeds. Bull Natl Res Inst Aquac 2:1–10

Gallardo WG, Hagiwara A, Tomita Y, Soyano K, Snell TW (1997) Effect of some vertebrate and invertebrate hormones on the population growth, mictic female production, and body size of the marine rotifer Brachionus plicatilis Mueller. Hydrobiologia 358:113–120

Gallardo WG, Hagiwara A, Tomita Y, Snell TW (1999) Effect of growth hormone and gamma - aminobutyric acid on Brachionus plicatilis (Rotifera) reproduction at low food or high ammonia levels. J Exp Mar Biol Ecol 240:179–191

Gallardo WG, Hagiwara A, Snell TW (2000) Effect of juvenile hormone and serotonin (5-HT) on mixis induction of the rotifer Brachionus plicatilis Muller. J Exp Mar Biol Ecol 252:97–107

Gerisch B, Rottiers V, Li D, Motola DL, Cummins CL, Lehrach H, Mangelsdorf DJ, Antebi A (2007) A bile-like steroid modulates C. elegans lifespan through nuclear receptor signaling. Proc Natl Acad Sci USA 104:5014–5019

Gilbert JJ (1974) Dormancy of rotifers. Trans Am Microsc Soc 93:490–513

Gilbert JJ (1989) Rorifera. In: Adiyodi KG, Adiyodi RG, Adiyodi RG (eds) Reproductive biology of invertebrates, vol 5, Sexual differentiation and behaviour. Oxford & IBH Publishing, New Delhi, pp 115–136

Gilbert JJ (2004a) Population density, sexual reproduction and diapause in monogonont rotifers: new data for *Brachionus* and a review. J Limnol 63:32–36

Gilbert JJ (2004b) Females from resting eggs and parthogenetic eggs in the rotifer Brachionus calyciflorus: lipi droplets, starvation resistance and reproduction. Freshw Biol 49:1505–1515

Gilbert JJ (2007a) Timing of diapause in monogonont rotifers: mechanisms and strategies. In: Alekseev VR, De Stasio B, Gilbert JJ (eds) Diapause in aquatic invertebrates: theory and human use. Springer, Dordrecht, pp 11–27

Gilbert JJ (2007b) Induction of mictic females in the rotifer Brachionus: oocytes of mictic females respond individually to population-density signal only during oogenesis shortly before oviposition. Freshw Biol 52:1417–1426

Gilbert JJ, Schröder T (2004) Rotifers from diapausing, fertilized eggs: unique features and emergence. Limnol Oceanogr 49:1341–1354

Gilbert JJ, Schröder T (2007) Intraclonal variation in propensity for mixis in several rotifers: variation among females and maternal age. Hydrobiologia 593:121–128

Gilbert JJ, Thompson GA Jr (1968) Alpha tocopherol control of sexuality and polymorphism in the rotifer *Asplanchna*. Science 159:734–736

Giraldez AJ, Mishima Y, Rihel J, Grocock RJ, Van Dongen S, Inoue K, Enright AJ, Schier AF (2006) Zebrafish MiR-430 promotes deadenylation and clearance of maternal mRNAs. Science 312:75–79

Giribert G, Distel DL, Plz M, Sterrer W, Wheeler WC (2000) Triploblastic relationships with emphasis on the acoelomates and the position of Gnathosmolulida Cycliophora, platyhelminthes, and Chaetognatha: a combined approach of 18S rDNA sequences and morphology. Syst Zool 49:539–562
Gómez A (2005) Molecular ecology of rotifers: from population differentiation to speciation. Hydrobiologia 546:83–99
Gómez A, Serra M (1996) Mate choice in male Brachionus plicatilis rotifers. Funct Ecol 10: 681–687
Hagiwara A (1996) Use of resting eggs for mass preservation of marine rotifers. Technical Reports of Japanese Sea Ranching Programs, 24:109–120. (In Japanese)
Hagiwara A, Hino A (1989) Effect of incubation and preservation on resting egg hatching and mixis in the derived clones of the rotifer Brachionus plicatilis. Hydrobiologia 186(187): 415–421
Hagiwara A, Hino A (1990) Feeding history and hatching of resting eggs in the marine rotifer Brachionus plicatilis. Nippon Suisan Gakkaishi 56:1965–1971
Hagiwara A, Hino A, Hirano R (1988a) Comparison of resting egg formation among five Japanese stocks of the rotifer Brachionus plicatilis. Nippon Suisan Gakkaishi 54:577–580
Hagiwara A, Hino A, Hirano R (1988b) Effects of temperature and chlorinity on resting egg formation in the rotifer Brachionus plicatilis. Nippon Suisan Gakkaishi 54:569–575
Hagiwara A, Lee C-S, Miyamoto G, Hino A (1989) Resting egg formation and hatching of the S-type rotifer Brachionus plicatilis at varying salinities. Mar Biol 103:327–332
Hagiwara A, Hamada K, Hori S, Hirayama K (1994) Increased sexual reproduction in *Brachionus plicatilis* (Rotifera) with the addition of bacterial and rotifer extracts. J Exp Mar Biol Ecol 181:1–8
Hagiwara A, Hoshi N, Kawahara F, Tominaga K, Hirayama K (1995a) Resting eggs of the marine rotifer Brachionus plicatilis Muller: development and effect of irradiation on hatching. Hydrobiologia 313(314):223–229
Hagiwara A, Kotani T, Snell TW, Assava-Aree M, Hirayama K (1995b) Morphology, reproduction, genetics, and mating behavior of small, tropical marine Brachionus strains (Rotifera). J Exp Mar Biol Ecol 194:25–37
Hagiwara A, Gallardo WG, Assavaaree M, Kotani T, de Araujo AB (2001) Live food production in Japan: recent progress and future aspects. Aquaculture 200:111–127
Hagiwara A, Kadota Y, Hino A (2005) Maternal effect by stem females in Brachionus plicatilis: effect of starvation on mixis induction in offspring. Hydrobiologia 546:275–279
Hagiwara A, Suga K, Akazawa A, Kotani T, Sakakura Y (2007) Development of rotifer strains with useful traits for rearing fish larvae. Aquaculture 268:44–52
Hino A, Hirano R (1977) Ecological studies on the mechanism of bisexual reproduction in the rotifer Brachionus plicatilis. II Effects of cumulative parthenogenetic generation on the frequency of bisexual reproduction. Nippon Suisan Gakkaishi 43:1147–1155
Hino A, Hirano R (1984) Relation between water temperature and bisexual reproduction rate in the rotifer Brachionus plicatilis. Nippon Suisan Gakkaishi 50:1481–1485
Hino A, Hirano R (1985) Relation between the temperature given at the time of fertilized egg formation and bisexual reproduction pattern in the deriving strains of the rotifer Brachionus plicatilis. Nippon Suisan Gakkaishi 51:511–514
Hino A, Hirano R (1988) Relationship between water chlorinity and bisexual reproduction rate in the rotifer Brachionus plicatilis. Nippon Suisan Gakkaishi 54:1329–1332
Hirata H (1980) Culture methods of the marine rotifer Brachionus plicatilis. Min Rev Data File Res 1:27–46
Holdsworth MJ, Soppe WS, Bentsink L (2008a) Molecular networks regulating Arabidopsis seed maturation, after-ripening, dormancy and germination. Invited Tansley Review. New Phytol 179:33–54
Holdsworth MJ, Finch-Savage WE, Grappin P, Job D (2008b) Post-genomics dissection of seed dormancy and germination. Trends Plant Sci 13:7–13

Ito T (1960) On the culture of mixohaline rotifer Brachionus plicatilis O.F. Muller in the sea water. Rep Fac Fish Pref Univ Mie 3:708–740, in Japanese with English abstract

Jeong P-Y, Kwon M-S, Joo H-J, Paik Y-J (2009) Molecular time-course and metabolic basis of entry in dauer in C. elegans. PLoS One 4:e4162

Joseph-Strauss D, Zenvirth D, Simchen G, Barkai N (2007) Spore germination in Saccharomyces cerevisiae: global gene expression patterns and cell cycle landmarks. Genome Biol 8:R241

Kaneko G, Kinoshita S, Yoshinaga T, Tsukamoto K, Watabe S (2002) Changes in expression pattern in stress protein genes during population growth of the rotifer Brachionus plicatilis. Fish Sci 68:1317–1323

Kaneko G, Yoshinaga T, Yanagawa Y, Kinoshita S, Tsukamoto K, Watabe S (2005) Molecular characterization of Mn-superoxide dismutase and gene expression studies in dietary restricted Brachionus plicatilis rotifers. Hydrobiologia 546:117–123

Kaneko G, Yoshinaga T, Tsukamoto K, Watabe S (2006) Insulin/IGF pathway possibly regulates population dynamics of rotifer. Growth Horm IGF Res 16:S41

Keilin D (1959) The problem of anabiosis or latent life: history and current concept. Proc R Soc Lond B 150:149–191

Kotani T, Ozaki M, Matsuoka K, Snell TW, Hagiwara A (2001) Reproductive isolation among geographically and temporally isolated marine Brachionus strains. Hydrobiologia 446–447: 283–290

Kranner I, Birtić S (2005) A modulating role for antioxidants in desiccation tolerance. Integr Comp Biol 45:734–740

Kubanek J, Snell TW (2008) Quorum sensing in rotifers. In: Winans SC, Bassler BL (eds) Chemical communication among bacteria. ASM Press, Washington DC, pp 453–461

Liberek K, Lewandowska A, Zietkiewicz S (2008) Chaperones in control of protein disaggregation. EMBO J 27:328–355

Lubzens E (1981) Rotifer resting eggs and their application to marine aquaculture. Spec Publ Eur Maricult Soc 6:163–179

Lubzens E (1987) Raising rotifers for use in aquaculture. Hydrobiologia 147:245–255

Lubzens E (1989) Possible use of rotifer resting eggs and preserved live rotifers (*Brachionus plicatilis*) in aquaculture. In: De Pauw N, Jaspers E, Ackefors H, Wilkins N (eds) Aquaculture – a biotechnology in progress. European Aquaculture, Bredene, pp 741–750

Lubzens E, Minkoff G (1988) Influence of the age of algae fed to rotifers (Brachionus plicatilis O. F. Muller) on the expression of mixis in their progenies. Oecologia 76:430–435

Lubzens E, Zmora O (2003) Production and nutritional values of rotifers. In: Støttrup JG, McEvoy LA (eds) Live foods in marine aquaculture. Blackwell, Oxford, UK, pp 17–64

Lubzens E, Minkoff G, Marom S (1985) Salinity dependence of sexual and asexual reproduction in the rotifer Brachionus plicatilis. Mar Biol 85:123–126

Mark Welch DB (2001) Early contributions of phylogenetics to understanding the evolution of Rotifera. Hydrobiologia 446(447):315–322

Mark Welch DB (2005) Bayesian and maximum likelihood analysis of rotifer–acanthocephalan relationships. Hydrobiologia 546:47–54

Mark Welch DB, Mark Welch JL (2005) Genomic approaches to rotifer ecology. Hydrobiologia 546:101–108

Mark Welch DB, Meselson M (2001) A survey of introns in three genes of rotifers. Hydrobiologia 446(447):333–336

Mark Welch DB, Cummings MP, Hillis DM, Meselson M (2004) Divergent gene copies in the asexual class Bdelloidea (Rotifera) separated before the bdelloid radiation or within bdelloid families. Proc Natl Acad Sci USA 101:1622–1625

Mark Welch DB, Mark Welch JL, Melselson M (2008) Evidence for degenerate tetraploidy in bdelloid rotifers. Proc Natl Acad Sci USA 105:5145–5149

Minkoff G, Lubzens E, Kahan D (1983) Environmental factors affecting hatching of rotifer (Brachionus plicatilis) resting eggs. Hydrobiologia 104:61–69

Morin P Jr, Duboc A, Storey KB (2008) Differential expression of microRNA species in organs of hibernating ground squirrels: a role in translational suppression during torpor. Biochem Biophys Acta 1779:628–633
Motola DL, Cummins CL, Rottiers V, Sharma KK, Li T, Li Y, Suino-Powell K, Xu HE, Auchus RJ, Antebi A, Mangelsdorf DJ (2006) Identification of ligands for DAF-12 that govern dauer formation and reproduction in C. elegans. Cell 124:1209–1223
Mtwisha L, Brandt W, McCready S, Lindsey GG (1998) HSP 12 is a LEA-like protein in Saccharomyces cerevisiae. Plant Mol Biol 37:513–521
Munuswamy N, Hagiwara A, Murugan G, Hirayama K, Dumont HJ (1996) Structural differences between the resting eggs of Brachionus plicatilis and Brachionus rotundiformis (Rotifera, Brachionidae): an electron microscopic study. Hydrobiologia 318:219–223
Örstan A (1998) Factors affecting long-term survival of dry bdelloid rotifers: a preliminary study. Hydrobiologia 387(388):327–331
Philippi T, Segers J (1989) Hedging one's evolutionary bets, revisited. Trends Ecol Evol 4:41–44
Pouchkina-Stantcheva NN, McGee BM, Boschetti C, Tolleter D, Chakrabortee S, Popova AV, Meersman F, Macherel D, Hincha DK, Tunnacliffe A (2007) Functional divergence of former alleles in an ancient asexual invertebrate. Science 318:268–271
Qiu Z, Bossier P, Wang X, Bojkova-Fournier S, MacRae TH (2006) Diversity, structure, and expression of the gene for p26, a small heat shock protein from Artemia. Genomics 88: 230–240
Qiu Z, Tsoi SCM, MacRae TH (2007) Gene expression in diapause-destined embryos of the crustacean Artemia francisacana. Mech Dev 124:856–867
Raz V, Bergervoet JHW, Koorneef M (2001) Sequential steps for developmental arrest in Arabidopsis seeds. Development 128:243–252
Ricci C (1998) Anhydrobiotic capabilities of bdelloid rotifers. Hydrobiologia 387(388):321–326
Ricci C (2001) Dormancy patterns in rotifers. Hydrobiologica 446(447):1–11
Ricci C, Caprioli M (2005) Anhydrobiosis in bdelloid species, population and individuals. Integr Comp Biol 45:759–763
Rumengan IFM, Kayano H, Hirayam K (1991) Karyotypes of S-type and L-type rotifers Brachionus plicatilis Muller, O.F. J Exp Mar Biol Ecol 154:171–176
Ruttner-Kolisko A (1974) Plankton rotifers: biology and taxonomy. Die Binnengewasser 26(Suppl 1):1–146
Schackwitz WS, Inoue T, Thomas JH (1996) Chemosensory neurons function in parallel to mediate a pheromone response in C. elegans. Neuron 17:719–728
Schröder T (2005) Diapause in monogonont rotifers. Hydrobiologia 546:291–306
Schröder T, Gilbert JJ (2004) Transgenerational plasticity for sexual reproduction and diapause in the life cycle of monogonont rotifers: intraclonal, intraspecific and interspecific variation in the response to crowding. Funct Ecol 18:458–466
Schurko AM, Logsdon M Jr (2008) Using a meiosis detection toolkit to investigate ancient asexual "scandals" and the evolution of sex. Bioessays 30:579–589
Schurko AM, Logsdon JM Jr, Eads BD (2009) Meiosis genes in Daphnia pulex and the role of parthenogenesis in genome evolution. BMC Evol Biol 9:78
Segers H (1995) Nomenclature consequences of some recent studies on Brachionus plicatilis (Rotifera, Brachionidae). Hydrobiologia 313:121–122
Segers H (2007) Annotated checklist of the rotifers (Phylum Rotifera) with notes on nomenclature, taxonomy and distribution. Zootaxa 1564:1–104
Segers J, Brockmann JH (1987) What is bet-hedging? Oxf Surv Evol Biol 4:182–211
Serra M, King C (1999) Optimal rates of bisexual reproduction in cyclical parthenogens with density-dependent growth. J Evol Biol 12:263–271
Serra M, Gomez A, Carmona MJ (1998) Ecological genetics of Brachionus. Hydrobiologia 387 (388):373–384
Serra M, Snell TW, Gilbert JJ (2005) Delayed mixis in rotifers: an adaptive response to the effects of density-dependent sex on population growth. J Plankton Res 27:37–45

Shaw WM, Luo S, Landis J, Ashraf J, Murphy CT (2007) The C. elegans TGF-β dauer pathway regulates longevity via insulin signaling. Curr Biol 17:1635–1645
Shearer TL, Snell TW (2007) Transfection of siRNA into Brachionus plicatilis (Rotifera). Hydrobiologia 593:141–150
Smith CA, Roeszler KN, Bowles J, Koopman P, Sinclair AH (2008) Onset of meiosis in the chicken embryo; evidence of a role for retinoic acid. BMC Dev Biol 8:85
Snell TW (1986) Effect of temperature, salinity and food level on sexual and asexual reproduction in Brachionus plicatilis (Rotifera). Mar Biol 92:157–162
Snell TW, Childress M (1987) The effect of age on male and female fertility in the rotifer Brachionus plicatilis. J Invertebr Reprod Dev 12:103–110
Snell TW, DesRosiers NJD (2008) Effect of progesterone on sexual reproduction of Brachionus manjavacas. J Exp Mar Biol Ecol 363:104–109
Snell TW, Hoff FH (1988) Recent advances in rotifer culture. Aquacult Mag 9(10):41–45
Snell TW, Kubanek J, Carter W, Oayne AB, Kim J, Hicks MK, Stelzer C-P (2006) A protein signal truggers sexual reproduction in Brachionus plicatilis (Rotifera). Mar Biol 149:763–773
Stelzer C-P, Snell TW (2006) Specificity of the crowding response in the Brachionus plicatilis species complex. Limnol Oceanog 5:125–130
Storey KB (2003) Mammalian hibernation. Transcriptional and translational controls. Adv Exp Med Biol 543:21–38
Suatoni E, Vicario S, Rice S, Snell T, Caccone A (2006) An analysis of species boundaries and biogeographical patterns in a cryptic species complex: the rotifer Brachionus plicatilis. Mol Phylogenet Evol 41:86–98
Suga K, Mark Welch DB, Tanaka Y, Sakakura Y, Hagiwara A (2007a) Analysis of expressed sequence tags of the cyclically parthenogenetic rotifer Brachionus plicatilis. PLoS One 2: e671
Suga K, Tanaka Y, Sakakura Y, Hagiwara A (2007b) Inheritance of mitochondrial DNA in the rotifer *Brachionus plicatilis*. Hydrobiologia 593:167–173
Suga K, Welch DB, Tanaka M, Sakakura Y, Hagiwara A (2008) Two circular chromosomes of unequal copy number make up the mitochondrial genome of the rotifer *Brachionus plicatilis*. Mol Biol Evol 25:1129–1137
Tunnacliffe A, Lapinski J (2003) Resurrecting Van Leeuwenhoek's rotifers: a reappraisal of the role of disaccharides in anhydrobiosis. Philos Trans R Soc Lond B Biol Sci 358:1755–1771
Tunnacliffe A, Wise MJ (2007) The contuing conundrum of the LEA proteins. Naturwissenschaften 94:791–812
Tunnacliffe A, Lapinski J, McGree B (2005) A putative LEA protein, but no trehalose, is present in anhydrobiotic bdelloid rotifers. Hydrobiologia 546:315–321
Van der Linden A, Blust R, Decleir W (1985) The influence of light on the hatching of *Artemia* cysts (Anostraca: Branchiopoda: Crustacea). J Exp Mar Biol Ecol 92:207–214
Van der Linden A, Vankerckhoven I, Caubergs R, Decleir W (1986) Action spectroscopy of light-induced hatching of *Artemia* cysts (Branchiopoda: Crustacea). Mar Biol 91:239–243
Wallace RL (2002) Rotifers: exquite metazoan. Integr Comp Biol 42:660–667
Wallace RL, Snell TW (1991) Rotifera. In: Thorp JH, Covish AP (eds) Ecology and classification of North American freshwater invertebrates. Academic Press, Inc., New York, pp 187–248
Wallace RL, Snell TW, Ricci C, Nogrady T (2006) Rotifera. In: Segerss H (ed) Biology, ecology and systematics, vol 1, 2nd edn, Guides to the identification of the microinvertebrates of the continental waters of the world 23. Backhuys Publishers, Leiden
Wang J, Kim SK (2003) Global analysis of dauer gene expression in Caenorhabditis elegans. Development 130:1621–1634
Watanabe T, Kitajima C, Fujita S (1983) Nutritional values of the live organisms used in Japan for mass propagation of fish: a review. Aquaculture 34:115–143
Wheelock CE, Wolfe MF, Olsen H, Tjeerdema RS, Sowby ML (1999) HSP60-induced tolerance in the rotifer *Brachionus plicatilis* exposed to multiple environmental contaminants. Arch Environ Contam Toxicol 36:281–287

Wurdak E, Gilbert JJ, Jagels R (1977) Resting egg ultrastructure and formation of the shell in Asplanchna sieboldi and Brachionus calyciflorus. Arch Hydrobiol Beih 8:298–302

Wurdak ES, Gilbert JJ, Jagels R (1978) Fine structure of the resting eggs of the rotifers *Brachionus calyciflorus* and *Asplanchna sieboldi*. Trans Am Microsc Soc 97:49–72

Yoshimura K, Usuki K, Yoshimatsu T, Kitajima C (1996) Culture technology of marine rotifers and implication for intensive culture of marine fish in Japana. Mar Freshw Res 47:217–222

Yoshinaga T, Kaneko G, Kinoshita S, Tsukamoto K, Watabe S (2003) The molecular mechanisms of life history alterations in a rotifer: a novel approach in population dynamics. Comp Biochem Physiol (B) 136:715–722

Yoshinaga T, Kaneko G, Kinoshita S, Furukawa S, Tsukamoto K, Watabe S (2005) Insulin-like signaling pathway involved in regulating longevity of the rotifer. Hydrobiologia 546:347–352

Chapter 8
Anhydrobiotic Abilities of Tardigrades

Ralph O Schill

Abstract Tardigrades have been discovered in 1773 and are found in a variety of habitats within marine, freshwater, and terrestrial ecosystems. To survive in habitats that are prone to occasional drought, they possess the ability to enter a reversible state known as anhydrobiosis. The desiccation tolerance allows them to cope with temporal variation of available water and extended lifespan in an anhydrobiotic state with up to 20 years by producing a time shift in the age of tardigrades. The period of anhydrobiosis is limited by cumulative DNA damage and the function of repair pathways during and after rehydration. The same pathways are probably responsible for the tolerance of high doses of radiation. Heat shock proteins serve as molecular chaperones to preserve or restore the protein integrity and late embryogenesis abundant (LEA) proteins LEA proteins play an important role as well. In several desiccated species glass transition has been detected, which support the vitrification hypothesis.

8.1 Discovery of the Tardigrades

In 1773, J. A. E. Goeze (1731–1793), a German pastor from Quedlingburg, Saxony-Anhalt, Germany, was the first to describe tardigrades as "small water bears" (Fig. 8.1). He wrote "…this animal is strange because of its extraordinary anatomy and at first glance its appearance has a strong resemblance to a little bear. It is because of this I will name them small water bears…" (Goeze 1773). However, another German, J.C. Eichhorn, was maybe the first to discover these animals in 1767 but did not publish his findings until 1781 when he published a description of a "water animal" (Eichhorn 1781), 8 years after Goeze's "Herrn Karl Bonnets

R.O. Schill
Department of Zoology, Biological Institute, University of Stuttgart, Stuttgart, Germany
e-mail: ralph.schill@bio.uni-stuttgart.de

E. Lubzens et al. (eds.), *Dormancy and Resistance in Harsh Environments*,
Topics in Current Genetics 21,
DOI 10.1007/978-3-642-12422-8_8,

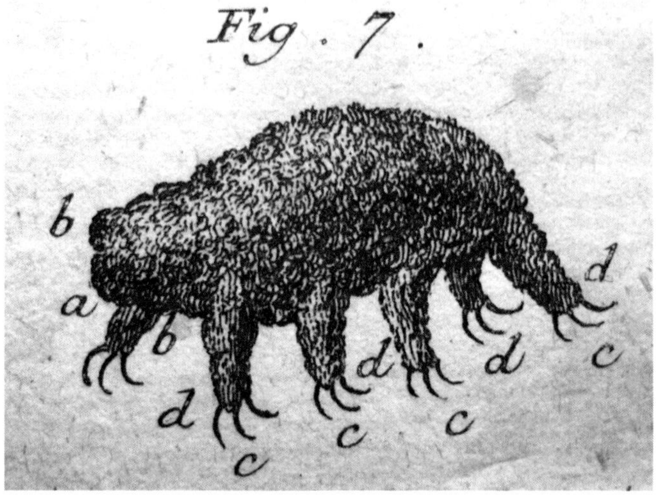

Fig. 8.1 The first drawing of a tardigrade, observed by J. A. E. Goeze and published in 1773

Abhandlungen aus der Insektologie." Shortly afterwards, L. Spallanzani (1729–1799), an Italian biologist and physiologist, also described the tardigrades and discovered their ability to desiccate and revive after "death" (Spallanzani 1776). He observed that these animals did not need a protective cover for this phenomenon and consequently suggested that these animals must slowly lose water to survive desiccation. The name "tardigrade" refers to the animal's way of movement (Lat. *tardus* – slow, *grado* – walker).

In 1834, the first tardigrade, *Macrobiotus hufelandi*, was described by C.A.S. Schultze (Schultze 1834). This species is probably the most well known of all tardigrades and was named after the German medical scientist C. W. Hufeland (1762–1836). He was the author of "Makrobiotik – Die Kunst das menschliche Leben zu verlängern" (The art of extending human life) (Hufeland 1817). In contrast to Spallanzani, Schultze spoke of a "reawakening" or "resurrection" of desiccated tardigrades after rehydration with water. His observations and interpretation were met with general disapproval. Chr. G. Ehrenberg (1795–1876), a German naturalist, zoologist, comparative anatomist, geologist, and microscopist, hypothesized that the animals secreted a medium during the dehydration process, in which they continued living and even reproduced. He believed that revived tardigrades which spent many years in desiccation were merely descendants of those formerly dehydrated (Ehrenberg 1834). Other scientists at the time also believed in abiogenesis (*generatio spontanea*).

The first substantive review of the field, "Mémoire sur les Tardigrades" (in 1840–1842), on the anatomy, systematics, and physiology of tardigrades was published by L. Doyère. His argument that the organism retained the integrity for revival during a slow desiccation process led to a quarrel with F.-A. Pouchet. Pouchet (1800–1872), a French naturalist and a leading proponent of spontaneous

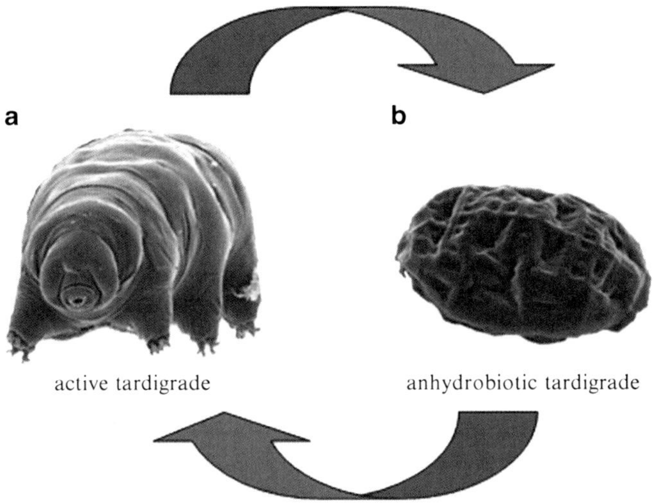

Fig. 8.2 A tardigrade in the active state (**a**) and in the anhydrobiotic, tun state (**b**)

generation of life from nonliving materials, was convinced that nothing could prevent the death of the entire organism after dehydration. This scientific dispute was settled in 1859 by the arbitral court of the Société de Biologie in Paris, France with Doyère as the victor. Since then, debates have focused more on whether tardigrade metabolism slows down (*vita minima*) or shuts down completely. At the start of the last century, the German Jesuit G. Rahm showed that desiccated tardigrades tolerated very low temperatures near absolute zero, and therefore metabolism seemed impossible (Rahm 1921). In contrast, H. Baumann, who studied the morphology and physiology of this desiccated stage worked on the assumption of *vita minima* (Baumann 1922). To this day, the ability of tardigrades to withstand unfavorable environmental conditions (in particular dryness) in a state of latent life or cryptobiosis remains a subject of particular interest (Fig. 8.2).

8.2 Cryptobiosis

A. v. Leeuwenhoek (1632–1723), the pioneering Dutch microscopist, was the first to describe the phenomenon of resurrection in desiccated animals (van Leeuwenhoek 1702). He discovered many small "animalcules" in the dry dust from a roof gutter after subsequent rehydration. These organisms were presumably bdelloid rotifers. Astonished, he confessed "I never thought that there could be any living creature in a substance so dried as this was" (van Leeuwenhoek 1702). Over the last 300 years different terms have been used to describe the resting stage of animals, until D. Keilin coined the term "cryptobiosis." This he defined as "the state of an organism when it shows no visible signs of life and when its metabolic rate becomes hardly measurable, or comes reversibly to a standstill" (Keilin 1959). According to Keilin's definition, there are four different types of cryptobiosis: states resulting from low temperature (cryobiosis), lack of oxygen (anoxybiosis), high concentrations of salt (osmobiosis), and desiccation (anhydrobiosis), or various combinations. Metabolite research on a variety of organisms suggests that there are similarities between the states of anhydrobiosis and cryobiosis owing to the reduction of free water in the cells; however, it has been suggested that freezing and dehydration are not equivalent phenomena, but involve different mechanisms (Crowe et al. 1992).

8.3 Longevity and Long-Term Anhydrobiotic Ability

8.3.1 Longevity in Tardigrades

Although there have been some studies on the biochemical modifications associated with attainment of an anhydrobiotic state in several organisms (Clegg 1967; Crowe and Clegg 1973, 1978; Tomos 1992), there have been a number of studies concerning the effects of desiccation on the life cycle, longevity, and aging of animals (Ricci 1998; Ricci and Caprioli 1998, 2005; Ricci and Covino 2005; Ricci and Pagani 1997; Ricci et al. 1987; Wharton 2002). Ricci and Pagani (1997) proposed three different hypothetical responses to the effects of anhydrobiosis on aging: The "Sleeping Beauty" model postulates a complete disregarding of the entire time spent in anhydrobiosis. Another model describes a partial discount of the time spent in the dry state, meaning that the physiological age is less than the anagraphical age, due to the slowing down of the metabolism and development. The third model assumes that the anhydrobiotic organism registers the exact time spent in the anhydrobiotic state, resulting in a nonextended longevity. For the tardigrade species *Milnesium tardigradum*, it has been shown that the "Sleeping Beauty" model is the appropriate (Hengherr et al. 2008a). This species shows a complete disregard of the time spent in anhydrobiosis and does not age during anhydrobiosis when compared with control cohorts.

8.3.2 Long-Term Anhydrobiotic Ability of Tardigrades

The ability to survive unfavorable environmental conditions in an anhydrobiotic state at all stages of the tardigrade life cycle is of great importance from an ecological and evolutionary point of view (Schill and Fritz 2008). This ability enables these organisms to inhabit extreme, harsh, and ephemeral environments as well as allowing them to disperse and colonize new habitats. Survival of newly founded populations is also facilitated by the widespread occurrence of parthenogenesis in tardigrades. Being able to survive desiccation in all embryonic states gives tardigrades an obvious advantage over predatory and parasitic organisms, which may be unable to survive such conditions (Schill and Fritz 2008). Despite the "Sleeping Beauty" model it is unlikely that natural habitats occupied by tardigrades stay dry for several years, decades, or even centuries and it is probable that there is an upper limit to the potential longevity in the anhydrobiotic state (Crowe 1975). Jönsson and Bertolani (Jönsson and Bertolani 2001) took a closer look at the long-term anhydrobiotic abilities and came to the conclusion that there is little support for the claim that tardigrades are capable of century-long survival, as suggested by Franceschi (1948). In fact, the data suggest that a decade is a more realistic estimate of the upper limit of anhydrobiotic survival in tardigrades. The longest known observation of an extended lifespan in an anhydrobiotic state was documented in the species *Echiniscus testudo* with up to 20 years survival (Jørgensen et al. 2007). However, most authors have reported an upper limit of around 10 years (Guidetti and Jönsson 2002; Rebecchi et al. 2006). This is presumably due to the cumulative damaging effects of free radical oxidation processes on cellular and molecular structures and the inactivation of DNA repair enzymes and free radical scavenging enzymes, e.g., superoxide dismutase (SOD) that possesses antioxidant effects (Crowe 1975; Örstan 1998; Wright et al. 1992). A particular problem is the damage to membranes caused by lipid peroxidation (Crowe and Madin 1975; Womersley 1981). The result is a chemical aging (Schöneich 1999) in the anhydrobiotic tun state, which limits the survival capacity of anhydrobiotic tardigrades.

8.4 DNA Damage and Repair Mechanisms

Cumulative damage incurring during cryptobiosis may have lead to selection for particularly efficient repair pathways. These are clearly essential upon rehydration, and DNA repair is presumably the secret of the tremendous ability to tolerate desiccation for a long time. This is evidenced by the fact that tardigrades tolerate high doses of radiation which should, of course, result in DNA damage and subsequently DNA repair. Jönsson et al. (2005) studied the radiation tolerance in the species *Richtersius coronifer*. Desiccated as well hydrated animals which were irradiated with 0.5 and 1 kGy did not deviate in survival from the control. He suggested that radiation tolerance in tardigrades is not due to cellular protection through metabolites or proteins, but to efficient repair mechanisms. For the species *M. tardigradum*, median lethal doses of 5,000 Gy (gamma-rays) and 6,200 Gy [high-linear energy transfer (LET) heavy ions, ^{4}He] in hydrated tardigrades, and 4,400 Gy (gamma-rays) and 5,200 Gy [high-linear energy transfer (LET) heavy ions, ^{4}He] in anhydrobiotic tardigrades at 48 h after irradiation, have been calculated (Horikawa et al. 2006). In *R. coronifer*, irradiation with $>$1,000 Gy prevents reproduction.

Gladyshev and Meselson Gladyshev and Meselson (2008) demonstrated the extraordinary resistance of bdelloid rotifers to ionizing radiation. The resulting double-strand DNA breakage was accompanied by enhanced either by systems that repair such damage, those that protect the repair systems, or both. They suggested that such breakage and repair systems may have maintained bdelloid chromosomes as colinear pairs and kept the load of transposable genetic elements low, because accurate repair of double-strand breaks requires the presence of a homologous template. However, the phenomenon seems to be species-specific and a consequence of the evolutionary adaptation of the bdelloid rotifers to survive the desiccation encountered in their characteristic habitats which periodically run dry and become wet after rainfall. In the tardigrade species *M. tardigradum*, the effect of anhydrobiosis on the integrity of DNA has been shown with the technique of microgel electrophoresis of single cells (DNA Comet Assay) (Neumann et al. 2009; Schill et al. 2008). The process of desiccation itself induces only minor DNA damage in cells, while the damage is clearly linked to the time the specimen was maintained in anhydrobiosis. A time-course experiment showed that the maximum level of DNA fragments in the comet tails was reached 90 min after rehydration. The formation and expansion of the comet tails is primarily not due to the damage itself, but instead reflects the excision repair system.

8.5 Protection and Repair with Proteins

8.5.1 Heat Shock Proteins

In the cellular stress response, several families of heat shock/stress proteins (HSPs) serve as molecular chaperones and can be induced by a wide variety of stressors that result in the accumulation of unfolded or misfolded proteins in cells, i.e., proteotoxicity (Hightower 1993). In general, during such proteotoxic conditions, there is a redirection of metabolism, leading to rapid up-regulation of HSP genes. One such family member, Hsp70 is a highly conserved and abundant protein, involved in facilitating protein maturation, stabilization of the conformation of folding intermediates, prevention of the formation of aberrant structures and leading polypeptides to biologically productive pathways (Gething and Sambrook 1992). In cysts of *Artemia* sp., Clegg et al. (1994) showed that protein p26 underwent an extensive stress-induced translocation to nuclei and other sites. These translocations exhibited strong pH dependence (Clegg et al. 1995) in a consistent trend with intracellular pH changes in vivo (Gnaiger et al. 2000; Hand 1998; Hand and Hardewig 1996; van Breukelen and Hand 2000; van Breukelen et al. 2000). Subsequent studies (Liang et al. 1997a, b) showed that p26 is a member of the small heat shock/α-crystallin family of proteins (de Jong et al. 1998; Sun and MacRae 2005) and demonstrated that this protein exhibited molecular chaperone activity in vitro (Liang et al. 1997b), and probably in vivo (Liang and MacRae 1999). In the tardigrade species *R. coronifer* when shifted from an active state to an anhydrobiotic state, Ramløv and Westh (2001) detected an induced protein with a molecular weight of approximately 71 kDa, which potentially belongs to the Hsp70 heat shock protein family. Three heat shock protein (*hsp70* family) genes are known in *M. tardigradum*, which are differently expressed during dehydration, rehydration, the active, and the anhydrobiotic state (Schill et al. 2004). One of these isoforms is significantly induced in the transitional stage between the active and the cryptobiotic state. The resulting comparatively high number of mRNAs may be stored for repair processes after rehydration. A further hint concerning general mechanisms of stabilization and repair at the protein level has been demonstrated in *R. Coronifer*, with elevated levels of stress proteins (Hsp70) are detected in rehydrated tardigrades after a period of desiccation (Jönsson and Schill 2007). This suggests that Hsp70 may be involved in the biochemical repair system and that it is connected to cellular repair processes after rehydration rather than to biochemical stabilization in the dry state.

8.5.2 LEA Proteins

In plants, and more recently in several animals, induction of high levels of hydrophilic proteins, in particular the late embryogenesis abundant (LEA) proteins, has been associated with water stress. LEA proteins were first identified 20 years ago in plants, where they are produced during seed development (Galau et al. 1986; Grzelezak et al. 1982). However, their precise function is poorly understood. To be classified as molecular chaperones, LEA proteins must not only prevent aggregation, they must additionally form transient, noncovalent complexes (Ellis 2004). Unfavorable protein–protein interactions, however, can lead to irreversible conformational changes and, in enzymes, a loss of catalytic activity (Carpenter et al. 1987). LEA proteins might simply function as "molecular shields", forming a physical barrier between partially unfolded neighboring proteins and preventing contact between them (Tunnacliffe and Wise 2007). LEA protein expression has more recently been demonstrated in other organisms, also linked to desiccation stress and the acquisition of desiccation tolerance. So far these organisms include microorganisms (Battista et al. 2001), nematodes (Browne et al. 2002; Goyal et al. 2003, 2005), rotifers (Tunnacliffe et al. 2005), a chironomid larvae (Kikawada et al. 2006), and Collembola (Bahrndorff et al. 2009). Research on LEA proteins has recently focussed on tradigrades because of the evidence that these proteins play an important role in protecting cellular proteins. In the tardigrade species *M. hufelandi*, putative LEA proteins have been isolated from the tun state (McGee et al. 2004); however, more results will be expected in the future.

8.6 Sugars and Vitrification

8.6.1 Hypothesis of Cell Stabilization

In the early sixties, S.J. Webb postulated the water replacement hypothesis in which accumulated hydrophilic molecules such as nonreducing sugars (e.g., trehalose and sucrose) interact with macromolecules through hydrogen bonds, leading to the replacement of water (Crowe et al. 1987, 1992; Crowe 2002; Webb 1964;

Webb et al. 1965). Over the last three decades considerable research has centered on the role of these sugars, particularly trehalose in animals (Yancey et al. 1982; Vertucci and Farrant 1995; Ingram and Bartels 1996; Chandler and Bartels 1999; Alpert 2000) and sucrose in higher plants (Crowe and Clegg 1973; Potts 2001; Crowe et al. 1997; Alpert 2000). These carbohydrates appeared to be ubiquitous in cryptobiotes and serve multiple additional roles (1) as compatible intracellular osmolytes during desiccation or freeze-dehydration; (2) as stabilizers of protein quaternary structure and lipid bilayer integrity with declining free water activity; and (3) as supercoolants. Trehalose is critical to desiccation tolerance in the cysts of the brine shrimp *Artemia* sp. Many studies have demonstrated its important role in the protection of cellular and macromolecular structures by replacing the water that is normally hydrogen-bonded to polar residues (Clegg 1986). Also of some importance is the vitrification hypothesis proposed by Crowe et al. (Crowe et al. 1998; Crowe 2002). This hypothesis suggests that hydrophilic molecules enter a glassy state during desiccation which prevents denaturation, aggregation, and disintegration because of immobilization. These two proposed hypotheses are not mutually exclusive.

8.6.2 The Role of Sugars in Tardigrades

There are many examples of trehalose accumulation in anhydrobiotic organisms, in adults, juveniles, and different developmental states. The desiccated embryo of brine shrimp *Artemia* sp. in the cyst contains substantial concentrations of trehalose. This can be as high as 13–18% of dry weight (Clegg 1965) and the anhydrobiotic nematode *Aphelenchus avenae* can produce 10–15% of its dry weight as trehalose when dried slowly (Madin and Crowe 1975). Similarly, a large amount of trehalose (18% of dry body mass) is rapidly accumulated in the African chironomid larvae *Polypedilum vanderplanki* (Okuda et al. 2004; Watanabe et al. 2002, 2003) and plays a role in water replacement and intracellular glass formation (Sakurai et al. 2008). Trehalose was also detected in anhydrobiotic states of the freshwater sponge *Trochospongilla* sp. (gemmules), the bryozoan *Cristatella mucedo* (statoblasts) and the mictic eggs of crustaceans *Daphnia magna, D. pulex*, *Triops longicaudatus,* and *T. cancriformis* (Hengherr et al. 2008b). The first confirmation of the presence of trehalose in tardigrades was in the species *Macrobiotus areolatus* (Crowe 1975); later Westh and Ramløv (1988) observed the production of trehalose in cryptobiosis in Arctic tardigrades. However, compared with the level of trehalose in *Artemia* sp., *A. avenae,* and *P. vanderplanki* very little trehalose was detected in the tun state of *R. coronifer* (0.1–2.3% of dry weight body mass) (Westh and Ramlov 1991). Hengherr et al. (2008c) measured the trehalose levels of six eutardigrade species (*M. tardigradum*, *Macrobiotus tonollii*, *Macrobiotus sapiens*, *Paramacrobiotus richtersi*, *Paramacrobiotus* "*richtersi* group" 1, and *Paramacrobiotus* "*richtersi* group" 2) and two heterotardigrade species (*E. testudo* and *E. granulatus*) during the induction of anhydrobiosis, the anhydrobiotic state, the rehydration phase, and after a 4-h active period. The data reveal great differences between the species. In *Macrobiotus* and *Paramacrobiotus* species, the increase in trehalose takes place during tun formation, and trehalose degradation is induced within 10–15 min following rehydration. *Echiniscus* showed the presence of trehalose but not desiccation-induced changes, while *Milnesium* showed no detectable trehalose. Lapinski and Tunnacliffe (2003) also demonstrated excellent desiccation tolerance of the bdelloid rotifers *Philodina roseola* and *Adineta vaga*, without measurable levels of trehalose and no expression of trehalose synthase genes. It seems that the apparent lack of trehalose in certain organisms does not preclude excellent desiccation tolerance, and that there are multiple strategies for desiccation tolerance in different organisms.

8.6.3 Vitrification in Tardigrades

A glass is a liquid of a quite high viscosity that it is capable of slowing down or inhibiting chemical reactions (see Sect. 6.1). Molecules embedded in a sugar and/or protein glass should gain substantial chemical stability compared with their counterparts diffusing freely in aqueous solutions. Glasses show temperature-dependent transitions during which they change from a glassy mechanical solid state to a state with decreased viscosity and the vitrification (glass transition) temperature and can be measured by differential scanning calorimetry (DSC). Viability studies on seeds resulted in predictions of the maximum temperature at which seeds will survive at a given water content and demonstrated the transition from a glassy mechanical solid to a state with a distinct decreased viscosity (Sun et al. 1984). Several other studies have been carried out on pollen (Hoekstra et al. 1992a, b) and animals (Clegg 1974; Sakurai et al. 2008). If vitrification is responsible for the tolerance in anhydrobiotic organisms, the thermal limits for the formation of amorphous glasses and survival in desiccated organisms should be dictated by the vitrification temperature and dependent on the final concentration of carbohydrates, proteins, and water. One of the first heat tolerance experiments demonstrated that anhydrobiotic tardigrades of the species *M. hufelandi* survived a short exposure to temperatures between 120 and 125°C (Doyère 1842). Later, Baumann (1927) reported that anhydrobiotic tardigrades could tolerate 100°C for 6 h and *R. coronifer* survived temperatures up to approximately 70°C for 60 min without any decrease in survival (Ramløv and Westh 2001). A systematic study with seven eutardigrade species (*M. sapiens*, *P "richtersi* group 1," *P*", *richtersi* group 2", *P.* "*richtersi* group 3", *M. tonollii*, *P. richtersi*, *M. tardigradum*) and two heterotardigrades (*E. granulatus* and *E. testudo*) has been completed recently (Hengherr et al. 2009). Exposure to temperatures up to 80°C for 1 h resulted in a moderate decrease in survival of the tardigrades. At 95°C all *Macrobiotus* and *Paramacrobiotus* species died. The heterotardigrade species *E. granulatus* and *E. testudo* survived, though with a low survival rate. However, *M. tardigradum* showed recovery rates of >90% after experiencing 100°C but followed by a steep decrease after exposure to higher temperatures. The temperature of 102°C for 1 h was the limit of the tolerance. Assuming the formation of an amorphous glass during desiccation by nonreducing carbohydrates and hydrophilic proteins as one of the major mechanisms to protect cellular structures from denaturation, coagulation, and disintegration, glass transition temperature is a limiting factor for high temperature tolerance in tardigrades. By using DSC, glass transition in the desiccated *Macrobiotus* and *Paramacrobiotus* species has been detected and therefore provides the first evidence for vitrification in tardigrades during anhydrobiosis (Hengherr et al. 2009). No glass transition could be detected in the two heterotardigrade species and *M. tardigradum*, which may indicate that these species use other mechanisms to tolerate desiccation and heat stress.

References

Alpert P (2000) The discovery, scope, and puzzle of desiccation tolerance in plants. Plant Ecol 151:5–17

Bahrndorff S, Tunnacliffe A, Wise MJ, McGee B, Holmstrup M, Loeschcke V (2009) Bioinformatics and protein expression analyses implicate LEA proteins in the drought response of Collembola. J Insect Physiol 55:210–217

Battista JR, Park MJ, McLemore AE (2001) Inactivation of two homologues of proteins presumed to be involved in the desiccation tolerance of plants sensitizes *Deinococcus radiodurans* R1 to desiccation. Cryobiology 43:133–139

Baumann H (1922) Die Anabiose der Tardigraden. Zool Jahrb 45:501–556

Baumann H (1927) Anabiosis of tardigrades. Zool Anz 72:175–179

Browne J, Tunnacliffe A, Burnell A (2002) Anhydrobiosis – plant desiccation gene found in a nematode. Nature 416:38

Carpenter JF, Crowe LM, Crowe JH (1987) Stabilization of phosphofructokinase with sugars during freeze-drying characterization of enhanced protection in the presence of divalent cations. Biochim Biophys Acta 923:109–115

Chandler J, Bartels D (1999) Plant desiccation. In: Lerner HR (ed) Plant Responses to Environmental Stresses: From Phytohormones to Genome Reorganization. Marcel Dekker, New York, pp 575–590

Clegg JS (1965) Origin of trehalose and its significance during formation of encysted dormant embryos of *Artemia salina*. Comp Biochem Physiol 14:135–143

Clegg JS (1967) Metabolic studies of crytobiosis in encysted embryos of *Artemia salina*. Comp Biochem Physiol 20:801–809

Clegg JS (1974) Biochemical adaptations associated with the embryonic dormancy of *Artemia salina*. Trans Am Microsc Soc 93:481–490

Clegg JS (1986) The physical properties and metabolic status of *Artemia* cysts at low water contents: the water replacement hypothesis. In: Leopold AC (ed) Membranes, Metabolism and Dry Organisms Cornell University Press, New York, pp 169–187

Clegg JS, Jackson SA, Warner AH (1994) Extensive intracellular translocations of a major protein accompany anoxia in embryos of *Artemia franciscana*. Exp Cell Res 212:77–83

Clegg JS, Jackson SA, Liang P, Macrae TH (1995) Nuclear–cytoplasmic translocations of protein p26 during aerobic–anoxic transitions in embryos of *Artemia franciscana*. Exp Cell Res 219:1–7

Crowe JH (1975) The physiology of cryptobiosis in tardigrades. Memorie dell'Istituto Italiano di Idrobiologia Dott Marco de Marchi 32:37–59

Crowe LM (2002) Lessons from nature: The role of sugars in anhydrobiosis. Comp Biochem Physiol A Mol Integr Physiol 131:505–513
Crowe JH, Clegg JS (1973) Anhydrobiosis dowden. Hutchinson and Ross, Stroudsburg, PA, p 477
Crowe JH, Clegg JS (eds) (1978) Dry biological systems. Academic, New York
Crowe JH, Madin KAC (1975) Anhydrobiosis in nematodes evaporative water loss and survival. J Exp Zool 193:323–334
Crowe JH, Crowe LM, Carpenter JF, Wistrom CA (1987) Stabilization of dry phospholipid-bilayers and proteins by sugars. Biochem J 242:1–10
Crowe JH, Carpenter JF, Crowe LM (1998) The role of vitrification in anhydrobiosis. Annu Rev Physiol 60:73–103
Crowe JH, Crowe LM, Petrelski S, Hoekstra FA, Araujo PD, Panek AD (1997) Anhydrobiosis: cellular adaptation to extreme dehydration. In: Dantzler WH (ed) Handbook of Physiology. Oxford University Press, New York
Crowe JH, Hoekstra FA, Crowe LM (1992) Anhydrobiosis. Annu Rev Physiol 54:579–599
de Jong WW, Caspers GJ, Leunissen JA (1998) Genealogy of the alpha-crystallin–small heat-shock protein superfamily. Int J Biol Macromol 22:151–162
Doyère PLN (1842) Memoires sur les Tardigrades. Sur le facilité que possedent les Tardigrades, les rotiferes, les anguillules des toits et quelques autres animalcules, de revenir à la vie après été completement desséchées. Ann Sci Nat Zool Biol Anim 2e:5–35
Ehrenberg CG (1834) Über das Wiederaufleben der Moosfauna. Okens Isis 6(7):710–713
Eichhorn JC (1781) Beyträge zur Naturgeschichte der kleinsten Wasserthiere, die mit blossem Auge nicht können gesehen werden und die sich in den Gewässern in und um Danzig befinden. F. Nicolai, Berlin and Stettin
Ellis JR (2004) From chloroplasts to chaperones: How one thing led to another. Photosynth Res 80:333–343
Franceschi T (1948) Anabiosi nei tardigradi. Boll Mus Ist Biol Univ Genova 22:47–49
Galau GA, Hughes DW, Dure L III (1986) Abscisic-acid induction of cloned cotton gossy-pium–hirsutum late embryogenesis-abundant lea messenger rna species. Plant Mol Biol 7:155–170
Gething MJ, Sambrook J (1992) Protein folding in the cell. Nature 355:33–45
Gladyshev E, Meselson M (2008) Extreme resistance of bdelloid rotifers to ionizing radiation. Proc Natl Acad Sci USA 105:5139–5144
Gnaiger E, Méndez G, Hand SC (2000) High phosphorylation efficiency and depression of uncoupled respiration in mitochondria under hypoxia. Proc Natl Acad Sci USA 97:11080–11085
Goeze JAE (1773) Herrn Karl Bonnets Abhandlungen aus der Insektologie aus d. Franz. übers. u. mit einigen Zusätzen hrsg. v. Joh. August Ephraim Goeze Gebauer, Halle. p 414
Goyal K, Tisi L, Basran A, Browne J, Burnell A, Zurdo J, Tunnacliffe A (2003) Transition from natively unfolded to folded state induced by desiccation in an anhydrobiotic nematode protein. J Biol Chem 278:12977–12984
Goyal K, Pinelli C, Maslen SL, Rastogi RK, Stephens E, Tunnacliffe A (2005) Dehydration-regulated processing of late embryogenesis abundant protein in a desiccation-tolerant nematode. FEBS Lett 579:4093–4098
Grzelezak ZF, Sattalo MH, Hanley-Bowdoin LK, Kennedy TD, Lane BG (1982) Synthesis and turnover of proteins and mRNA in germinating wheat embryos. Can J Biochem Physiol 60(3): 389–397
Guidetti R, Jönsson KI (2002) Long-term anhydrobiotic survival in semi-terrestrial micrometazonas. J Zool (1987) 257:181–187
Hand SC (1998) Quiescence in *Artemia franciscana* embryos: reversible arrest of metabolism and gene expression at low oxygen levels. J Exp Biol 201:1233–1242
Hand SC, Hardewig I (1996) Downregulation of cellular metabolism during environmental stress: Mechanisms and implications. Annu Rev Physiol 58:539–563
Hengherr S, Brümmer F, Schill RO (2008a) Anhydrobiosis in tardigrades and its effects on longevity traits. J Zool (1987) 275:216–220

Hengherr S, Heyer AG, Brümmer F, Schill RO (2008b) Trehalose as protecting agent in aquatic invertebrates during dormancy induced by desiccation. Comp Biochem Physiol A Comp Physiol 151:S34

Hengherr S, Heyer AG, Köhler HR, Schill RO (2008c) Trehalose and anhydrobiosis in tardigrades – evidence for divergence in responses to dehydration. FEBS J 275:281–288

Hengherr S, Worland MR, Reuner A, Brümmer F, Schill RO (2009) High-temperature tolerance in anhydrobiotic tardigrades is limited by glass transition. Physiol Biochem Zool 82:749–755

Hightower LEH (1993) A brief perspective on the heat-shock response and stress proteins. Mar Environ Res 35:79–83

Hoekstra FA, Crowe JH, Crowe LM (1992a) Germination and ion leakage are linked with phase transitions of membrane lipids during imbibition of *Typha latifolia* pollen. Physiol Plant 84:29–34

Hoekstra FA, Crowe JH, Crowe LM, van Roekel T, Vermeer E (1992b) Do phospholipids and sucrose determine membrane phase transitions in dehydrating pollen species? Plant Cell Environ 15(5):601–606

Horikawa DD, Sakashita T, Katagiri C, Watanabe M, Kikawada T, Nakahara Y, Hamada N, Wada S, Funayama T, Higashi S, Kobayashi Y, Okuda T, Kuwabara M (2006) Radiation tolerance in the tardigrade *Milnesium tardigradum*. Int J Radiat Biol 82:843–848

Hufeland CW (1817) Makrobiotik – Die Kunst das menschliche Leben zu verlängern. Fleischhauer, Reutlingen

Ingram J, Bartels D (1996) The molecular basis of dehydration tolerance in plants. Annu Rev Plant Physiol Plant Mol Biol 47:377–403

Jönsson KI, Bertolani R (2001) Facts and fiction about long-term survival in tardigrades. J Zool (Lond) 255:121–123

Jönsson KI, Schill RO (2007) Induction of Hsp70 by desiccation, ionising radiation and heat-shock in the eutardigrade *Richtersius coronifer*. J Comp Physiol B 146:456–460

Jönsson KI, Harms-Ringdahl M, Torudd J (2005) Radiation tolerance in the eutardigrade Richtersius coronifer. Int J Radiat Biol 81:649–656

Jørgensen A, Møbjerg N, Kristensen RM (2007) A molecular study of the tardigrade *Echiniscus testudo* (Echiniscidae) reveals low DNA sequence diversity over a large geographical area. J Limnol 66:77–83

Keilin D (1959) The Leeuwenhoek Lecture. The problem of anabiosis or latent life: History and current concept. Proc R Soc Lond B Biol Sci 150:149–191

Kikawada T, Nakahara Y, Kanamori Y, Iwata K, Watanabe M, McGee B, Tunnacliffe A, Okuda T (2006) Dehydration-induced expression of LEA proteins in an anhydrobiotic chironomid. Biochem Biophys Res Commun 348:56–61

Lapinski J, Tunnacliffe A (2003) Anhydrobiosis without trehalose in bdelloid rotifers. FEBS Lett 553:387–390

Liang P, MacRae TH (1999) The synthesis of a small heat shock/alpha-crystallin protein in *Artemia* and its relationship to stress tolerance during development. Dev Biol 207:445–456

Liang P, Amons R, Clegg JS, MacRae TH (1997a) Molecular characterization of a small heat shock alpha-crystallin protein in encysted *Artemia* embryos. J Biol Chem 272:19051–19058

Liang P, Amons R, Macrae TH, Clegg JS (1997b) Purification, structure and in vitro molecular-chaperone activity of Artemia p26, a small heat-shock/alpha-crystallin protein. Eur J Biochem 243:225–232

Madin KAC, Crowe JH (1975) Anhydrobiosis in nematodes – carbohydrate and lipid-metabolism during dehydration. J Exp Zool 193:335–342

McGee B, Schill RO, Tunnacliffe A (2004) Hydrophilic proteins in invertebrate anhydrobiosis. Integr Comp Biol 44:679–679

Neumann S, Reuner A, Brümmer F, Schill RO (2009) DNA damage in storage cells of anhydrobiotic tardigrades. Comp Biochem Physiol A Mol Integr Physiol 153(4):425–429

Okuda T, Watanabe M, Kikawada T, Fujita A, Forczek E (2004) Cryptobiosis in the african chironomid: Physiological mechanism to survive complete dehydration. Proc Arthropod Embryol Soc Jpn 39:1–7

Potts M (2001) Desiccation tolerance: a simple process? Trends Microbiol 9:553–559
Örstan A (1998) Factors affecting long-term survival of dry bdelloid rotifers: a preliminary study. Hydrobiologia 387/388:327–331
Rahm P (1921) Effect of very low temperatures on the fauna of moss. Proc K Ned Akad Wet Ser C Biol Med Sci 23:235–248
Ramløv H, Westh P (2001) Cryptobiosis in the eutardigrade *Adorybiotus* (*Richtersius*) *coronifer*: Tolerance to alcohols, temperature and de novo protein synthesis. Zool Anz 240:517–523
Rebecchi L, Guidetti R, Borsari S, Altiero T, Bertolani R (2006) Dynamics of long-term anhydrobiotic survival of lichen-dwelling tardigrades. Hydrobiologia 558:23–30
Ricci C (1998) Anhydrobiotic capabilities of bdelloid rotifers. Hydrobiologia 387(388):321–326
Ricci C, Caprioli M (1998) Stress during dormancy: Effect on recovery rates and life-history traits of anhydrobiotic animals. Aquat Ecol 32:353–359
Ricci C, Caprioli M (2005) Anhydrobiosis in bdelloid species, populations and individuals. Integr Comp Biol 45:759–763
Ricci C, Covino C (2005) Anhydrobiosis of *Adineta ricciae*: Costs and Benefits. Hydrobiologia 546:307–314
Ricci C, Pagani M (1997) Desiccation of Panagrolaimus rigidus (nematoda): Survival, reproduction and the influence on the internal clock. Hydrobiologia 347:1–13
Ricci C, Vaghi L, Manzini ML (1987) Desiccation of rotifers (Macrotrachela quadricornifera): Survival and reproduction. Ecology 68:1488–1494
Sakurai M, Furuki T, K-i A, Tanaka D, Nakahara Y, Kikawada T, Watanabe M, Okuda T (2008) Vitrification is essential for anhydrobiosis in an African chironomid, *Polypedilum vanderplanki*. Proc Natl Acad Sci USA 105:5093–5098
Schill RO, Fritz GB (2008) Desiccation tolerance in embryonic stages of the tardigrade *Milnesium tardigradum*. J Zool (Lond) 276:103–107
Schill R, Steinbrück G, Köhler H (2004) Stress gene (*hsp70*) sequences and quantitative expression in *Milnesium tardigradum* (Tardigrada) during active and cryptobiotic stages. J Exp Biol 207:1607–1613
Schill RO, Neumann S, Reuner A, Brümmer F (2008) Detection of DNA damage with single-cell gel electrophoresis in anhydrobiotic tardigrades. Comp Biochem Physiol A Comp Physiol 151:S32
Schöneich C (1999) Reactive oxygen species and biological aging: amechanistical approach. Exp Gerontol 34:19–34
Schultze CAS (1834) *Macrobiotus hufelandii*, animale e crustaceorum classe novum, Berlin, p 8
Spallanzani L (1776) Opuscoli di Fisica Animale e Vegetabile. Modena: Società Tipografica, pp 203–285
Sun Y, MacRae TH (2005) Small heat shock proteins: molecular structure and chaperone function. Cell Mol Life Sci 62:2460–2476
Sun WQ, Irving TC, Leopold AC (1984) The role of sugar, vitrification and membrane phase transition in seed desiccation tolerance. Physiol Plant 90(4):621–628
Tomos D (1992) Life without water. Curr Biol 2:594–596
Tunnacliffe A, Wise MJ (2007) The continuing conundrum of the LEA proteins. Naturwissenschaften 94:791–812
Tunnacliffe A, Lapinski J, McGee B (2005) A putative LEA protein, but no trehalose, is present in anhydrobiotic bdelloid rotifers. Hydrobiologia 546:315–321
van Breukelen F, Hand SC (2000) Characterization of ATP-dependent proteolysis in embryos of the brine shrimp, *Artemia franciscana*. J Comp Physiol B 170:125–133
van Breukelen F, Maier R, Hand SC (2000) Depression of nuclear transcription and extension of mRNA half-life under anoxia in *Artemia franciscana* embryos. J Exp Biol 203:1123–1130
van Leeuwenhoek A (1702) On certain animalcules found in the sediments in gutters of the roofs of houses. Letter 144. The Selected Works of Anton van Leeuwenhoek, London, pp 207–213
Vertucci CW, Farrant JM (1995) Acquisition and loss of desiccation Marcel Dekker: New York

Watanabe M, Kikawada T, Minagawa N, Yukuhiro F, Okuda T (2002) Mechanism allowing an insect to survive complete dehydration and extreme temperatures. J Exp Biol 205:2799–2802

Watanabe M, Kikawada T, Okuda T (2003) Increase of internal ion concentration triggers trehalose synthesis associated with cryptobiosis in larvae of *Polypedilum vanderplanki*. J Exp Biol 206:2281–2286

Webb SJ (1964) Bound water, metabolites and genetic continuity. Nature 203:374–377

Webb SJ, Dumasia MD, Bhorjee JS (1965) Bound Water, Inositol, and the Biosynthesis of Temperate and Virulent Bacteriophages by Air-Dried *Escherichia coli*. Can J Microbiol 11:141–150

Westh P, Ramlov H (1991) Trehalose accumulation in the tardigrade *Adorybiotus coronifer* during anhydrobiosis. J Exp Zool 258:303–311

Westh P, Ramløv H (1988) Cryptobiosis in Arctic tardigrades with special attention to the appearance of trehalose. In: Höpner Petersen G (ed) Grönland Exkursion 2.–25., August 1987. Institut für Polarökologie, Kiel, pp 227–245

Wharton DA (2002) Nematode survival strategies. In: Lee DL (ed) The biology of nematodes. Taylor and Francis, London, pp 389–411

Womersley C (1981) Biochemical and physiological aspects of anhydrobiosis. Comp Biochem Physiol B Biochem Mol Biol 70:669–678

Wright JC, Westh P, Ramlov H (1992) Cryptobiosis in Tardigrada. Biol Rev 67:1–29

Yancey PH, Clark ME, Hand SC, Bowlus RD, Somero GN (1982) Living with water stress: evolution of osmolyte systems. Science 217:1214–1222

Chapter 9
Cryoprotective Dehydration: Clues from an Insect

M. Roger Worland, Gordana Grubor-Lajšić, Jelena Purać, Michael A.S. Thorne, and Melody S. Clark

Abstract Arthropods have evolved a number of different adaptations to survive extreme environmental temperatures including, in some regions, over-wintering temperatures well below 0°C. One of the less common adaptations to surviving cold is that of cryoprotective dehydration, where the animal becomes almost anhydrobiotic with the loss of virtually all osmotically active water. In this chapter, we describe integrated studies utilising physiology, biochemistry and molecular biology to understand this phenomenon in the Arctic springtail (*Megaphorura arctica*) (formerly *Onychiurus arcticus*). These studies concentrate on the action of trehalose as a cryoprotectant, the production of antioxidants to reduce cell damage and changes in membrane composition.

9.1 Introduction

Terrestrial arthropods have evolved a complex array of adaptations to survive in some of the coldest and driest environments on earth. In fact, the stresses placed upon insects by sub-zero temperatures and low humidity are very similar; both result in an increase in osmolarity of cellular fluids resulting in cell shrinkage and

M. Roger Worland, M.A.S. Thorne, and M.S. Clark (✉)
Biological Sciences Division, British Antarctic Survey, Natural Environment Research Council, High Cross, Madingley Road, Cambridge, CB3 0ET UK
e-mail: mscl@bas.ac.uk

Grubor-Lajšić
University of Novi-Sad, Faculty of Sciences, Trg Dositeja Obradovića 3, 21000 Novi Sad, Republic of Serbia

J. Purać
Biological Sciences Division, British Antarctic Survey, Natural Environment Research Council, High Cross, Madingley Road, Cambridge CB3 0ET, UK
University of Novi-ad, Faculty of Sciences, Trg Dositeja Obradovic´a 3, 21000 Novi Sad, Republic of Serbia

E. Lubzens et al. (eds.), *Dormancy and Resistance in Harsh Environments*, Topics in Current Genetics 21,
DOI 10.1007/978-3-642-12422-8_9,

potentially death (Ring and Danks 1994). The ability of insects to survive sub-zero temperatures is one of the most researched of their extraordinary abilities. In fact full credit should be accorded to R.W. Salt who pioneered this work in the 1930s, describing the fundamental principles of insect cryobiology and providing a comprehensive list of working hypotheses (see Ring and Riegert 1991 for a bibliography, plus Salt 1957, 1961).

Insects possess three main strategies to survive the cold: freeze tolerance, freeze avoidance and cryoprotective dehydration (Block 1990; Sinclair et al. 2003; Worland 1996). The majority of micro-arthropods living in polar regions avoid freezing by lowering the freezing point of their body fluids to below that of their environment and survive in a supercooled state. This is achieved by increasing the concentration of osmolytes in their body fluids by accumulating cryoprotective compounds and/or by removing ice nucleators from their bodies. The most common cryoprotective compounds utilised are low molecular weight sugars and sugar alcohols, such as glycerol, sorbitol, ribitol, mannitol and trehalose (Zachariassen 1985; Cannon and Block 1988). These compounds reduce the melting point of body fluids by their colligative properties. Their effect is often enhanced by a reduction in body water content. Whereas many polar arthropods reduce the levels of body water during winter (Worland and Block 2003), a very few become almost anhydrobiotic under stressful environmental conditions. This response has been termed cryoprotective dehydration (Holmstrup and Westh 1994). Once all the osmotically active water has been removed from the animal's body, mechanical damage to cell structure by ice formation is avoided. However, survival of the resultant high osmotic concentration requires specialised biological responses and novel compounds. The most common cryoprotectant found in cold tolerant insects is glycerol (Storey and Storey 1991), which readily penetrates cell membranes through trans-membrane protein channels. Animals that tolerate both cold and desiccation usually synthesise trehalose, which, like glycerol, is non-toxic even in high concentrations. The cryoprotective properties of glycerol are mainly related to its colligative properties (Zachariassen 1979), whereas trehalose acts as a water replacement molecule to maintain proteins and membranes in their native state during desiccation (Crowe and Crowe 1986).

Very few terrestrial arthropods are known to utilise the cryoprotective dehydration mechanism to survive low winter temperatures. This may be because a prerequisite of living on land is resistance to water loss, which limits their ability to make use of this particular strategy. In general, rates of water loss in arthropods are reduced at low temperatures (Sømme 1995; Worland and Block 1986); however, as the vapour pressure of the haemolymph of a supercooled insect is higher than that of ice in its surroundings at the same temperature, water will inevitably continue to be lost at sub-zero temperatures until the increase in solute concentration reduces the diffusion gradient and the system reaches equilibrium (Lundheim and Zachariassen 1993). For this reason, only terrestrial taxa susceptible to water loss such as the Annelida and Diptera are likely to utilise the protective dehydration strategy. Freeze tolerant species are not affected by this desiccating force once their extra cellular body fluids are frozen (Worland and Block 2003). One of the first examples of protective dehydration to be described was that of the egg capsules of the earthworm *Dendrobaena octaedra*, which have a very permeable membrane and desiccate in frozen soil (Holmstrup 1992). Despite further studies (Worland and Block 2003) only a few other examples of animals using this survival mechanism have been identified. These include a nematode (Wharton et al. 2003), an enchytraied worm (Pedersen and Holmstrup 2003) and the larvae of the Antarctic midge *Belgica antarctica* (Elnitsky et al. 2008). Perhaps one of the best documented examples is that of the Arctic springtail *M. arctica* (Worland 1996; Holmstrup and Sømme 1998; Worland et al. 1998). In many ways, this arthropod can be viewed as a model organism for understanding the process of cryoprotective dehydration. There is a history of physiological and biochemical studies in this organism (Worland 1996; Holmstrup and Sømme 1998; Worland et al. 1998; Bahrndorff et al. 2007), which is being increasingly supplemented with molecular data (Clark et al. 2007, 2009; Bahrndorff et al. 2008). Such multi-disciplinary investigations are essential to gain a comprehensive understanding of not only specialised cellular adaptations but also to understand how this species interacts with the environment.

M. arctica (Collembola, Onychiuridae) is an apterygote (wingless) arthropod belonging to the Collembola class that have no metamorphosis and moult continuously throughout their entire life cycle. This pale yellow springtail is relatively large weighing up to approximately 1 mg and growing up to 3 mm in length (Fig. 1a). During the desiccation process the total water content of the animal reduces from 70 to 40% of fresh weight (Fig. 1b) (Worland et al. 1998). *M. Arctica* is common in coastal areas of the Palaearctic region and often found in the high nutrient, vegetated areas beneath bird cliffs. Samples for studies discussed in this chapter were collected from beneath stones lying on grass and moss below the bird cliffs at Krykkjefjellet and Stuphallet near the Ny-Ålesund base, West Spitsbergen, Svalbard (Fig. 2).

9.2 Laboratory Induced Cold Tolerance in *M. arctica*

Summer populations of *M. arctica* are freezing intolerant with a mean SCP of −6.1°C. This relatively high SCP is stable throughout the summer and is not affected by acclimation at 0°C (Block et al. 1994). This is in contrast with some other species of polar springtails (e.g. *Cryptopygus antarcticus*), which can be induced to lower their SCPs by culturing them for 2–3 weeks without food at 0°C (Block and Worland 2001). The soil temperature at the main collection site at Stuphallet typically varies between 3 and 6°C during the summer, but during winter it can fall to below −20°C (Coulson et al. 1995a, 1995b). This suggests that changes in cold hardiness should occur to enable overwinter survival. Preliminary studies by Worland (1996) found that when *M. arctica* was acclimated at sub-zero temperatures (−2.2°C) in the presence of ice, the springtails became partially desiccated and had lower SCPs. The samples regained their normal water content and recovered within a few hours after warming and melting of the surrounding ice.

a

b

Fig. 1 (**a**) *M. arctica* in native state; (**b**) *M. arctica* in desiccated state as viewed by SEM. Photographs courtesy of K. Robinson and M. R. Worland. Animals are approximately 3 mm in length

More recent laboratory studies have shown that *M. arctica* is capable of surviving temperatures down to −14°C for periods of several weeks. Detailed acclimation experiments, where the temperature was slowly reduced from 5 to −14°C over a period of several weeks, have shown that the supercooling point (SCP) of the springtails is reduced from approximately −5°C to below −30°C (Fig. 3a). At the same time the freezeable water content is reduced by almost 97% from approximately 2.7 g/g dry weight at 5°C to 0.08 g/g dry weight at −14°C (Fig. 3b). Freezeable water content was calculated from the exotherm produced by the animal's body water freezing, measured by Differential Scanning Calorimetry (DSC) (see Worland et al. 1998 for details). During the treatment period the amount of water remaining unfrozen (osmotically inactive) remained fairly constant at

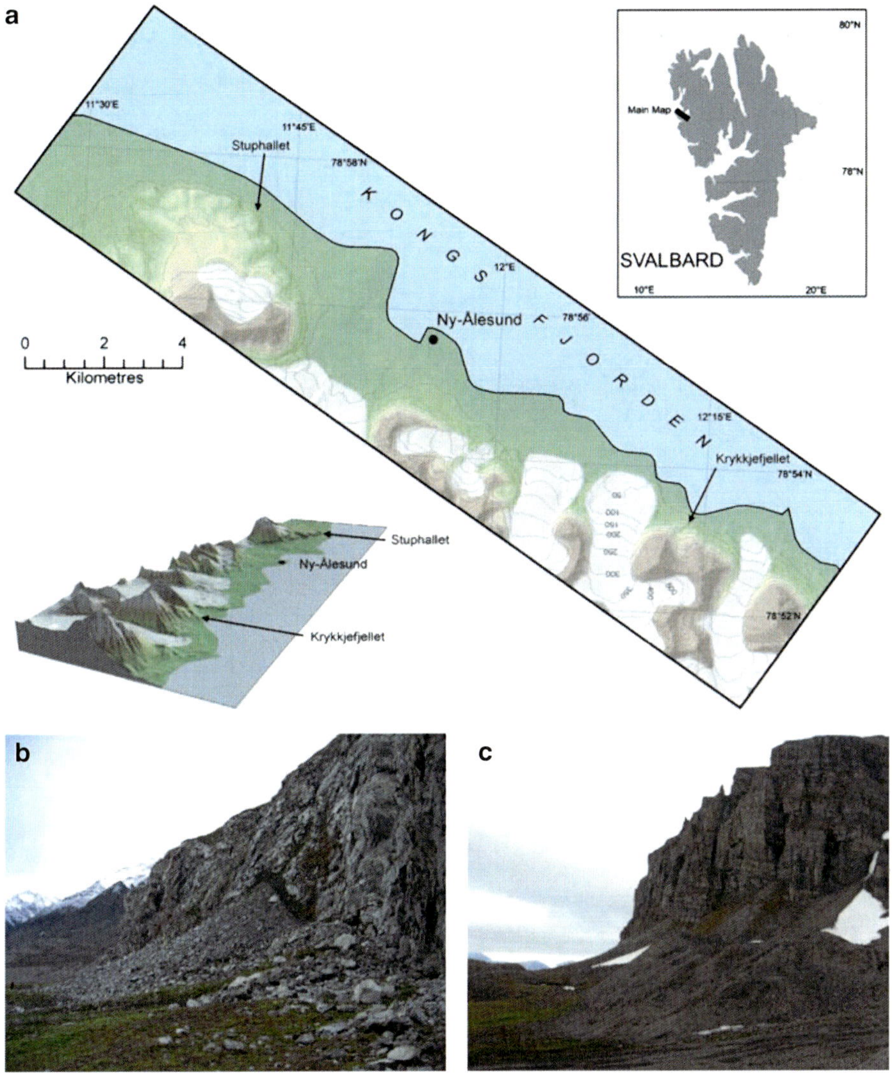

Fig. 2 Arctic field sites on Svalbard near the Ny-Ålesund Research Station. (**a**) Map data courtesy of the Norsk Polar Institute; (**b**) Photographs of Krykkjefjellet; (**c**) Stuphallet courtesy of M.S. Clark and G. Hillyard respectively

0.36 ± 0.14 g/g dry weight (Fig. 3b). Recovery time from the semi-desiccated state varied according to the level of desiccation attained (Fig. 3c) but even the most desiccated animals, treated to −14°C, recovered within 9 h at 5°C.

These physiological and biochemical studies have now been combined with gene and protein expression work in a functional genomics approach to further our understanding of the cellular mechanisms which *M. arctica* uses to combat

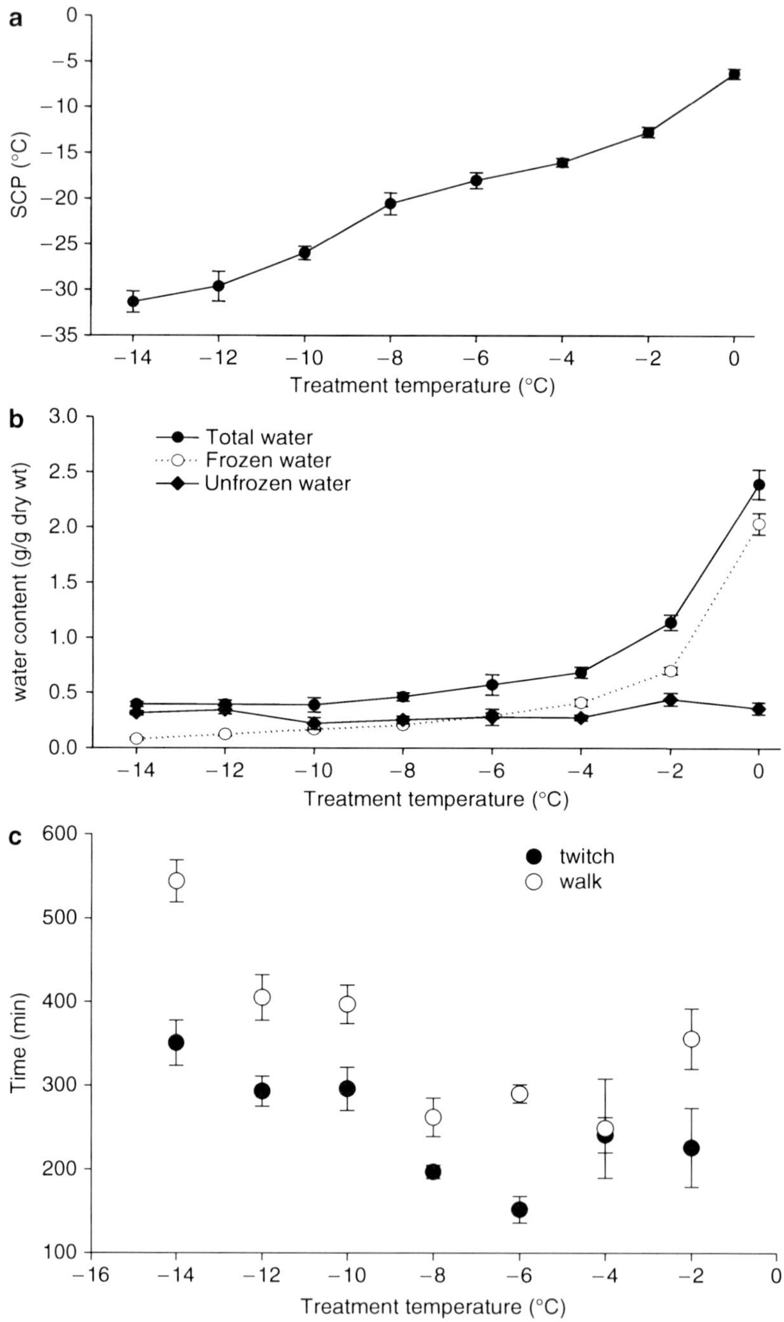

Fig. 3 (**a**) Graph showing reduction in the super cooling point (SCP) of *M. arctica* with treatment temperature; (**b**) Graph showing reduction in water content of *M. arctica* with reduction in treatment temperature; (**c**) Graph showing variation in recovery times of *M. arctica* with treatment temperature. Data courtesy of M.R. Worland and J. Purać (unpublished)

desiccation stress during this remarkable cold hardening process. EST (expressed sequence tag or gene transcription) analysis identified a number of candidate genes putatively involved in desiccation, such as the aquaporins (transmembrane proteins involved in water and glycerol transport) and heat shock proteins (Clark et al. 2007). This has now been expanded with more comprehensive microarray analyses (Clark et al. 2009). However, this chapter will concentrate on three main areas of study, where the integration between the physiology, biochemical and molecular studies are strongest, namely trehalose as a cryoprotectant, antioxidant enzymes and phospholipid fatty acid composition. These examples also demonstrate the added value obtained by utilising such interdisciplinary approaches.

9.3 Trehalose as a Cryo/Anhydro Protectant

Apart from insects, as previously discussed, trehalose has also been found in microorganisms, plants and invertebrates (nematodes, embryonic cysts of crustaceans and tardigrades), but so far not in mammals (Benaroudj et al. 2001). The mechanisms by which trehalose acts as an anhydroprotectant, preserving the functionality of biomolecules, involve water replacement, glass formation and chemical stability (Crowe et al. 1988). These are due to the presence of hydroxyl groups and the chemical properties of trehalose. It is a non-reducing sugar with a low crystallisation tendency. Therefore, trehalose can be accumulated in very high concentrations because it has no capacity to reduce and damage other compounds. This is particularly important during dehydration when water loss additionally increases trehalose concentrations. Thus a colligative effect of a high trehalose concentration is also present, lowering the freezing point of the body fluids allowing the animals to supercool to as low as −35 to −40°C (Storey and Storey 1992). Trehalose has a specific function in stabilising macromolecules and the membrane bi-layer structure and this is achieved via hydrogen bonds formed between trehalose and the proteins or polar head groups of membrane lipids. These interactions have been extensively described in anhydrobiotic systems (Crowe et al. 1987) and have been confirmed for freezing preservation in studies with isolated membranes (Rudolf and Crowe 1985). However, trehalose is also important during rehydration, when the membrane changes from the gel to the liquid crystalline state. Such changes cause membranes to become transiently leaky with a fatal loss of contents, but trehalose prevents this transition occurring in the first place by acting as a water replacement (Crowe et al. 1992).

Trehalose is commonly synthesised from glycogen, which in insects is stored in the fat body (Avonce et al. 2006) in a process that is under hormonal control (Steele 1999). Previously, we showed that the induced desiccation of *M. arctica* is combined with a rapid breakdown of glycogen, an increase in trehalose 6-phosphate synthase activity and a concomitant accumulation of trehalose (Worland et al. 1998). Glycogen is a storage compound accumulated during the active feeding stage in the summer and is converted to trehalose via glucose under dehydrating conditions (Fig. 4a). Trehalose concentration then increases as body water is lost reaching a maximum of 60–80 μg/mg fresh weight at temperatures below −6°C (Fig. 4b).

The pathway for trehalose synthesis includes several enzymes and input of high energy phosphate from ATP. The formation of trehalose occurs in a two-step

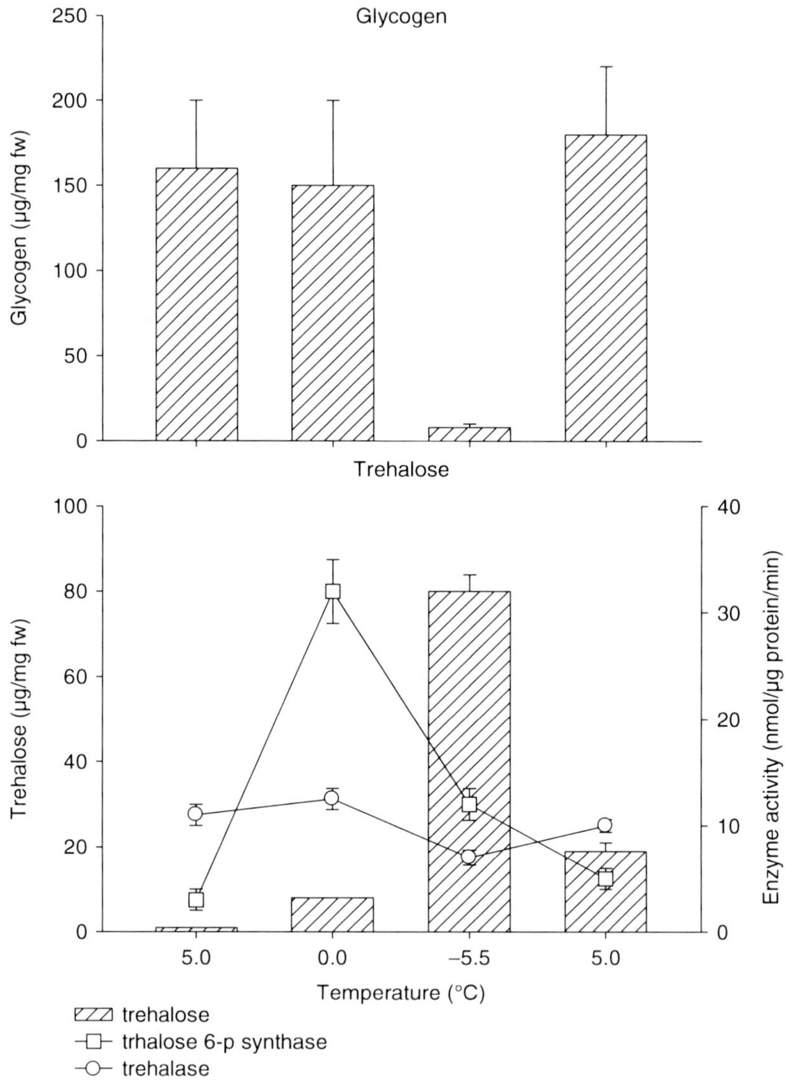

Fig. 4 (**a**) Graph showing depletion of glycogen stores with treatment temperature from +5°C through 0°C, down to −5.5°C and back through recovery to +5°C. (**b**) Graph showing production of trehalose with treatment temperature, alongside activity of TPS and trehalase. Data re-drawn from Worland et al. (1998)

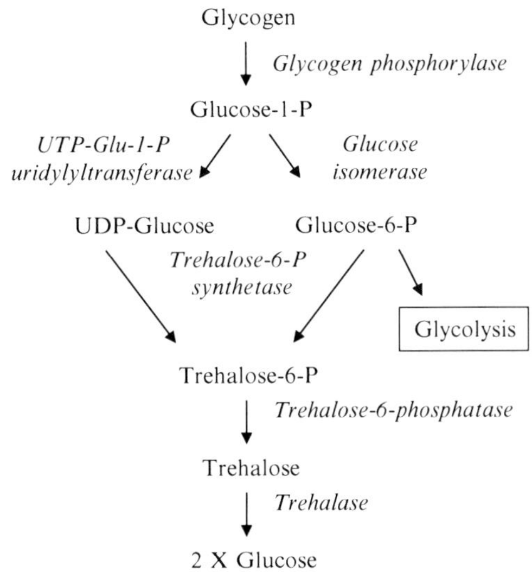

Fig. 5 Trehalose biochemical pathway

process from glucose-6-phosphate and UDP-glucose. Two enzymes are directly involved in the synthesis of trehalose: trehalose-6-phosphate synthase and trehalose 6-phosphatase. Trehalose-6-phosphate synthase and UTP-Glu-1-phosphate uridylyltransferase are thought to be one enzyme with two catalytic activities. Trehalose is broken down by the enzyme trehalase into two molecules of glucose (Fig. 5). Molecular studies in *M. arctica* revealed a further complexity to this pathway, with the discovery of duplicated forms of trehalose-6-phosphate synthase (TPS) (Clark et al. 2009).

9.3.1 Duplication of TPS Genes in M. arctica

Six EST clones were identified as showing sequence similarity to TPS using database searches (Clark et al. 2009). However, further analysis showed that these sequences differentiated into two distinct groupings (called contigs) (Fig. 6a). Contig 1 showed most sequence similarity to TPS from the Oriental migratory locust, whilst contig 2 more closely matched TPS from a *Drosophila* species (Fig. 6b). In-depth comparison of the two paralogues was difficult due to the fragmentary nature of the EST data; however, where both contigs overlapped the same region of 100 amino acids, percentage identity was calculated as 42.1%. Gene duplication is a common mechanism for increasing genome complexity and expansion of functionality (Ohno 1970). The fate of duplicated genes within a genome is complex, but retention tends to be accompanied by acquisition or partitioning of function between the paralogues (commonly known as sub-functionalisation) (Force et al. 1999). Indeed there is evidence of such in the TPS genes of *M. arctica*

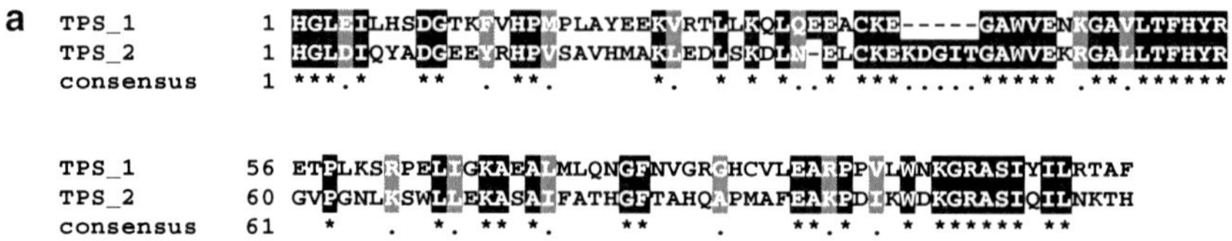

b

	Best database match	Organism	% amino acid identity
TPS_1	A8D372	*Locusta migratoria malinensis* Oriental migratory locust	86
TPS_2	B4I383	*Drosophila sechellia* *Drosophila* species	66

Fig. 6 (**a**) Sequence alignment of the two TPS contigs identified in *M. Arctica*. *Asterisks* in consensus line and *dark shading* of sequence indicate conserved amino acids with *dots* and *pale shading* indicating semi-conservative changes. (**b**) Closest database matches to the two *M. arctica* TPS contigs annotated with amino acid percentage identities

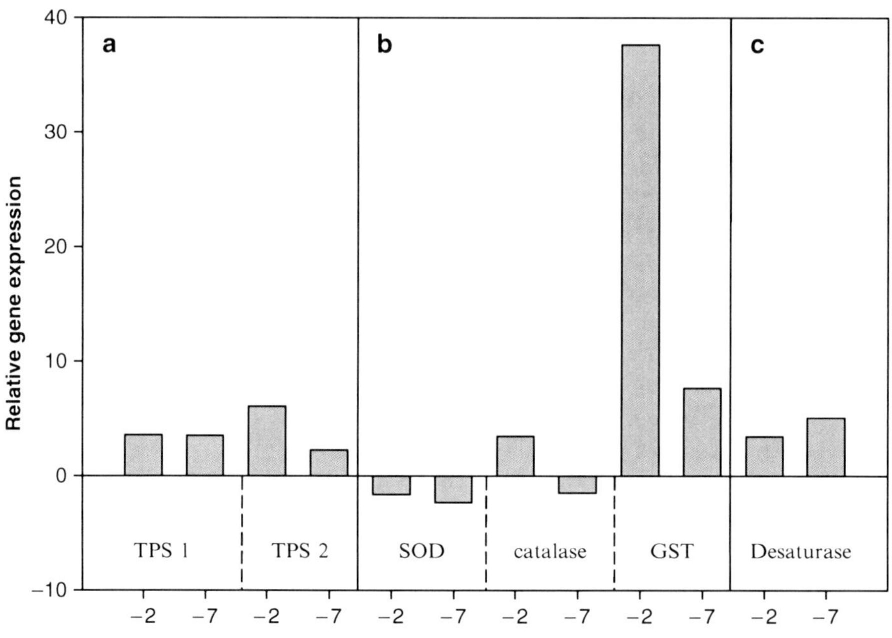

Fig. 7 Graphs showing relative gene expression levels in *M. arctica* during desiccation treatments of (**a**) duplicated TPS; (**b**) antioxidants: superoxide dismutase (SOD); catalase and glutathione-s-transferase (GST); (**c**) desaturase genes

with differing expression levels during the desiccation process (Fig. 7). Comparison with other organisms highlighted a similar gene duplication in the nematode *C. elegans*, an organism that accumulates trehalose in response to stressful conditions

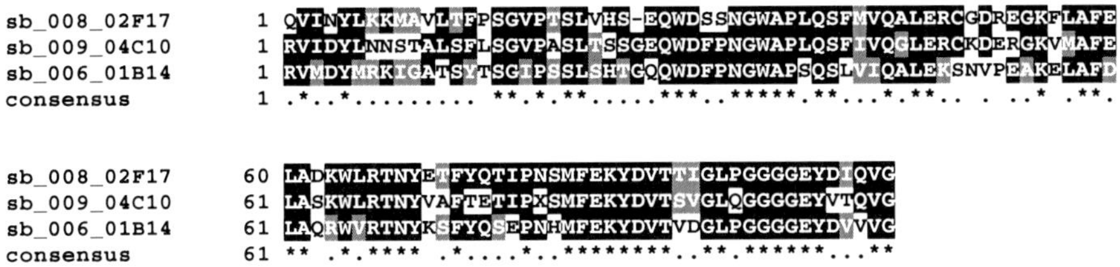

Fig. 8 Sequence alignment of the three trehalase sequences identified in *M. arctica*. *Asterisks* in consensus line and *dark shading* of sequence indicate conserved amino acids with *dots* and *pale shading* indicating semi-conservative changes

(Pellerone et al. 2003). *C. elegans* also processes multiple trehalase genes. Investigation of the *M. arctica* ESTs produced three gene fragments with sufficient difference between them (between 54.4 and 68.3% identity at the amino acid level) to define them as potentially different paralogues (Fig. 8).

Although both organisms accumulate trehalose in response to stressful conditions, why the duplication events have occurred and persisted is not clear from RNAi knock-out experiments in *C. elegans*. No obvious short-term loss of function was identified in the TPS engineered nematodes despite trehalose levels being reduced to 7% of normal in the double knock-out experiments (Pellerone et al. 2003). A further puzzle is the recent identification of duplicated TPS genes in the rotifer *Brachionus plicatilis* (Denekamp et al. 2009). This organism regularly survives stressful conditions via the production of resting eggs, but does not appear to accumulate trehalose under such conditions (Denekamp et al. 2009).

9.4 Reactive Oxygen Species and Antioxidant Enzymes

The molecular and biochemical basis of the cold hardiness strategies involve not only the production of anti-freeze agents, but also the control of pathways for their synthesis. Thus, some well-known biochemical regulatory mechanisms, such as the change of enzyme levels and the action of allosteric effectors and covalent modification of enzymes, are part of these adaptive processes. In recent years, the study of cell signalling pathways clearly suggests that beside the classical hormones and growth factors, the oxidants are the important determinants of well-adapted organisms.

Respiration and oxygen consumption in aerobic cells is accompanied by generation of reactive oxygen species (ROS) such as hydrogen peroxide (H_2O_2), superoxide anion radical (O_2^-) lipid peroxides and hydroxyl radicals. If not eliminated, ROS propagate further oxidative processes leading to damage of cellular molecules and resulting in disturbed homeostasis and cellular death (Hermes-Lima and Zenteno-Savin 2002). Like other eukaryotes, insect and other arthropods possess a suite of antioxidant enzymes, which protect their cells from the damaging effects of ROS (Ahmad 1992). Antioxidant enzymes such as superoxide dismutase (SOD) participate in dismutating the O_2^- radical to the peroxide and oxygen ($2O_2^- + H^+ \rightarrow H_2O_2 + O_2$, Fridovich 1978). Hydrogen peroxide is then converted to water and oxygen ($2H_2O_2 \rightarrow 2H_2O + O_2$) by catalase (CAT). In insects, ascorbate peroxidase (APOX) also scavenges hydrogen peroxide, but only at low concentrations (Clavaron-Mathews et al. 1997), which are not normally scavenged by CAT (which has a high Km). In addition, a glutathione s-transferase (GST) (a detoxification enzyme with peroxidase-like activity) decomposes hydroperoxides to lipid peroxides, using reduced glutathione as a substrate ($ROOH + 2GSH \rightarrow ROH + H_2O + GSSG$) (Ahmad et al. 1991; Ahmad 1992). Oxidised glutathione is then reduced by glutathione reductase (GR).

Water stress increases the formation of ROS (Hansen et al. 2006). This may be minimised to a certain extent in *M. arctica* as trehalose has the added advantage of being an antioxidant (Oku et al. 2003). Hence the accumulation of trehalose during the cryoprotective dehydration process in *M. arctica* could be additionally important in protecting proteins and unsaturated fatty acids from oxidation during dehydration. On the other hand, the cellular antioxidant system mediates the production of low to moderate concentrations of ROS as key regulators of many intracellular pathways (Valko et al. 2007). ROS are involved in diverse physiological responses affecting numerous cellular signalling pathways including gene expression and physiological regulation (Dalton et al. 1999; Suzuki et al. 1997; Halliwell and Gutteridge 2007). On the basis of the existing data that the H_2O_2 exposure of *Candida albicans* cells induces the intracellular accumulation of trehalose (Benaroudj et al. 2001), we can speculate that this ROS could be involved in the metabolically adjusted pathways for the cold resistance of *M. arctica*. Our preliminary analyses of differentially expressed genes in *M. arctica* using Q-PCR indicate the initial up-regulation of CAT in response to cold treatment, followed by down regulation at $-7°C$ (the latter result is mirrored in biochemical experiments on animals maintained down to $-20°C$ (Grubor-Lajšić, pers comm)). GST would appear to be the major antioxidant of the three tested with strong up-regulation during the desiccation process (Fig. 7). Thus revealing a role of antioxidants in establishing redox homeostasis and potentially at a time point and level that may trigger trehalose synthesis, as the main anhydro-protectant of *M. arctica*.

9.5 Phospholipid Fatty Acid Composition

In addition to accumulation of sugars and polyols, membrane phospholipids adjustments to cold and drought are also recorded in insects (Bennett et al. 1997; Bayley et al. 2001; Holmstrup et al. 2002; Bahrndorff et al. 2007; Michaud and Denlinger 2007). Bayley et al. (2001) proposed that acclimation of soil collembolans (*Folsomia candida*) to drought stress can induce changes in membrane lipid composition that are typical for low temperature acclimation. This suggests that the mechanisms of cold and drought have many similarities. These changes include the introduction of unsaturated bonds into the fatty acids of membrane phospholipids decreasing the temperature for the transition from gel to liquid crystalline state (Hazel 1995). This is part of the homeoviscous adaptation (HVA) of cellular membranes (Hazel 1995). It is well known that in fully hydrated cells, membrane lipids are in a liquid crystalline state (Luzzati and Husson 1962). The removal of water from the hydration shell of phospholipid polar head groups of desiccation-sensitive cells results in the compaction of the phospholipid molecules. This leads to a phase transition into the gel phase, resulting in serious damage to the integrity of cell membranes (Crowe and Crowe 1982, 1984). Therefore, HVA of cellular membranes with the degree of fatty acid desaturation is crucial to enhance the cold hardiness of organisms.

Our unpublished data on the composition of fatty acids of non-polar and polar lipids of *M. arctica* in control (maintained at $+5^{\circ}$C) and cold-exposed (-2°C) animals suggest changes that could be related to adaptive response to low temperature and lack of water. In response to cold the content of 16:1ω7 (palmitoleic acid), 18:1ω 9 (oleic acids) and 18:3ω3 (α-linolenic acid) in the polar pool were significantly increased. This resulted in an increase in the UFA/SFA (unsaturated fatty acid/saturated fatty acid) ratio in cold exposed springtails compared with control animals. The proportion of unsaturated fatty acids increasing as growth temperature decreased was first reported in *Escherichia coli* (Marr and Ingraham 1962) and has long been considered as an adaptive response to maintain the liquid crystalline phase at low temperatures.

The enzyme Δ9-acyl-CoA desaturase plays an essential role in HVA by increasing the ratio of unsaturated to saturated fatty acids in cell membranes and is expressed at low temperatures in fish (Tiku et al. 1996), bacteria (Sakamoto and Bryant 1997), plants (Vega et al. 2004) and insects (Kayukawa et al. 2007). Our finding of up-regulation of a desaturase gene in cold exposed springtails suggests that it could contribute to enhanced cold hardiness in *M. arctica* through the production of these unsaturated fatty acids (Fig. 7). Indeed, Kayukawa et al. (2007) have already demonstrated the role of this gene in cold hardiness, with an increase in the expression of the *Delia antiqua* Δ9-acyl-CoA desaturase with cold hardiness treatments. This was accompanied by a concomitant increase in the production of palmitoleic and oleic acids (the presumed products of Δ9-acyl-CoA desaturase).

9.6 Summary

Integrated physiological, biochemical and molecular studies can dramatically improve our understanding of organism environmental adaptations. They also enable us to draw wider conclusions across whole taxa as to the cellular changes needed to withstand environmental change. In the example of *M. arctica*, presented here, these encompass cryoprotectants, antioxidant enzymes and membrane phospholipids, all of which have been identified in other species undergoing similar desiccation-type processes. However, to expand our knowledge of the desiccation gene repertoire, high throughput genomics techniques, such as the use of microarrays, are required. These enable the identification of many genes at any one time, including putative novel candidate sequences, which can then be subjected to functional analyses. These studies are currently in progress in our laboratory, not only in *M. arctica*, but also in other springtails, such as *Cryptopygus antarcticus*, which survive extreme cold by other physiological mechanisms. Our aim is to understand how insects survive in the harsh polar environments across a range of scales from single genes through whole animal physiology to ecosystem interactions using multi-disciplinary approaches.

Acknowledgements This paper was produced within the BAS GSAC BIOREACH/BIOFLAME core programmes and also contributes to the SCAR EBA programme. JP was sponsored by the EU Sleeping Beauty Consortium: Specific Targeted Research Project, Contract no 012674 (NEST). JP and GG-L are also funded by the MSTD grant 143034, awarded by the Republic of Serbia. The authors would like to thank NERC for access to the NERC Arctic Research Station (Harland Huset) at Ny-Ålesund and Nick Cox, the Arctic base commander. We would also like to thank Pete Convey for critical reading of the manuscript and Barbara Worland and Guy Hillyard for their help with animal collection in the 2007 and 2008 field seasons respectively and Zeljko Popovic for his help with the Q-PCR.

References

Ahmad S (1992) Biochemical defence of pro-oxidant plant allelochemicals by herbivorous insects. Biochem Syst Ecol 20:269–296

Ahmad S, Duval DL, Weinhold LC, Pardini RS (1991) Cabbage looper antioxidant enzymes: tissue specificity. Insect Biochem 21:563–572

Avonce N, Mendoza-Vargas A, Morett E, Iturriage G (2006) Insights on the evolution of trehalose biosynthesis. BMC Evol Biol 6:109

Bahrndorff S, Petersen SO, Loeschcke V, Overgaard J, Holmstrup M (2007) Differences in cold and drought tolerance of high arctic and sub-arctic populations of *Megaphorura arctica* Tullberg 1876 (Onychiuridae: Collembola). Cryobiology 55:315–323

Bahrndorff S, Tunnacliffe A, Wise MJ, McGee B, Holmstrup M, Loeschke V (2008) Bioinformatics and protein expression analyses implicate LEA proteins in the drought response of colemmbola. J Insect Physiol 55:210–217

Bayley M, Petersen SO, Knigge T, Köhler HR, Holmstrup M (2001) Drought acclimation confers cold tolerance in the soil collembolan *Folsomia candida*. J Insect Physiol 47:1197–1204

Benaroudj N, Lee DH, Goldberg AL (2001) Trehalose accumulation during cellular stress protects cells and cellular proteins from damage by oxygen radicals. J Biol Chem 276:24261–24267

Bennett VA, Pruitt NL, Lee RE (1997) Seasonal changes in fatty acid composition associated with cold-hardening in third larvae of *Eurosta solidaginis*. J Comp Physiol B 167:249–255

Block W (1990) Cold tolerance of insects and other arthropods. Philos Trans R Soc Lond B Biol Sci 326:613–633

Block W, Worland MR (2001) Experimental studies of ice nucleation in an antarctic springtail (Collembola, Isotomidae). Cryobiology 42:170–181

Block W, Webb NR, Coulson S, Hodkinson ID, Worland MR (1994) Thermal adaptation in the arctic collembolan *Onychirus arcticus* (Tullberg). J Insect Physiol 40:715–722

Cannon RJC, Block W (1988) Cold tolerance of microarthropods. Biol Rev 63:23–77

Clark MS, Thorne MAS, Purać J, Grubor-Lajšić G, Kube M, Reinhardt R, Worland MR (2007) Surviving extreme polar winters by desiccation: clues from arctic springtail (*Onychiurus arcticus*) EST libraries. BMC Genomics 8:475

Clark MS, Thorne MAS, Purać J, Burns G, Hillyard G, Popovic Z, Grubor-Lajšić G, Worland MR (2009) Surviving the cold: molecular analyses of insect cryoprotective dehydration in the arctic springtail (*Megaphorura arctica*). BMC Genomics 10:328

Clavaron-Mathews M, Summers CB, Felton GW (1997) Ascorbate peroxidase: a novel antioxidant enzyme in insects. Arch Insect Biochem Physiol 34:57–68

Coulson SJ, Hodkinson ID, Block W, Webb NR, Worland MR (1995a) Low summer temperatures – a potential mortality factor for high arctic soil microarthropods. J Insect Physiol 41:783–792

Coulson SJ, Hodkinson ID, Strathdee AT, Block W, Webb NR, Bale JS, Worland MR (1995b) Thermal environments of arctic soil microorganisms during winter. Arctic Alpine Res 27:364–370

Crowe JH, Crowe LM (1982) Induction of anhydrobiosis-membrane changes during drying. Cryobiology 19:317–328

Crowe JH, Crowe LM (1984) Effects of dehydration on membranes and membrane stabilization at low water activities. In: Chapman D (ed) Biological Membranes 5. Academic Press, London, pp 57–103

Crowe JH, Crowe LM (1986) Water and carbohydrate interactions with membranes – studies with infrared-spectroscopy and differential scanning calorimetry methods. Methods Enzymol 127:696–703

Crowe JH, Crowe LM, Carpenter JF, Aurell Wistrom C (1987) Stabilisation of dry phospholipid bilayers and proteins by sugars. Biochem J 242:1–10

Crowe JH, Crowe LM, Carpenter JF, Rudolph AS, Wistrom CA, Spargo BJ, Anchordoguy TJ (1988) Interactions of sugars with membranes. Biochim Biophys Acta 947:367–384

Crowe JH, Hoekstra F, Crowe LM (1992) Anhydrobiosis. Annu Rev Physiol 54:579–599

Dalton TP, Shertzer HG, Puga A (1999) Regulation of gene expression by reactive oxygen. Annu Rev Pharmacol Toxicol 39:67–101

Denekamp NY, Thorne MAS, Clark MS, Kube M, Reinhardt R, Lubzens E (2009) Discovering genes associated with dormancy in the monogonont rotifer *Brachionus plicatilis*. BMC Genomics 10:108

Elnitsky MA, Hayward SAL, Rinehart JP, Denlinger DL, Lee RE (2008) Cryoprotective dehydration and the resistance to inoculative freezing in the antarctic midge, *Belgica antarctica*. J Exp Biol 211:524–530

Force A, Lynch M, Pickett FB, Amores A, Yan YL, Postlethwait J (1999) Preservation of duplicate genes by complementary, degenerative mutations. Genetics 151:1531–1545

Fridovich I (1978) Biology of oxygen radicals. Science 201:875–880

Halliwell B, Gutteridge JMC (2007) Free radicals in biology and medicine. Oxford University Press, Oxford

Hansen BH, Romma S, Garmo OA, Olsvik PA, Andersen RA (2006) Antioxidative stress proteins and their gene expression in brown trout (*Salmo trutta*) from three rivers with different heavy metal levels. Comp Biochem Physiol C 143:263–274

Hazel JR (1995) Thermal adaptation in biological membranes: is homeoviscous adaptation the explanation? Annu Rev Physiol 57:19–42

Hermes-Lima M, Zenteno-Savin T (2002) Animal response to drastic changes in oxygen availability and physiological oxidative stress. Comp Biochem Physiol C 133:537–556

Holmstrup M (1992) Cold hardiness strategy in cocoons of the lumbricoid earthworm *Dendrobaena octaedra* (Savigny). Comp Biochem Physiol B 102:49–54

Holmstrup M, Sømme L (1998) Dehydration and cold hardiness in the arctic collembolan *Onychiurus arcticus* Tullberg 1876. J Comp Physiol B 168:197–203

Holmstrup M, Westh P (1994) Dehydration of earthworm cocoons exposed to cold: a novel cold hardiness mechanism. J Comp Physiol B 164:312–315

Holmstrup M, Bayley M, Ramløv H (2002) Supercoolor dehydrate? An experimental analysis of overwintering strategies in small permeable arctic invertebrates. Proc Natl Acad Sci USA 99:5716–5720

Kayukawa T, Chen B, Hoshizaki S, Ishikawa Y (2007) Upregulation of a desaturase is associated with the enhancement of cold hardiness in the onion maggot, *Delia antiqua*. Insect Biochem Mol Biol 37:1160–1167

Lundheim R, Zachariassen KE (1993) Water balance in over-wintering beetles in relation to strategies for cold tolerance. J Comp Biochem Physiol 163:1–4

Luzzati V, Husson F (1962) Structure and liquid-crystalline phases of lipid-water systems. J Cell Biol 12:207–219

Marr AG, Ingraham JL (1962) Effect of temperature on composition of fatty acids in *Escherichia coli*. J Bacteriol 84:1260–1267

Michaud MR, Denlinger DL (2007) Shifts in the carbohydrate, polyol, and amino acid pools during rapid cold-hardening and diapause-associated cold-hardening in flesh fies (*Sarcophaga crassipalpis*): a metabolomic comparison. J Comp Physiol B 177:753–763

Ohno S (1970) Evolution by gene duplication. Springer, Berlin

Oku K, Watanabe H, Kubota FS, Kurimoto M, Tsujisaka Y, Komori M, Inoue Y, Sakurai M (2003) NMR and quantum chemical study on the OH$\cdots\pi$ and CH$\cdots$O Interactions between Trehalose and Unsaturated Fatty Acids: Implication for the Mechanism of Antioxidant Function of Trehalose. J Am Chem Soc 125:12739–12748

Pedersen PG, Holmstrup M (2003) Freeze or dehydrate: only two options for the survival of subzero temperatures in the arctic Enchytraeid *Fridericia ratzeli*. J Comp Physiol B 173:601–609

Pellerone FI, Archer SK, Behm CA, Grant WN, Lacey MJ, Somerville AC (2003) Trehalose metabolism genes in *Caenorhabditis elegans* and filarial nematodes. Int J Parasitol 33:1195–1206

Ring RA, Danks HV (1994) Desiccation and cryoprotection – overlapping adaptations. Cryoletters 15:181–190

Ring RA, Riegert PW (1991) A tribute to RW Salt. In: Lee RE, Denlinger DL (eds) Insects at low temperature. Chapman and Hall, London, pp 3–16

Rudolf AS, Crowe JH (1985) Membrane stabilisation during freezing: the role of two natural cryoprotectants, trehalose and proline. Cryobiology 22:367–377

Sakamoto T, Bryant DA (1997) Temperature-regulated mRNA accumulation and stabilization for fatty acid desaturase genes in the cyanobacterium *Synechococcus* sp. strain PCC 7002. Mol Microbiol 23:1281–1290

Salt RW (1957) Natural occurrence of glycerol in insects and its relation to their ability to survive freezing. Can Entomol 89:491–494

Salt RW (1961) Principles of insect cold hardiness. Annu Rev Entomol 6:55–74

Sinclair BJ, Vernon P, Klok CJ (2003) Insects at low temperature: an ecological perspective. Trends Ecol Evol 18:257–262

Sømme L (1995) Invertebrates in hot and cold arid environments. Springer, Berlin

Steele JE (1999) Activation of fat body in *Periplaneta americana* (Blattoptera: Blattidae) by hypertrehalosemic hormones (HTH): new insights into the mechanism of cell signalling. Eur J Entomol 96:317–322

Storey KB, Storey JM (1991) Glucose-6-phosphate-dehydrogenase in cold hardy insects – kinetic-properties, freezing stabilization, and control of hexose-monophosphate shunt activity. Insect Biochem 21:157–164

Storey KB, Storey JM (1992) Natural freeze tolerance in ectothermic vertebrates. Annu Rev Physiol 54:619–637

Suzuki YJ, Forman HJ, Sevanian A (1997) Oxidants as stimulators of signal transduction. Free Radic Biol Med 22:269–285

Tiku PE, Gracey AY, Macartney AI, Benyon RJ, Cossins AR (1996) Cold-induced expression of Delta(9)-desaturase in carp by transcriptional and post-translational mechanisms. Science 271:815–818

Valko M, Leibfrit D, Moncol J, Cronin MTD, Mazur M, Telser J (2007) Free radicals and antioxidants in normal physiological functions and human disease. Int J Biochem Cell Biol 39:44–84

Vega SE, del Rio AH, Bamberg JB, Palta JP (2004) Evidence for the up-regulation of stearoyl-ACP (A9) desaturase gene expression during cold acclimation. Am J Potato Res 81:125–135

Wharton DA, Goodall G, Marshall CJ (2003) Freezing survival and cryoprotective dehydration as cold tolerance mechanisms in the antarctic nematode (*Panagrolaimus davidi*). J Expt Biol 206:215–222

Worland MR (1996) The relationship between water content and cold tolerance in the arctic collembolan *Onychiurus arcticus* (Collembola: Onychiuridae). Eur J Entomol 93:341–348

Worland MR, Block W (1986) Survival and water loss in some antarctic arthropods. J Insect Physiol 32:579–584

Worland MR, Block W (2003) Desiccation stress at sub-zero temperatures in polar terrestrial arthropods. J Insect Physiol 49:193–203

Worland MR, Grubor-Lajsic G, Montiel PO (1998) Partial desiccation induced by sub-zero temperatures as a component of the survival strategy of the arctic collembolan *Onychiurus arcticus* (Tullberg). J Insect Physiol 44:211–219

Zachariassen KE (1979) Mechanism of the cryoprotective effect of glycerol in beetles tolerant to freezing. J Insect Physiol 25:29–32

Zachariassen KE (1985) Physiology of cold tolerance in insects. Physiol Rev 4:799–832

Chapter 10
A Molecular Overview of Diapause in Embryos of the Crustacean, *Artemia franciscana*

Zhijun Qiu and Thomas H. MacRae

Abstract Diapause-destined embryos of *Artemia franciscana* arrest as gastrulae and are enclosed within a chitinous shell to form cysts. Upon release from females encysted embryos enter diapause, becoming essentially ametabolic and very stress tolerant. An objective of our research is to characterize the molecular mechanisms of diapause initiation and maintenance. Subtractive hybridization was therefore used to identify up-regulated genes in diapause-destined *Artemia* embryos and these were divided into functional categories including cellular growth and stress tolerance, metabolism, and genetic and environmental information processing. Of particular interest mRNA for p8, a stress-inducible transcription co-factor, increased early in diapause-destined *Artemia* embryos, representing one of the few transcription co-factors known to be up-regulated during diapause. mRNAs encoding the three small heat shock proteins p26, ArHsp21, and ArHsp22 were also up-regulated and these proteins may promote diapause maintenance by enhancing stress tolerance. The work has revealed proteins potentially crucial to diapause, a physiological condition of fundamental and applied significance.

10.1 Introduction

10.1.1 Diapause

The ability to reversibly arrest growth by entering a dormant or semidormant state while increasing innate tolerance to environmental and physiological stress is known as diapause (Hairston 1998; MacRae 2005). Insects undergo diapause

---

Z. Qiu
Department of Biology, Dalhousie University, Halifax, NS B3H 4J1, Canada

T.H. MacRae (✉)
Department of Biology, Dalhousie University, Halifax, NS B3H 4J1, Canada
e-mail: tmacrae@dal.ca

E. Lubzens et al. (eds.), *Dormancy and Resistance in Harsh Environments*,
Topics in Current Genetics 21,
DOI 10.1007/978-3-642-12422-8_10,

(Denlinger 2002; Koštál 2006; Hahn and Denlinger 2007; Koštál et al. 2008; Sim and Denlinger 2008), as do crustaceans (Hairston 1998; MacRae 2003, 2005; Tarrant et al. 2008), other invertebrates such as *Caenorhabditis elegans* (Baumeister et al. 2006; Fielenbach and Antebi 2008) and tardigrades (Guidetti et al. 2008), fish (Podrabsky and Somero 2007), and mammals, the latter as delayed embryo implantation (Hamatani et al. 2004; Lopes et al. 2004). Diapause is divided into phases termed initiation, maintenance, and termination (Koštál 2006) and it is defined by several characteristics not all of which pertain to every organism that enters this condition. As described by Hairston (1998), two primary characteristics of diapause are neurohormonal control, or in broader terms, initiation by signals originating within organisms, and the persistence of reduced metabolism in situations that normally promote growth. Other properties of diapause include diminished changes in morphology, greater stress tolerance, behavioral modifications, occurrence at genetically predetermined developmental times, and complete development of requisite diapause characteristics in advance of exposure to adverse conditions (Hairston 1998). Achieving this altered physiological state is an active process governed by specific patterns of gene expression yielding changes in key cellular properties including metabolic activity, growth, and stress tolerance (Fig. 10.1).

Organisms in diapause and quiescence possess similar characteristics with the major difference being that quiescence terminates immediately upon return to a favorable environment without the need for an activation step such as desiccation or low temperature, as required to end diapause (MacRae 2003; Guidetti et al. 2008). Interestingly, the initiation of both quiescence and diapause involves the up-regulation of p8, a stress-inducible transcription co-factor, and proteins associated with BRCA2, a suppressor of cell growth (Qiu et al. 2007; Qiu and MacRae 2007; Sambasivan et al. 2008). Such observations suggest there are mechanistic similarities between these two events at the molecular level with entry into and maintenance of quiescence seen as active processes which need reprogramming of gene expression and adjustment of cell protein composition, just as for diapause.

10.1.2 **Artemia franciscana** ***Life History***

The brine shrimp *A. franciscana* is a Branchiopod crustacean found in extremely saline environments where fluctuations in food, oxygen, temperature, irradiation, and water availability occur routinely. *Artemia* survive unpredictable habitats because they possess a complex life history with embryos developing either ovoviviparously or oviparously (Liang and MacRae 1999; Clegg and Trotman 2002; MacRae 2003). Subsequent to fertilization ovoviviparous embryos undergo uninterrupted development and they are released from females in approximately 5 days as swimming nauplii or larvae. In contrast, oviparous embryos cease development at gastrulation, a chitinous shell permeable only to water and gasses surrounds the embryo forming a cyst that exits the female, a process that also takes about 5 days

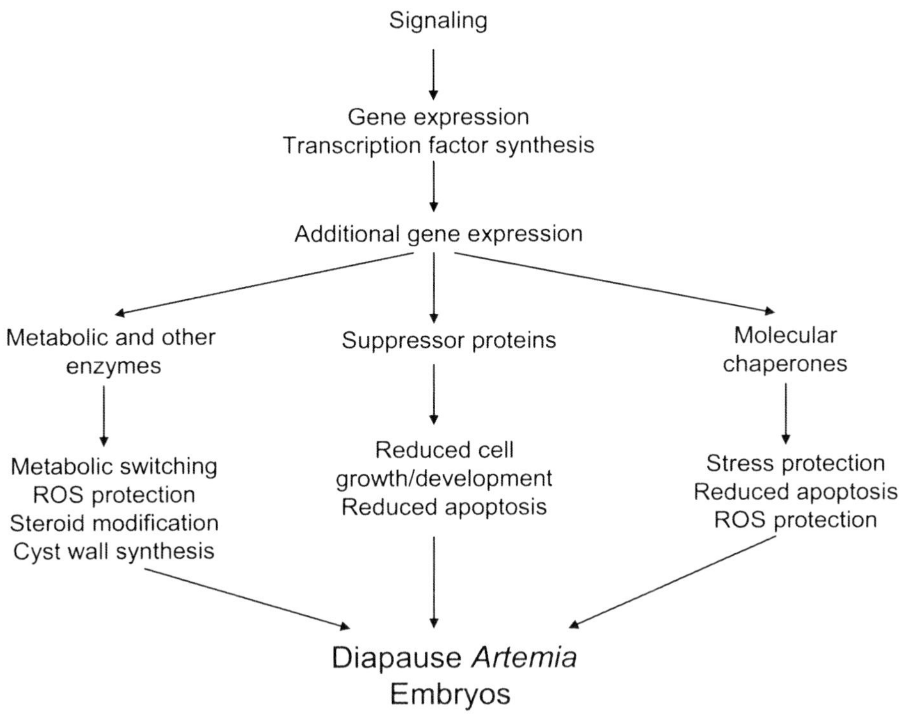

Fig. 10.1 A model of diapause initiation and maintenance. Biotic and/or abiotic signals activate transcription factor gene expression during diapause initiation and maintenance thereby rendering proteins that modulate the expression of other genes, some of which encode transcription factors. Consequently, the expression of other genes is either up- or down-regulated leading to protein changes that affect growth and development, apoptosis, metabolic shifts, and stress tolerance. Control is likely to involve kinase/phosphatase cascades, but little evidence to support this proposal is available. *ROS* reactive oxygen species. From Qiu et al. (2007)

(de Chaffoy et al. 1978; MacRae 2003; Tanguay et al. 2004). Within several days of discharge, cyst metabolic activity declines to a level difficult to detect experimentally (Clegg 1997; Clegg and Jackson 1998). Thus, for *Artemia*, attaining profound dormancy associated with diapause requires suppression of growth and cell division followed by inhibition of metabolic activity, related processes that for convenience can be considered respectively as developmental and metabolic diapause. Contingent with achieving diapause, and during postdiapause growth, encysted *Artemia* embryos are exceptionally resistant to environmental and physiological stress a capability undoubtedly critical for survival upon exposure to harsh habitats. As one example, completely hydrated, encysted, postdiapause *Artemia* embryos held at ambient temperature endure anoxia for as long as 4–6 years, a condition that kills most metazoans (Clegg 1997, 2007; Clegg et al. 2000). Possession of such characteristics suggests *Artemia* is well suited as a model system for the study of molecular mechanisms that regulate entry into and maintenance of diapause and quiescence. Consequently, we have used subtractive hybridization to examine gene expression in *Artemia* embryos 2 days postfertilization (Qiu et al. 2007), the purpose being to recognize proteins that influence diapause initiation and maintenance.

10.2 Gene Expression During Diapause

10.2.1 Identification by Subtractive Hybridization of Differentially Regulated Genes in Diapause-Destined Artemia *Embryos*

The molecular signals that initiate diapause are unknown for the most part, but encysting embryos of *A. franciscana* undergo a developmental program where cell division and metabolic activity are inhibited. Moreover, cysts acquire stress tolerance which contributes to diapause maintenance. The realization of mechanistic processes that promote diapause initiation in *Artemia* embryos is likely to depend on the expression of genes not transcribed in other life history stages. Moreover, the protein products encoded by up-regulated genes, including transcription factors or signaling molecules, may persist in cells beyond initiation and mediate diapause maintenance and termination. This possibility does not, of course, preclude the differential expression of genes responsible for maintenance and termination at other times in the diapause developmental program.

The objective of the work to be described was to characterize differential gene expression in *A. franciscana* during diapause initiation and maintenance. Subtractive hybridization was used to examine mRNA populations in oviparous (diapause-destined) vs. ovoviviparous (nauplii-destined) embryos (Qiu et al. 2007), an experimental approach employed successfully to describe gene expression in diapausing *Calanus finmarchicus* (Tarrant et al. 2008) and *Culex pipiens* L. (Robich and Denlinger 2005; Robich et al. 2007). Subtractive hybridization yielded 85 up-regulated cDNAs (genes/mRNAs) in diapause-destined *Artemia* embryos identifiable by comparison to sequences in GenBank (Table 1 in Qiu et al. 2007). The cDNAs represented 55 different deduced proteins which were divided into four categories by application of KEGG PATHWAY found at www.genome.ad.jp/keg/pathway.html (Fig. 10.2). These included proteins that participate in metabolic activities, cellular processes such as growth and stress tolerance, genetic information processing which focused predominantly on protein synthesis, and environmental information processing or gene regulation. Of the 1,128 sequenced isolates obtained by subtractive hybridization 424 (36.7%) encoded unidentified proteins and 299 (26.5%) represented hypothetical proteins. The large reservoir of unidentified proteins implies that the diapause program is mechanistically complex and that study of this process will provide interesting insights into gene expression and physiological adaptation.

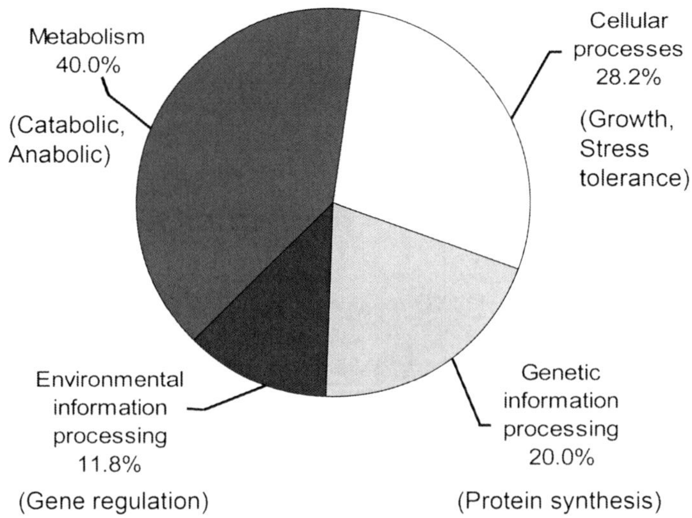

Fig. 10.2 cDNAs and proteins from diapause-destined *Artemia* embryos. Proteins identified by subtractive hybridization in diapause-destined *Artemia* embryos were divided into functional categories using KEGG PATHWAY at www.genome.ad.jp/keg/pathway.html. Adapted from Qiu et al. (2007)

The up-regulation of selected mRNAs in diapause-destined *Artemia* embryos was validated by semiquantitative PCR (Fig. 10.3). Of those mRNAs tested, 17 were shown to be up-regulated but 4 were down-regulated (Table 10.1), demonstrating that false positives are obtained by subtractive hybridization and showing the need for verification of results by alternative experimental approaches. Transcripts represented by 12 cDNAs failed to amplify by PCR (Qiu et al. 2007). The probable reason these mRNAs did not amplify is that primers were inappropriate for PCR conditions, but this has yet to be investigated.

Several of the mRNAs identified by subtractive hybridization are of particular relevance within the scheme (Fig. 10.1) proposed for diapause initiation and maintenance and some have been studied in our laboratory. p8, also known as nuclear protein 1 (NUPR1) (Chowdhury et al. 2009) or candidate of metastasis-1 (com 1) (Ree et al. 1999), is a stress-inducible basic helix–loop–helix transcription co-factor (Mallo et al 1997; Qiu and MacRae 2007) with the potential to differentially control gene expression. Three small heat shock proteins (sHSPs), molecular chaperones known for preventing stress-induced irreversible protein denaturation and cell death by apoptosis (Sun and MacRae 2005a; Haslbeck et al. 2005), were recognized and they are likely to function in diapause maintenance. The *Artemia* sHSPs include p26-like, renamed ArHsp21 (Qiu and MacRae 2008a), Hsp16-like, renamed ArHsp22 (Qiu and MacRae 2008b) and p26, an *Artemia* cyst-specific sHSP (Jackson and Clegg 1996; Liang and MacRae 1999; Sun et al. 2004, 2006; Qiu et al. 2006; Villeneuve et al. 2006).

Other mRNAs were distinguished by subtractive hybridization and characterized by semiquantitative PCR, but unlike the mRNAs just mentioned their protein

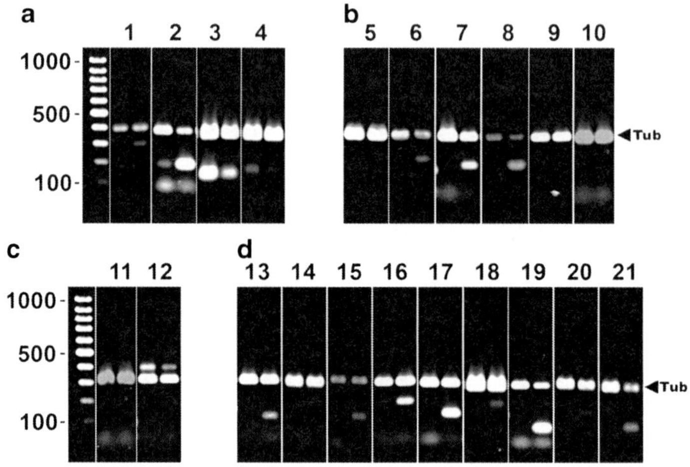

Fig. 10.3 Gene expression in diapause-destined *Artemia* embryos. RT-PCR products obtained by amplification of RNA from nauplii- (*left lane* in each pair) and diapause- (*right lane* in each pair) destined *Artemia* embryos at 2 days postfertilization were electrophoresed in 1.5% agarose gels and stained with Gel Star. (**a**) Environmental information processing: 1, p8; 2, Takeout; 3, N33 protein; 4, FLI-LRR associated protein 1. (**b**) Cellular processes: 5, BRCA1 associated protein-1; 6, p26-like; 7, neuralized-like protein 2; 8, Hsp16-like; 9, CDK5 activator binding protein (a weak band was visible in the *right lane*); 10, RNA binding protein (a weak band was visible in the *left lane*). (**c**) Genetic information processing: 11, Erv1-like growth factor (a weak band was visible in the *right lane*); 12, QM protein, ribosomal protein L10. (**d**) Metabolism: 13, glucose-6-phosphate isomerase; 14, aldehyde dehydrogenase; 15, steroid dehydrogenase; 16, dimeric dihydrodiol dehydrogenase; 17, cytochrome P450; 18, uncoupling protein; 19, chitin binding protein (DW678180); 20, chitin binding peritrophin-A precursor (DW678202); 21, chitin binding protein (DW678209). Accession numbers differentiate chitin binding proteins. Size markers in base pairs are on the *left*. Tub, tubulin. From Qiu et al. (2007)

products remain to be examined. The up-regulation of mRNA encoding Takeout, a hemolymph juvenile hormone binding protein able to modulate insect development by interacting with juvenile hormone (Noriega et al. 2006; Saito et al. 2006), suggests hormonal influences on *Artemia* diapause. BRAC1 associated protein-1, an ubiquitin carboxy-terminal hydrolase with tumor suppressor activity (Jensen and Rauscher III 1999; Mullan et al. 2006), potentially inhibits cell growth and division, thus contributing to cessation of embryogenesis, an early step in the *Artemia* diapause developmental program. Unexpectedly, the putative tumor suppressors N33 protein (MacGrogan et al. 1996; Ahuja et al. 1998) and the QM protein/ribosomal protein L10 (Oh et al. 2002; Park and Jeong 2006) are down-regulated in *Artemia*, not what is expected if they promote cessation of embryo development. However, QM protein/ribosomal L10 protein binds with the SH3 domain of Src kinases such as c-Yes and decreases their activity. Consequently, reduced amounts of QM protein/ribosomal protein L10 may lead to increased Src-like kinase activity thereby directing signaling pathways that shape diapause initiation and maintenance. Neuralized-like protein 2, an E3 ubiquitin ligase similar to BRAC1, regulates the Notch signaling pathway in insects (Hu et al. 2005; Song et al. 2006).

Table 10.1 Expression of selected genes in diapause-destined *Artemia* embryos

Functional category	Gene name/ accession number	Function	Expression up-regulated	Expression down-regulated
Environmental information processing	p8 DW678166	Co-transcription factor	Yes	
	Takeout DW678196	Hemolymph juvenile hormone binding protein	Yes	
	N33 protein DW678157	Putative tumor suppressor		Yes
	FLI–LRR associated protein-1 DW678205	Actin organization		Yes
Cellular processes	BRCA1 associated protein-1 DW678176	Tumor suppressor Ubiquitin carboxy-terminal hydrolase	Yes	
	p26-like DW678179	Small heat shock protein, molecular chaperone	Yes	
	Neuralized-like protein 2 DW678200	Ubiquitin ligase, protein degradation	Yes	
	Hsp16-like DW678191	Small heat shock protein, molecular chaperone	Yes	
	CDK5 activator binding protein DW678172	Binds activating subunit of Cdc2-like kinase	Yes	
	RNA binding protein DW678193	Translation, splicing, hnRNP formation		Yes
Genetic information processing	Erv1-like growth factor DW678207	Sulfhydryl oxidase, augmenter of liver regeneration ALRp	Yes	
	QM protein, ribosomal protein L10 DW678182	Tumor suppressor, signal regulator		Yes
Metabolism	Glucose-6-phosphate isomerase DW678186	Glycolysis	Yes	
	Aldehyde dehydrogenase DW678194	Aldehyde oxidation	Yes	

(*continued*)

Table 10.1 (continued)

Functional category	Gene name/ accession number	Function	Expression up-regulated	Expression down-regulated
	Steroid dehydrogenase DW678203	Steroid oxidation	Yes	
	Dimeric dihydrodiol Dehydrogenase DW678175	Oxidation of aromatic hydrocarbon transdihydrodiols to catechols	Yes	
	Cytochrome P450 DW678185	Hydroxylation of organics including steroids	Yes	
	Uncoupling protein DW678167	Proton channel, uncouples oxidative phosphorylation and ATP production	Yes	
	Chitin binding protein DW678180	Peritrophic matrix	Yes	
	Chitin binding peritrophin-A precursor DW678202	Chitin binding, junction protein cej-1	Yes	
	Chitin binding protein DW678209	Peritrophic matrix	Yes	

From Qiu et al. (2007)

The relationship to Notch evokes a role for neuralized-like protein 2 in suppression of macromolecular synthesis and cell division in diapause-destined *Artemia* embryos.

The largest cluster of up-regulated mRNAs detected by subtractive hybridization is housed in the metabolism category (Table 1 in Qiu et al. 2007). The up-regulation of nine mRNAs in this group was confirmed by PCR (Table 10.1), but their protein products have not been studied. Diapause-destined *Artemia* embryos remain metabolically active in the brood sac and for several days after release from females, with changes in metabolism occurring when development plays out. For example, the up-regulation of glucose-6-phosphate isomerase mRNA indicates greater reliance on glycolysis as cysts mature and enter diapause. By comparison, the northern house mosquito *Culex pipiens* undergoes a metabolic switch as adult females change from blood feeding to sugar utilization and lipid accumulation during diapause (Robich and Denlinger 2005). mRNAs for steroid dehydrogenase and cytochrome P450, enzymes capable of steroid modification (Bishop et al. 2006; Lukacik et al. 2007), increase in diapause-destined *Artemia* embryos. Therefore, hormonal regulators which have prominent roles in insect diapause (Denlinger 2002; MacRae 2005; Jing et al. 2007; Singtripop et al. 2008) could mold diapause in *Artemia* embryos, although there is no direct evidence to support this proposal. The up-regulation of mRNAs encoding dimeric dihydrodiol dehydrogenase and chitin binding proteins was observed and they may mediate formation of the chitin-enriched cyst shell. Changes in metabolic enzymes have the potential to shift catabolic and anabolic activities in *Artemia* embryos thus governing diapause initiation and maintenance.

10.2.2 Other Examples of Diapause-Dependent Gene Regulation in Arthropods

The mosquito *C. pipiens* undergoes adult diapause in response to decreasing temperature, day length shortening, and reduced juvenile hormone (Robich et al. 2007). Except for a sHSP mRNA, there is little similarity between modified transcripts identified during *A. franciscana* embryo diapause (Table 10.1, Fig. 10.3) (Qiu et al. 2007) and *C. pipiens* adult diapause (Robich and Denlinger 2005; Robich et al. 2007). There are parallels between functional categories, but this is difficult to escape due to the general nature of the groups used to partition mRNAs. Additionally, even though the most extensive analyses of differential gene expression during arthropod diapause have been accomplished for *A. franciscana* and *C. pipiens*, the number of mRNAs (genes) identified, especially when the complexity of changes experienced by diapausing organisms is considered, is limited. Diapause in the flesh fly *Sarcophaga crassipalpis* has also been extensively characterized with much of the data describing modifications of heat shock proteins (Denlinger 2002; Hayward et al. 2005; Fujiwara and Denlinger 2007; Li et al. 2007; Rinehart et al. 2007).

Diapause in the copepod *C. finmarchicus* involves a developmental delay where molting ceases and preadult C5 copepodids move deeper into the ocean, limit their metabolic activity, and consume stored lipids (Tarrant et al. 2008). As determined by suppressive subtractive hybridization and quantitative real-time PCR the genes encoding ferritin and the ecdysteroid receptor are up-regulated in diapausing *Calanus*. Up-regulation of ferritin, a metal chelating protein, was proposed to shield against oxidative damage and perhaps delay development by binding iron, whereas increased ecdysteroid receptor, a transcription factor activated by ecdysteroid binding, implies endocrine control (Johnson 2003; Tarrant et al. 2008). The inability to induce copepod diapause in the laboratory limits molecular analysis at this time, but the study of natural populations has produced useful data, and as for *Artemia* the results suggest a regulatory role for steroid hormones. A ferritin homologue termed artemin is produced in diapausing *Artemia*. Artemin does not bind metals and it is thought to be a molecular chaperone, an activity shared by ferritin (Chen et al. 2003, 2007). It will be interesting to learn if *C. finmarchicus* ferritin is a chaperone.

mRNA characterization, whether by subtractive hybridization or other methodologies, indicates genes with potential roles in diapause initiation and maintenance, thereby suggesting avenues for further work. For example, the presence within cells of protein products encoded by identified genes/mRNAs can be studied revealing if the mRNA has been translated. It is possible to search for proteins after their corresponding mRNAs have disappeared revealing aspects of protein stability and influence not possible to see by examining mRNA only. Moreover, studies focusing on mRNA detection and quantification do not provide a definitive demonstration of gene expression at a particular time, only that the gene was expressed. Thus, in studies designed to elucidate mechanistic aspects of diapause it is important to consider the proteome, but even this is limited unless protein function is demonstrated at the time and intracellular location of interest. There are effective ways to address issues of protein function, including knock down by RNAi, gene knock out, and analysis of mutations. The method of choice will vary and unpublished data generated in my laboratory by the application of RNAi technology is proving useful for the functional analysis of proteins encoded by mRNAs up-regulated in diapause-destined embryos of *Artemia*.

10.3 p8, a Transcription Co-factor Up-regulated Early in *Artemia* Diapause

10.3.1 Characterization of p8

Protein products of mRNAs up-regulated in diapause-destined *Artemia* embryos at 2 days postfertilization have been characterized. The first of these is p8, a stress-inducible, basic helix–loop–helix transcription co-factor with a bipartite nuclear localization signal indicative of proteins that enter nuclei (Fig. 10.4) (Qiu and MacRae 2007). Similar transcription co-factors occur in mammalian cells (Mallo et al. 1997; Vasseur et al. 1999; Jones 2004; Valacco et al. 2006). The deduced amino acid sequence of *Artemia* p8 consists of 66 residues and in comparison to mammalian p8 it, as well as other p8 proteins from invertebrates, exhibits an

a

```
                                               M  S  E  D
CATTTGAAAGGAAGAGTTTCAGGTAGTAGGATCAGGTGTGACGATGTCAGAAGATC 56
H  F  D  R  F  E  H  F  N  F  D  M  D  K  H  L  F  S  G  Q
ATTTTGATAGATTTGAACACTTTAACTTTGATATGGACAAGCATCTCTTCTCCGGGCAA 115
 S  G  R  L  R  T  K  Q  E  A  A  L  H  T  N  R  F  D  P  S
AGTGGAAGGCTGCGGACGAAACAGGAAGCTGCTCTTCATACAAACCGATTTGACCCTAG 174
  G  H  S  R  K  L  V  T  K  M  K  N  T  E  I  K  K  R  A
TGGACACTCAAGAAAGCTAGTAACGAAAATGAAAAACACGGAAATTAAAAAACGTGCTG 233
D  K  K
ACAAAAAATGACTTCTCCAGGGTTGTGAATTCTATAGGTCCCGGACTTCCGATCTCGTT 292
TTATGGCCTTCGTAATTGCAGTGGGACTGATTTTTTTTCAATTAATTGTATTAATTGTG 351
GACAATGGTGCTGGTAATTTAGCAACATGTAGGGCTCTTGGTGACAAAAGTGATAAAAA 410
TAGTTCCAAAAATGTGATATTTTGGGTTTTATATTTACTCCTTTCTTTTGCTGGCAACA 469
CTGTCTGAACATGTTTTTTTTTAATATTGTACTCGGTTGTGGTTTTCTTGCGTATACTG 528
TCTTTTGTATTCATTTTATCTGATAATAAACCTTTTGTAAAAAAAAAAA          577
```

b

```
                                                           Basic  Helix I  Loop    Helix II
Mosquito  ------------------------DDYDKYDFEMEKHIHSGHSGKQRSKKEASDHTNHYDPSGHSRKIVTKLKNTESNKK------- 56
Fly       ------------------MSEAHFDEYEHYNFDHDKHIFSGHSGKQRNKREANEHTNHFDPSGHSRKILTKLMNTNNNNKKAAACKN 69
Shrimp    ------------------MSEDHFDRFEHFNFDMDKHLFSGQSGRLRTKQEAALHTNRFDPSGHSRKLVTKMKNTEIKKRADKK--- 66
Rat       MATLPPTAHTSQQPVNIEDEDGILDEYDQYSLAQSY-VVGGGR-KGRTKREAAANTNRPSPGGHERKLLTKFQNSE-RKK--AWR-- 80
Mouse     MATLPPTANPSQQPLNLEDEDGILDEYDQYSLAHPC-VVGGGR-KGRTKREAAANTNRPSPGGHERKLLTKFQNSE-RKK--AWR-- 80
Human     MATFPPATSAPQQPPGPEDEDSSLDESDLYSLAHSY-LGGGGR-KGRTKREAAANTNRPSPGGHERKLVTKLQNSE-RKKRGARR-- 82
Frog      MKTSYIEAN-KVQPTDFEVQ--YFDEYEYYNLTDRYSLPTAAR-KGRTKKEAEANTNRESPCGHERKISSKLQRSECKKKLKVAKV- 82
                                  *   : :.:      :   .    : *.*:**   :**:  .*  **.**:  :*:  .::  .::
                                                                          |____NLS____|
                                                                          R48  Artemia  D64
```

Fig. 10.4 Sequence of *Artemia* p8. (**a**) The complete *Artemia* p8 cDNA and deduced amino acid sequences. *Boxed* C and A, transcription start sites; *boxed* ATG, translation start site; *boxed* TGA, stop codon; *boxed* AATAAA, polyadenylation signal; *shaded* ATTTA, sequences indicative of rapidly degraded mRNAs. (**b**) p8 amino acid sequences from mosquito (XP_315506), fly (NP_609539), rat (AAB94673), mouse (NP_062712), human (AAC19384), and frog (BAB33387) were compared by ClustalW to *Artemia* (shrimp) p8 (ABD19714). *Asterisk* identical residue; *colon*, conserved substitution; *dot*, semi-conserved substitution. Basic, helix I, loop, and helix II motifs are *boxed*; *NLS* nuclear localization signal with key basic residues *shaded* in *Artemia* p8. Adapted from Qiu and MacRae (2007)

amino-terminal deletion (Fig. 10.4). The missing residues encompass a PEST motif thought to enhance ubiquitin-dependent protein degradation (Goruppi and Kyriakis 2004), and its absence suggests stabilization of *Artemia* p8.

10.3.2 Developmental Regulation of p8 in Artemia *Embryos*

The amount of p8 mRNA in *Artemia* embryos was measured by quantitative PCR demonstrating an abrupt rise at day 1 postfertilization in diapause-destined embryos followed by a gradual decline until the mRNA had all but disappeared prior to cyst release (Fig. 10.5). In contrast, p8 mRNA did not increase in *Artemia* embryos developing directly into nauplii (Fig. 10.5). The disappearance of p8 mRNA

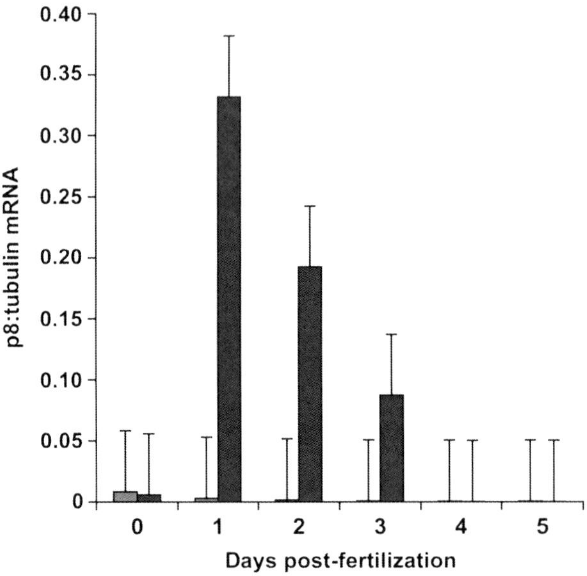

Fig. 10.5 p8 mRNA during *Artemia* embryo development. p8 mRNA from *Artemia* embryos was measured at daily intervals postfertilization by quantitative PCR using *Artemia* α-tubulin mRNA as standard. *Light* and *dark shaded bars*, p8 mRNA in nauplii- and diapause-destined embryos, respectively. From Qiu and MacRae (2007)

indicates that this transcription co-factor regulates gene expression early in diapause and has less influence later in development. To test this idea antibody raised to p8 purified from transformed bacteria was used to probe cell-free extracts of *Artemia* embryos collected at daily intervals postfertilization. p8 was observed only in diapause-destined embryos, but unlike p8 mRNA, the protein remained in embryos until their discharge from females (Fig. 10.6). Persistence in embryos extends the potential influence of p8 to diapause maintenance and termination, a conclusion that would not have been possible if only mRNA were examined. Immunofluorescent staining disclosed p8 in nuclei from cysts but not in nuclei from second instar larvae arising from cysts (Qiu and MacRae 2007). This result, in concert with the absence of p8 in nauplii-destined embryos, indicates that p8 regulates the expression of genes within the diapause developmental program, but perhaps not beyond this point.

10.3.3 Diapause-Related Transcription Factors in Organisms Other than Artemia

The best studied diapause-associated transcription factors are DAF-12 and DAF-16/FOXO (forkhead transcription factor) in *C. elegans* (Baumeister et al. 2006;

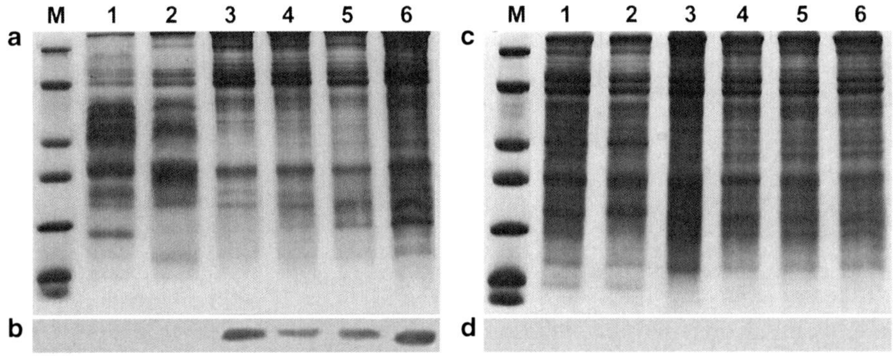

Fig. 10.6 p8 in *Artemia* embryos. Protein extracts prepared from diapause-destined (**a**, **b**) and nauplii-destined (**c**, **d**) *Artemia* embryos at daily intervals postfertilization were electrophoresed in SDS polyacrylamide gels and either stained with Coomassie blue (**a**, **c**) or transferred to nitrocellulose and probed with antibody to p8 (**b**, **d**). Lanes 1–6, day 0 (fertilization) to day 5 postfertilization. All lanes received 30 μg of protein. p8 had a molecular mass of approximately 8.0 kDa. M, molecular mass markers of 116.0, 66.2, 45.9, 35.0, 25.0, 18.4, and 14.4 kDa. Adapted from Qiu and MacRae (2007)

Fielenbach and Antebi 2008), an organism that enters diapause (dauer) in response to crowding and/or nutrient limitation. Additionally, as indicated by RNAi, the transcription factor FOXO regulates lipid sequestering in *C. pipiens*, a process essential to diapause in this organism (Sim and Denlinger 2008). Other transcription factors observed in diapausing insects are POU, which may modulate synthesis of pheromone biosynthesis-activating neuropeptide and diapause hormone (Zhang et al. 2004), and an ETS homologue (Suzuki et al. 1999), both in *Bombyx mori*. A basic helix–loop–helix protein named Har-DHMBP-3, thought to influence production of the same proteins as POU, was identified in *Helicoverpa armigera*. Transcription factors affecting delayed implantation were also tentatively recognized by differential gene expression (Hamatani et al. 2004; Lopes et al. 2004) but their functions are unknown.

p8 is the first transcription factor proposed to have a role in crustacean diapause. In other biological systems p8 promotes expression of genes that suppress cell growth (Malicet et al. 2003; Jiang et al. 2006) and apoptosis (Giroux et al. 2006), although opposite effects have been reported (Vasseur et al. 2002; Malicet et al. 2006). Inhibition of cell growth and apoptosis by p8-dependent gene regulation, if it were to occur, is important to *Artemia* diapause because embryo development must halt at gastrulation and prevention of death offers additional protection to cells. Additionally, the gene encoding p26, a diapause specific, abundant sHSP has a basic helix–loop–helix transcription factor interaction site (Qiu et al. 2006) suggesting p8 regulates mRNA production and ultimately, synthesis of proteins important in diapause maintenance.

10.4 sHSPs and Diapause Maintenance in *A. franciscana*

10.4.1 Artemia *sHSPs*

mRNAs encoding the sHSPs p26, ArHsp21, and ArHsp22 are up-regulated during development of diapause-destined *Artemia* embryos (Table 10.1; p26, a well characterized sHSP, was identified by subtractive hybridization but not included in the table). Deduced amino acid sequences of the *Artemia* sHSPs, each obtained by cloning of cDNAs, have conserved α-crystallin domains but divergent flanking amino- and carboxy-terminal regions (Fig. 10.7). p26 possesses 29% identity and 51% similarity to ArHsp21 and 38% identity and 50% similarity to ArHsp22, whereas ArHsp21 and ArHsp22 share 27% identity and 52% similarity. Several other *Artemia* sHSPs properties are compared in Table 10.2 (Liang et al. 1997; Liang and MacRae 1999; Sun et al. 2004, 2006; Qiu and MacRae 2008a, b). Of note, each *Artemia* sHSP possesses a highly conserved arginine within the α-crystallin domain, a distinction of most sHSPs (Sun and MacRae 2005a). Substitution of the conserved arginine with another residue is associated with several mammalian diseases and generally has a significant effect on sHSP structure and function (Sun and MacRae 2005a, b), as is true for p26 (Sun et al. 2006). The *Artemia* sHSPs have conserved amino-terminal WXDPF motifs which are duplicated only in ArHsp22 and each sHSP has a carboxy-terminal I/VXI/V motif thought to facilitate monomer–monomer association. All *Artemia* sHSPs oligomerize (Crack et al. 2002; Sun et al. 2004; Qiu and MacRae 2008a, b) as is common for

```
p26             -MALNPWYGGFGGMTDPWSDPFG------FGG-FGGGMDLDIDRPF--------RRRMMR  44
ArHsp21         -MSGMRLARSLLLLGRPQSRHLF------WGRRTWDPFEELRMIMREMENQFQNINQNVF  53
ArHsp22         MTTLVPWTDQW---TDPWEDPFADLPVETFTGRWRDPFAADVYKPYGLPRTHLHRRRRRR  57
                  :             * .  : . . .::     . :                 .:

p26             RGPDTSRALKELATPGSLRDTADEFQVQLDVGHFLPNEITVKTTDDDILVHGKHDERSD-  103
ArHsp21         KALPSSFKEETAVPVISSKGDDNMYRLVLDLSGFKPEDVKIDLMDRNLRVTGKCEQKTS-  112
ArHsp22         RIRTVQRVFSRKGTDVRTREDDKEWEITMQLPGFLPSDITVNSTDKEIIVHGVHKERPDY  117
                :    .   .   .    :   . :.: :::  * *.::.:.  * :: * *  .::..

p26             --EYGHVQREFRRRYRLPEHVKPESVSSTLSSDGVLTIHAPKTALSSPTERIVPITPAPA  161
ArHsp21         --DGCRMYHETQREYLLPENVNLNELKSAFTDSGYLTIEAPMPEGMKPKE--IPINRGAQ  168
ArHsp22         EGEEGYVSREIRRRFVPPKTINPGELSSTFSSDGELRIHAPKAIPGEPRQRRIQIMPAPI  177
                ..:   : :* :*.:  *: ::  .:.*:::..* * *.** .   .* :  : * : :

p26             VGRIEGGTTGTTTGSTASSTPARTTRSGGAA  192
ArHsp21         QIESESKE--------SKRED----------  181
ArHsp22         GSRFEG--------ENEEEWP----------  190
                    *
```

Fig. 10.7 Alignment of *Artemia* sHSPs. The deduced amino acid sequences of p26 (AF031367), ArHSP21 (ABD19712), and ArHsp22 (ABD19713) were aligned by CLUSTALW. *Asterisk* identical residues; :, conserved substitution; ., semi-conserved substitution. The α-crystallin domain is *shaded* as are amino-terminal arginines in ArHsp22 and p26. Amino-terminal WXDPF-like motifs, highly conserved arginines in α-crystallin domains, and carboxy-terminal I/VXI/V motifs are *boxed*. Adapted from Qiu and MacRae (2008b)

Table 10.2 Comparison of *Artemia* sHSPs

	Artemia sHSPs		
Characteristic	p26	ArHsp21	ArHsp22
Length (amino acids)	192	181	190
Molecular mass (kDa)	20.8	21.1	22.4
α-Crystallin domain	+(61–152)	+(70–162)	+(74–168)
Conserved arginine	+(114)	+(123)	+(130)
Arginine enriched amino-terminus	+	−	+
WXDPF motif	+ (17–21)	+ (27–31)	+ (15–19) (31–35)
I/VXI/V motif	+(154–156)	+(161–163)	+(170–172)
Oligomerization	+	+	+
Chaperone activity	+	+	+
Nuclear localization	+	−	+
Heat Induced	−	−	+(Adults)
Developmentally regulated	+	+	+

Numbers in brackets positions of amino acid residues; (*Adults*) heat induced expression of ArHsp22 occurred only in adults; + characteristic is observed; − characteristic is not observed

sHSPs, and all exhibit molecular chaperone activity as demonstrated in vitro by inhibiting heat-induced denaturation of citrate synthase (Fig. 10.8a) and reduction-induced denaturation of insulin at room temperature (Fig. 10.8b). p26 confers thermotolerance on transformed bacteria (Liang and MacRae 1999; Crack et al. 2002) and transfected mammalian cells (Villeneuve et al. 2006) but this has not been tested for ArHsp21 and ArHsp22. Overall, the *Artemia* sHSPs are similar to one another and to sHSPs from other organisms and they are likely to protect proteins within embryos from irreversible stress-induced denaturation.

10.4.2 sHSP Synthesis and Localization in Diapause-Destined Artemia *Embryos*

The synthesis and accumulation of p26, which represents about 10% of soluble protein in *Artemia* cysts, have been well documented (Jackson and Clegg 1996; Liang and MacRae 1999; MacRae 2003; Qiu et al. 2006). Probing of northern blots revealed p26 mRNA at day 2 postfertilization in diapause-destined embryos but not in embryos developing directly into nauplii (Liang and MacRae 1999). Analysis by real-time PCR, a more sensitive technique, disclosed a small quantity of p26 mRNA in nauplii-destined embryos at days 2 and 3 postfertilization but none later in development (Qiu et al. 2006). p26 first appeared at day 3 postfertilization in diapause-destined *Artemia* embryos but the protein was never observed in ovoviviparously developing embryos (Jackson and Clegg 1996; Liang and MacRae 1999). p26 localizes to cyst nuclei and cytoplasm under stress and nonstress conditions implying that proteins in both cell compartments are clients of this chaperone (Clegg et al. 1995; Liang and MacRae 1999; Willsie and Clegg 2001).

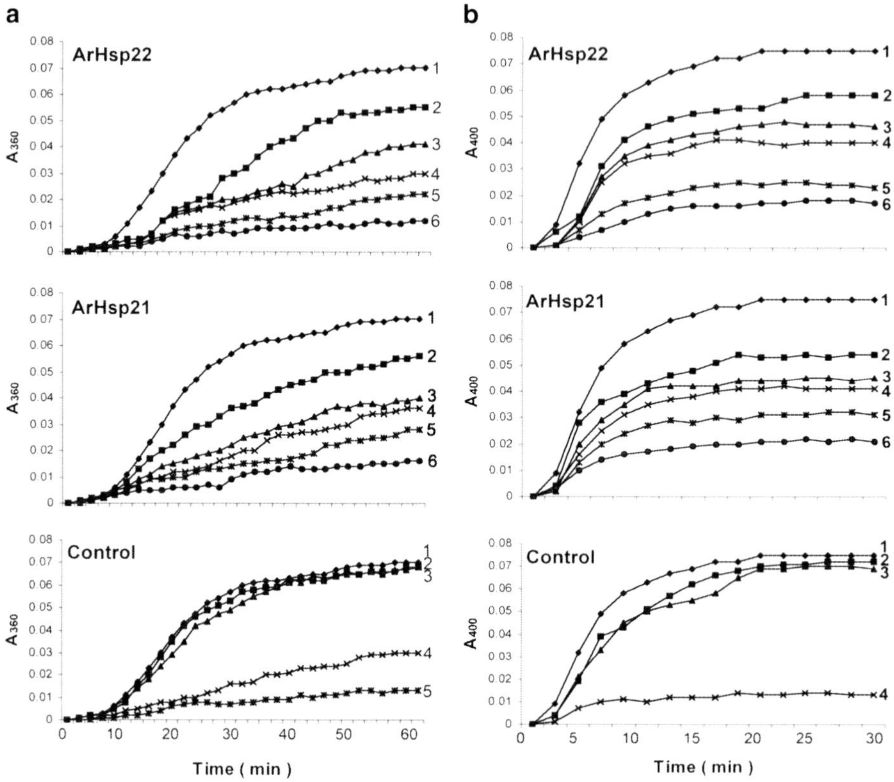

Fig. 10.8 Chaperone activity of *Artemia* sHSPs. (**a**) Purified *Artemia* sHSPs produced in transformed bacteria were incubated with 150 nM citrate synthase at 43°C and turbidity was measured at A_{360}. The curves in graphs labeled ArHsp22 and ArHsp21 represent: 1, no sHSP added; 2, 37.5 nM sHSP; 3, 75 nM sHSP; 4, 150 nM sHSP; 5, 300 nM sHSP; 6, 600 nM sHSP. The *curves* in the graph labeled Control represent: 1, no sHSP added; 2, 600 nM bovine immunoglobulin; 3, 600 nM bovine serum albumin; 4, 300 nM p26; 5, 600 nM p26. (**b**) Purified *Artemia* sHSPs produced in transformed bacteria were incubated with 4 μM insulin at 25°C and turbidity increases were measured at A_{400} upon addition of dithiothreitol. The *curves* in graphs labeled ArHsp22 and ArHsp21 represent: 1, no sHSP added; 2, 0.1 μM sHSP; 3, 0.2 μM sHSP; 4, 0.4 μM sHSP; 5, 0.8 μM sHSP; 6, 1.6 μM sHSP. *Curves* in the graph labeled Control represent: 1, no sHSPs added; 2, 1.6 μM bovine immunoglobulin; 3, 1.6 μM bovine serum albumin; 4, 1.6 μM p26. Adapted from Qiu and MacRae (2008a, b)

ArHsp21 and ArHsp22 accumulation in developing embryos was initially investigated by using real-time PCR to quantify their respective mRNAs (Qiu and MacRae 2008a, b). During development of diapause-destined embryos, ArHsp21 mRNA increased from day 2 postfertilization onward, peaking at day 5, whereas in nauplii-destined embryos ArHsp21 mRNA rose marginally by day 3 and then decreased (Fig. 10.9). ArHsp22 mRNA also increased substantially in postfertilization diapause-destined embryos, but during nauplii-destined development only a

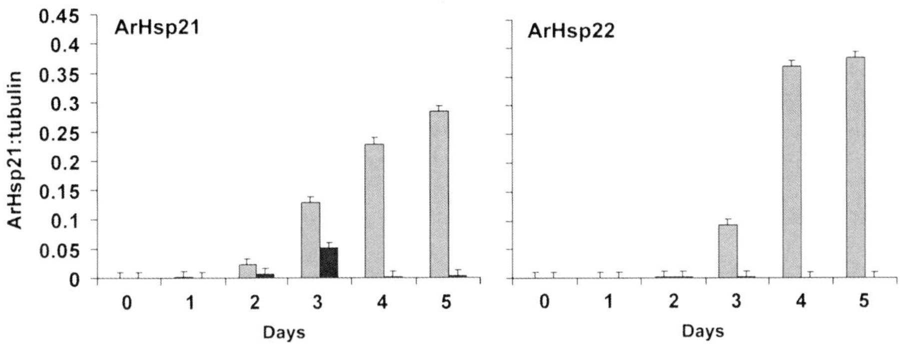

Fig. 10.9 ArHsp21 and ArHsp22 mRNA accumulation during *Artemia* embryo development. *Artemia* embryo RNA prepared daily postfertilization was reverse-transcribed and ArHsp21 and ArHsp22 mRNAs were quantified by real-time PCR using α-tubulin as the internal standard. *Gray bars*, diapause-destined embryos; *black bars*, nauplii-destined embryos. Adapted from Qiu and MacRae (2008a, b)

very small amount of ArHsp22 mRNA was observed at days 2 and 3 postfertilization (Fig. 10.9). ArHsp22 mRNA was about 35% more abundant than ArHsp21 mRNA in diapause-destined *Artemia* embryos.

Detection of sHSPs in embryos was made possible by preparation of antibodies that specifically recognize each of the *Artemia* sHSPs (Qiu and MacRae 2008a, b). ArHsp21 was first noted in diapause-destined *Artemia* embryos at day 3 postfertilization by immunoprobing of western blots. The protein bands representing ArHsp21 at days 3 and 4 postfertilization were similar in staining intensity, but weaker than at day 5, whereas the protein was not detected in embryos developing directly into nauplii (Fig. 10.10a, b, d, e). ArHsp22 initially appeared in cell-free extracts of diapause-destined embryos prepared at day 3 postfertilization with the staining of protein bands equally strong at days 4 and 5, but the protein was not seen in nauplii-destined embryos (Fig. 10.10a, c, d, f). ArHsp21 and ArHsp22 disappeared early in postdiapause development, presumably due to proteolytic degradation (Fig. 10.10g–i) and suggesting, as for p26, that these proteins accumulate specifically for diapause-related functions. ArHsp22 synthesis is, however, induced by heat in adult males, but not in other *Artemia* life history stages tested, making this the only known example of a crustacean sHSP whose production is both developmentally regulated and stress inducible (Qiu and MacRae 2008b). Accumulation in cysts but not nauplii corroborates the proposal that p26, ArHsp21, and ArHsp22 contribute to diapause maintenance, perhaps protecting cells by preventing irreversible protein denaturation upon exposure to detrimental environmental circumstances. In addition to this shared activity, the *Artemia* sHSPs may each have novel beneficial functions that contribute to maintenance of the organism during diapause, with inhibition of apoptosis by p26 as an example (Villeneuve et al. 2006).

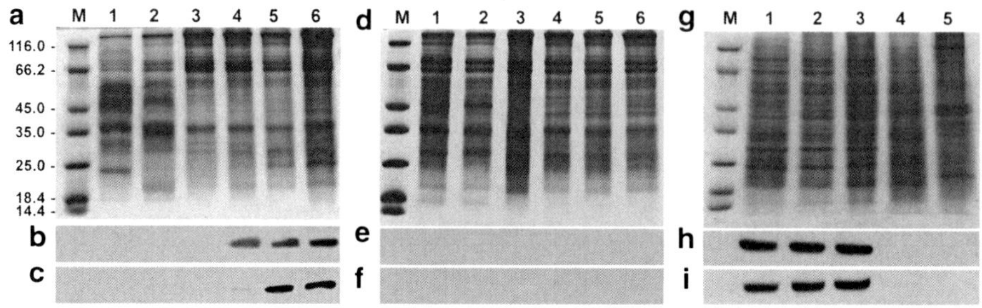

Fig. 10.10 ArHsp21 and ArHsp22 during *Artemia* development. Protein extracts prepared from *Artemia* embryos at daily intervals postfertilization were electrophoresed in SDS polyacrylamide gels and either stained with Coomassie blue (**a**, **d**) or blotted to nitrocellulose and stained with antibodies to ArHsp21 (**b**, **e**) and ArHsp22 (**c**, **f**). (**a–c**), diapause-destined embryos; (**d–f**), nauplii-destined embryos. Lane 1, day 0; 2, day 1; 3, day 2; 4, day 3; 5, day 4; 6, day 5. 30 μg of protein was loaded in each lane. (**g–i**) Protein extracts were prepared from postdiapause *Artemia* in the presence of protease inhibitors and electrophoresed in SDS polyacrylamide gels. The gels were either stained with Coomassie blue (**g**) or transferred to nitrocellulose and probed with antibodies to ArHsp21 (**h**) and ArHsp22 (**i**). Lane 1, undeveloped cysts; 2, emerged larvae; 3, nauplii; 4, instar II larvae; 5, adult males. Fifty μg of protein was loaded in each lane. ArHsp21 and ArHsp22 respectively had molecular masses of approximately 21.0 and 22.0 kDa. M, molecular mass markers in kDa. Adapted from Qiu and MacRae (2008a, b)

10.4.3 Diapause-Related sHSPs in Organisms Other than Artemia

sHSP gene expression is activated in several organisms during diapause and mRNA encoding Hsp12.6 is the most highly up-regulated transcript during *C. elegans* dauer (Jones et al. 2001). The flesh fly *S. crassipalpis* undergoes pupal diapause and accrues a number of sHSPs along with their corresponding transcripts (Denlinger 2002; Li et al. 2007; Rinehart et al. 2007). The sHSPs build up in the absence of stress induction, as occurs in diapause-destined *Artemia* embryos, and they confer survival to cold temperatures, an asset for insects which over-winter in diapause. In contrast to these findings, diapause in other insects is not accompanied by sHSP synthesis. For example, adult diapause in *Drosophila triauraria* (Goto and Kimura 2004) and larval diapause in *Lucilia sericata* (Tachibana et al. 2005) do not entail synthesis and accumulation of sHSPs, and in *C. pipiens* augmentation of sHSPs during diapause is low (Robich et al. 2007). The reasons for organismal variation in sHSP synthesis during diapause are unknown, but potentially reflect the life history stage at which diapause occurs, the degree of dormancy attained, and the level of stress tolerance. Because *Artemia* enters profound dormancy early in development, exhibits exceptional metabolic decline, and has a level of stress tolerance unequaled in most metazoans, it could require several different sHSPs, some in abundance. On the other hand, the survival of organisms that undergo diapause later in development and remain partially active requires smaller amounts of sHSPs.

10.5 Conclusions

Artemia embryos cease development as gastrulae and enter diapause, a resting stage important to the survival of many different organisms. In addition to enhanced survival, which is commonly observed, attainment and maintenance of diapause may share over-riding regulatory and mechanistic features. *Artemia* diapause is characterized by extreme levels of physiological adaptation, representing a novel model system for study that informs our more general understanding of this developmental process. Diapause involves specific patterns of gene expression not otherwise seen and this necessitates generation of new transcription factors and/or reshuffling of their associations with one another. In this regard, the up-regulation of p8, a transcription co-factor that potentially influences diapause initiation, maintenance, and termination, is restricted to oviparously developing *Artemia* embryos. Identifying more transcription factors and target genes, in addition to signals that promote their synthesis, is key to understanding diapause. Morphological, cellular, and biochemical adaptations characterize diapause and in *Artemia* these changes take on several forms including those associated with production of protective structures such as the cyst shell, inhibition of cell growth, and division, reduction of metabolism, and enhancement of stress tolerance. Molecular chaperones such as the sHSPs are likely to have important roles in maintaining *Artemia* diapause, although this is not the case for all organisms, nor are diapause-related sHSP functions inevitably identical in all organisms. Deriving definitive sHSP functions during *Artemia* diapause requires functional analysis in vitro, as described herein, and determining how embryos respond upon elimination of these proteins, experiments currently underway. The molecular analysis of diapause promises to provide important fundamental information with practical applications. As examples, large amounts of *Artemia* cysts are consumed in the aquaculture industry and understanding diapause will make their use more efficient. Moreover, the ability to enter diapause enhances the survival of several pest insects in agriculture and forestry and increases the severity of their effects. Disruption of diapause in these organisms therefore has enormous social and economical implications.

Acknowledgment Research described in this chapter was supported by a Natural Sciences and Engineering Research Council of Canada Discovery Grant to T.H.M.

References

Ahuja N, Li Q, Mohan AL, Baylin SB, Issa J-PJ (1998) Aging and DNA methylation in colorectal mucosa and cancer. Cancer Res 58:5489–5494

Baumeister R, Schaffitzel E, Hertweck M (2006) Endocrine signaling in *Caenorhabditis elegans* controls stress response and longevity. J Endocrinol 190:191–202

Bishop G, Nomura T, Yokota T, Montoya T, Castle J, Harrison K, Kushiro T, Kamiya Y, Yamaguchi S, Bancos S, Szatmári A-M, Szekeres M (2006) Dwarfism and cytochrome P450-mediated C-6 oxidation of plant steroid hormones. Biochem Soc Trans 34:1199–1201

Chen T, Amons R, Clegg JS, Warner AH, MacRae TH (2003) Molecular characterization of artemin and ferritin from *Artemia franciscana*. Eur J Biochem 270:137–145
Chen T, Villeneuve TS, Garant KA, Amons R, MacRae TH (2007) Functional characterization of artemin, a ferritin homolog synthesized in *Artemia* embryos during encystment and diapause. FEBS J 274:1093–1101
Chowdhury UR, Samant RS, Fodstad O, Shevde LA (2009) Emerging role of nuclear protein 1 (NUPR1) in cancer biology. Cancer Metastasis Rev 28:225–232
Clegg JS (1997) Embryos of *Artemia franciscana* survive four years of continuous anoxia: the case for complete metabolic rate depression. J Exp Biol 200:467–475
Clegg JS (2007) Protein stability in *Artemia* embryos during prolonged anoxia. Biol Bull 212:74–81
Clegg JS, Jackson SA (1998) The metabolic status of quiescent and diapause embryos of *Artemia franciscana* (Kellogg). Arch Hydrobiol Spec Issues Adv Limnol 52:425–439
Clegg JS, Trotman CNA (2002) Physiological and biochemical aspects of *Artemia* ecology. In: Abatzopoulos THJ, Beardmore JA, Clegg JS, Sorgeloos P (eds) Artemia basic and applied biology. Kluwer Academic Publishers, Dordrecht, pp 129–170
Clegg JS, Jackson SA, Liang P, MacRae TH (1995) Nuclear–cytoplasmic translocations of protein p26 during aerobic–anoxic transitions in embryos of *Artemia franciscana*. Exp Cell Res 219:1–7
Clegg JS, Jackson SA, Popov VI (2000) Long-term anoxia in encysted embryos of the crustacean, *Artemia franciscana*: viability, ultrastructure, and stress proteins. Cell Tissue Res 301:433–446
Crack JA, Mansour M, Sun Y, MacRae TH (2002) Functional analysis of a small heat shock/α-crystallin protein from *Artemia franciscana*: oligomerization and thermotolerance. Eur J Biochem 269:933–942
de Chaffoy D, de Maeyer-Criel G, Kondo M (1978) On the permeability and formation of the embryonic cuticle during development in vivo and in vitro of *Artemia salina* embryos. Differentiation 12:99–109
Denlinger DL (2002) Regulation of diapause. Annu Rev Entomol 47:93–122
Fielenbach N, Antebi A (2008) *C. elegans* dauer formation and the molecular basis of plasticity. Genes Dev 22:2149–2165
Fujiwara Y, Denlinger DL (2007) High temperature and hexane break pupal diapause in the flesh fly, *Sarcophaga crassipalpis*, by activating ERK/MAPK. J Insect Physiol 53:1276–1282
Giroux V, Malicet C, Barthet M, Gironella M, Archange C, Dagorn J-C, Vasseur S, Iovanna JL (2006) p8 is a new target of gemcitabine in pancreatic cancer cells. Clin Cancer Res 12:235–241
Goruppi S, Kyriakis JM (2004) The pro-hypertropic basic helix–loop–helix protein p8 is degraded by the ubiquitin/proteasome system in a protein kinase B/Akt- and glycogen synthase kinase-3-dependent manner, whereas endothelin induction of p8 mRNA and renal mesangial cell hypertrophy require NFAT4. J Biol Chem 279:20950–20958
Goto SG, Kimura MT (2004) Heat shock-responsive genes are not involved in the adult diapause of *Drosophila triauraria*. Gene 326:117–122
Guidetti R, Boschini D, Altiero T, Bertolani R, Rebecchi L (2008) Diapause in tardigrades: a study of factors involved in encystment. J Exp Biol 211:2296–2302
Hahn DA, Denlinger DL (2007) Meeting the energetic demands of insect diapause: nutrient storage and utilization. J Insect Physiol 53:760–773
Hairston NG Jr (1998) Time travelers: what's timely in diapause research? Arch Hydrobiol Spec Issues Adv Limnol 52:1–15
Hamatani T, Daikoku T, Wang H, Matsumoto H, Carter MG, Ko MSH, Dey SK (2004) Global gene expression analysis identifies molecular pathways distinguishing blastocyst dormancy and activation. Proc Natl Acad Sci USA 101:10326–10331
Haslbeck M, Franzmann T, Weinfurtner D, Buchner J (2005) Some like it hot: the structure and function of small heat shock proteins. Nat Struct Mol Biol 12:842–846

Hayward SAL, Pavlides SC, Tammariello SP, Rinehart JP, Denlinger DL (2005) Temporal expression patterns of diapause-associated genes in flesh fly pupae from the onset of diapause through post-diapause quiescence. J Insect Physiol 51:631–640
Hu Y, Nguyen TT, Bui KC, deMello DE, Smith JB (2005) A novel inflammation-induced ubiquitin E3 ligase in alveolar type II cells. Biochem Biophys Res Commun 333:253–263
Jackson SA, Clegg JS (1996) Ontogeny of low molecular weight stress protein p26 during early development of the brine shrimp, *Artemia franciscana*. Dev Growth Differ 38:153–160
Jensen DE, Rauscher FJ III (1999) Defining biochemical functions for the BRAC1 tumor suppressor protein: analysis of the BRCA1 binding protein BAP1. Cancer Lett 143 (Suppl):S13–S17
Jiang WG, Davies G, Martin TA, Kynaston H, Mason MD, Fodstad O (2006) Com-1/p8 acts as a putative tumour suppressor in prostate cancer. Int J Mol Med 18:981–986
Jing T-Z, Wang Z-Y, Qi F-H, Liu K-Y (2007) Molecular characterization of diapause hormone and pheromone biosynthesis activating neuropeptide from the black-back prominent moth, *Clostera anastomosis* (L.) (Lepidoptera, Notodontidae). Insect Biochem Mol Biol 37: 1262–1271
Johnson CL (2003) Ecdysteroids in the oceanic copepod *Calanus pacificus*: variation during molt cycle and change associated with diapause. Mar Ecol Prog Ser 257:159–165
Jones S (2004) An overview of the basic helix–loop–helix proteins. Genome Biol 5(6):226
Jones SJM, Riddle DL, Pouzyrev AT, Velculescu VE, Hillier L, Eddy SR, Stricklin SL, Baillie DL, Waterston R, Marra MA (2001) Changes in gene expression associated with developmental arrest and longevity in *Caenorhabditis elegans*. Genome Res 11:1346–1352
Koštál V (2006) Eco-physiological phases of insect diapause. J Insect Physiol 52:113–127
Koštál V, Tollarová M, Doležel D (2008) Dynamism in physiology and gene transcription during reproductive diapause in a heteropteran bug, *Pyrrhocoris apterus*. J Insect Physiol 54:77–88
Li AQ, Popova-Butler A, Dean DH, Denlinger DL (2007) Proteomics of the flesh fly brain reveals an abundance of upregulated heat shock proteins during pupal diapause. J Insect Physiol 53:385–391
Liang P, MacRae TH (1999) The synthesis of a small heat shock/α-crystallin protein in *Artemia* and its relationship to stress tolerance during development. Dev Biol 207:445–456
Liang L, Amons R, Clegg JS, MacRae TH (1997) Molecular characterization of a small heat shock/α-crystallin protein in encysted *Artemia* embryos. J Biol Chem 272:19051–19058
Lopes FL, Desmarais JA, Murphy BD (2004) Embryonic diapause and its regulation. Reproduction 128:669–678
Lukacik P, Keller B, Bunkoczi G, Kavanagh K, Lee WHWA, Adamski J, Oppermann U (2007) Structural and biochemical characterization of human orphan DHRS10 reveals a novel cytosolic enzyme with steroid dehydrogenase activity. Biochem J 402:419–427
MacGrogan D, Levy A, Bova GS, Isaacs WB, Bookstein R (1996) Structure and methylation-associated silencing of a gene within a homozygously deleted region of human chromosome band 8p22. Genomics 35:55–65
MacRae TH (2003) Molecular chaperones, stress resistance and development in *Artemia franciscana*. Semin Cell Dev Biol 14:251–258
MacRae TH (2005) Diapause: diverse states of developmental and metabolic arrest. J Biol Res 3:3–14
Malicet C, Lesavre N, Vasseur S, Iovanna JL (2003) p8 inhibits the growth of human pancreatic cancer cells and its expression is induced through pathways involved in growth inhibition and repressed by factors promoting cell growth. Mol Cancer 2:37. http://www.molecular-cancer.com/content/2/1/37
Malicet C, Hoffmeister A, Moreno S, Closa D, Dagorn J-C, Vasseur S, Iovanna JL (2006) Interaction of the stress protein p8 with Jab1 is required for Jab1-dependent p27 nuclear-to-cytoplasm translocation. Biochem Biophys Res Commun 339:284–289
Mallo GV, Fiedler F, Calvo EL, Ortiz EM, Vasseur S, Keim V, Morisset J, Iovanna JL (1997) Cloning and expression of the rat p8 cDNA, a new gene activated in pancreas during the acute

phase of pancreatitis, pancreatic development, and regeneration, and which promotes cellular growth. J Biol Chem 272:32360–32369
Mullan PB, Quinn JE, Harkin DP (2006) The role of BRCA1 in transcriptional regulation and cell cycle control. Oncogene 25:5854–5863
Noriega FG, Ribeiro JMC, Koener JF, Valenzuela JG, Hernan-dez-Martinez S, Pham VM, Feyereisen R (2006) Comparative genomics of insect juvenile hormone biosynthesis. Insect Biochem Mol Biol 36:366–374
Oh HS, Kwon H, Sun SK, Yang C-H (2002) QM, a putative tumor suppressor, regulates proto-oncogene c-yes. J Biol Chem 277:36489–36498
Park S, Jeong D-G (2006) Ribosomal protein L10 interacts with the SH3 domain and regulates GDNF-induced neurite growth in SH–SY–5y cells. J Cell Biochem 99:624–634
Podrabsky JE, Somero GN (2007) An inducible 70 kDa-class heat shock protein is constitutively expressed during early development and diapause in the annual killifish *Austrofundulus limnaeus*. Cell Stress Chaperones 12:199–204
Qiu Z, MacRae TH (2007) Developmentally regulated synthesis of p8, a stress-associated transcription cofactor, in diapause-destined embryos of *Artemia franciscana*. Cell Stress Chaperones 12:255–264
Qiu Z, MacRae TH (2008a) ArHsp21, a developmentally regulated small heat shock protein synthesized in diapausing embryos of *Artemia franciscana*. Biochem J 411:605–611
Qiu Z, MacRae TH (2008b) ArHsp22, a developmentally regulated small heat shock protein produced in diapause-destined *Artemia* embryos, is stress inducible in adults. FEBS J 275:3556–3566
Qiu Z, Bossier P, Wang X, Bojikova-Fournier S, MacRae TH (2006) Diversity, structure, and expression of the gene for p26, a small heat shock protein from *Artemia*. Genomics 88:230–240
Qiu Z, Tsoi SCM, MacRae TH (2007) Gene expression in diapause-destined embryos of the crustacean, *Artemia franciscana*. Mech Dev 124:856–867
Ree AH, Tvermyr M, Engebraaten O, Rooman M, Røsok Ø, Hovig E, Meza-Zepeda LA, Bruland ØS, Fodstad Ø (1999) Expression of a novel factor in human breast cancer cells with metastatic potential. Cancer Res 59:4675–4680
Rinehart JP, Li A, Yocum GD, Robich RM, Hayward SAL, Denlinger DL (2007) Up-regulation of heat shock proteins is essential for cold survival during insect diapause. Proc Natl Acad Sci USA 104:11130–11137
Robich RM, Denlinger DL (2005) Diapause in the mosquito *Culex pipiens* evokes a metabolic switch from blood feeding to sugar gluttony. Proc Natl Acad Sci USA 102:15912–15917
Robich RM, Rinehart JP, Kitchen LJ, Denlinger DL (2007) Diapause-specific gene expression in the northern house mosquito, *Culex pipiens* L., identified by suppressive subtractive hybridization. J Insect Physiol 53:235–245
Saito K, Su Z-H, Emi A, Mita K, Takeda M, Fujiwara Y (2006) Cloning and expression analysis of *takeout/JHBP* family genes of silkworm, *Bombyx mori*. Insect Mol Biol 15:245–251
Sambasivan R, Pavlath GR, Dhawan J (2008) A gene-trap strategy identifies quiescence-induced genes in synchronized myoblasts. J Biosci 33:27–44
Sim C, Denlinger DL (2008) Insulin signaling and FOXO regulate the overwintering diapause of the mosquito *Culex pipiens*. Proc Natl Acad Sci USA 105:6777–6781
Singtripop T, Manaboon M, Tatun N, Kaneko Y, Sakurai S (2008) Hormonal mechanisms underlying termination of larval diapause by juvenile hormone in the bamboo borer, *Omphisa fuscidentalis*. J Insect Physiol 54:137–145
Song R, Koo B-K, Yoon K-J, Yoon M-J, Yoo K-W, Kim H-T, Oh H-J, Kim Y-Y, Han J-K, Kim C-H, Kong Y-Y (2006) Neuralized-2 regulates a Notch ligand in cooperation with Mind Bomb-1. J Biol Chem 281:36391–36400
Sun Y, MacRae TH (2005a) Small heat shock proteins: molecular structure and chaperone activity. Cell Mol Life Sci 62:2460–2476
Sun Y, MacRae TH (2005b) The small heat shock proteins and their role in human disease. FEBS J 272:2613–2627

Sun Y, Mansour M, Crack JA, Gass GL, MacRae TH (2004) Oligomerization, chaperone activity, and nuclear localization of p26, a small heat shock protein from *Artemia franciscana*. J Biol Chem 279:39999–40006

Sun Y, Bojikova-Fournier S, MacRae TH (2006) Structural and functional roles for β-strand 7 in the α-crystallin domain of p26, a polydisperse small heat shock protein from *Artemia franciscana*. FEBS J 273:1020–1034

Suzuki MG, Terada T, Kobayashi M, Shimadu T (1999) Diapause-associated transcription of *BmEts*, a gene encoding an ETS transcription factor homolog in *Bombyx mori*. Insect Biochem Mol Biol 29:339–347

Tachibana S-I, Numata H, Goto SG (2005) Gene expression of heat shock proteins (*Hsp23, Hsp70 and Hsp90*) during and after larval diapause in the blow fly *Lucilia sericata*. J Insect Physiol 51:641–647

Tanguay JA, Reyes RC, Clegg JS (2004) Habitat diversity and adaptation to environmental stress in encysted embryos of the crustacean *Artemia*. J Biosci 29:489–501

Tarrant AM, Baumgartner MF, Verslycke T, Johnson CL (2008) Differential gene expression in diapausing and active *Calanus finmarchicus* (Copepoda). Mar Ecol Prog Ser 355:193–207

Valacco MP, Varone C, Malicet C, Cánepa E, Iovanna JL, Moreno S (2006) Cell growth-dependent subcellular localization of p8. J Cell Biochem 97:1066–1079

Vasseur S, Mallo GV, Fiedler F, Bödeker H, Cánepa E, Moreno S, Iovanna JL (1999) Cloning and expression of the human p8, a nuclear protein with mitogenic activity. Eur J Biochem 259:670–675

Vasseur S, Hoffmeister A, Garcia-Montero A, Mallo GV, Feil R, Kühbandner S, Dagorn J-C, Iovanna JL (2002) p8-deficient fibroblasts grow more rapidly and are more resistant to adriamycin-induced apoptosis. Oncogene 21:1685–1694

Villeneuve TS, Ma X, Sun Y, Oulton MM, Oliver AE, MacRae TH (2006) Inhibition of apoptosis by p26: implications for small heat shock protein function during *Artemia* development. Cell Stress Chaperones 11:71–80

Willsie JK, Clegg JS (2001) Nuclear p26, a small heat shock/α-crystallin protein, and its relationship to stress resistance in *Artemia franciscana* embryos. J Exp Biol 204:2339–2350

Zhang T-Y, Kang L, Zhang Z-F, Xu W-H (2004) Identification of a POU factor involved in regulating the neuron-specific expression of the gene encoding diapause hormone and pheromone biosynthesis-activating neuropeptide in *Bombyx mori*. Biochem J 380:255–263

Chapter 11
An Exploratory Review on the Molecular Mechanisms of Diapause Termination in the Waterflea, *Daphnia*

Evelyne Vanvlasselaer and Luc De Meester

Abstract The production of dormant stages is a key trait in many organisms, including those animals that inhabit freshwater systems. Dormancy has important ecological and evolutionary consequences, and traits related to induction of dormancy, survival during the dormant state, and termination of dormancy are likely to be under strong selection and may profoundly impact the dynamics of natural populations. Insight into the biochemical pathways that lead to diapause termination may allow more efficient hatching of dormant stages, which is currently a major bottleneck in many experimental studies and applications. We review the current knowledge on the biochemical pathways leading to the termination of dormancy in the waterflea, *Daphnia*. Even though *Daphnia* is a key model organism in ecology, evolution, and ecotoxicology, there is surprisingly little known about the mechanistic underpinnings of diapause termination. We provide a general but still hypothetical scheme on potential biochemical pathways that may be involved in diapause termination in *Daphnia*, largely based on studies that were carried out on related organisms, and suggest approaches for future research.

11.1 Introduction

The production of dormant stages, like seeds, eggs, and cysts, is an important trait in plants, many microorganisms and animal taxa that inhabit freshwater systems, and is viewed as a strategy to cope with habitat uncertainty in space and time (Brendonck and De Meester 2003). In this review, we focus on dormancy in the freshwater crustacean zooplankter *Daphnia*, the waterflea. *Daphnia* is an important model organism in ecotoxicology, ecology, and evolutionary biology (De Coen and

E. Vanvlasselaer (✉) and L. De Meester
Laboratory of Aquatic Ecology and Evolutionary Biology, Katholieke Universiteit Leuven, Ch. Deberiotstraat 32, 3000 Leuven, Belgium

E. Lubzens et al. (eds.), *Dormancy and Resistance in Harsh Environments*, Topics in Current Genetics 21,
DOI 10.1007/978-3-642-12422-8_11,

Janssens 1998; Feder and Mitchell-Olds 2003; Lampert 2006), and many studies have documented the importance of dormant egg production for ecological and evolutionary dynamics, as well as the use of dormant eggs in research (e.g., resurrection ecology, Kerfoot et al. 1999) and various applications (e.g., ecotoxicology, Persoone et al. 1992). Yet, despite the strong interest of the scientific community in *Daphnia* and the importance of dormancy to the ecology of organisms, surprisingly few studies have focused on the mechanistic underpinnings of diapause termination in *Daphnia*. Many studies have reported on hatching characteristics under different conditions (Pancella and Stross 1963; Davison 1969; De Meester and De Jager 1993b; Dos Santos and Persoone 1998) and have designed hatching protocols (Doma 1979; Schwartz and Hebert 1987), but the overall pattern remains that variability in hatching success is very high and overall hatching rates are sometimes disappointingly low. These low and variable hatching rates constrain the efficient use of dormant eggs in research and applications. Part of the variability is no doubt instrinsic and related to bet-hedging strategies that result in a spread of hatching over time (Philippi and Seger 1989; Cáceres and Tessier 2003), whereas other variation, e.g., the reduction in hatching rates as eggs age, is likely to have different causes. It is our hope that insight into the biochemical pathways leading to dormancy termination in *Daphnia* may lead to the identification of shortcuts in the hatching cascade and improved protocols that result in increased and more reliable hatching rates. In this paper, we start with giving a short overview on the ecological and evolutionary implications of dormancy in *Daphnia* and point to some applications that would benefit from a better understanding of the biochemical pathways leading to dormancy termination. We then attempt to build a quite general and still largely hypothetical scheme of biochemical pathways that may lead to termination of dormancy in *Daphnia*. Our aim is to provide an overview of the current knowledge, to inspire research that specifically tests these hypothetical pathways, and to stimulate the development of protocols that break dormancy. Given that very little biochemical work has been carried out on dormant eggs of *Daphnia* (Pauwels et al. 2007), our scheme is largely based on studies on ecologically or phylogenetically related organisms such as rotifers and anostracans, including the brine shrimp *Artemia*, a branchiopod occupying a widely different habitat, namely saline lakes (Abatzopoulos et al. 2002).

11.2 Ecological and Evolutionary Implications of Diapause

The reproduction cycle of the waterflea *Daphnia* typically features the production of dormant eggs. All species, both the majority of species that are cyclically parthenogenetic as well as the taxa that have become obligately parthenogenetic, share the capacity to produce dormant eggs that are protected by an envelope, called the ephippium. The ephippia are produced by the dorsal part of the carapax of the mother. In cyclic parthenogenetic taxa, the production of dormant eggs is associated with sexual reproduction. The ecological and evolutionary implications of dormant eggs have been reviewed by Brendonck and De Meester (2003). Summarizing, dormant eggs are important for local survival in habitats that vary in quality through time and become temporarily unfavorable to survival of active *Daphnia*. Clear cut examples are habitats that freeze solid or dry out, or of which abiotic conditions such as temperature or pH temporarily fall outside the tolerance limits of active animals. But dormant eggs also allow escape from periods of low food abundance, high predator abundance, or parasite outbreaks. In addition, dormant stages are the most important way by which *Daphnia* can disperse from one habitat to the other. *Daphnia* are good colonizers of new habitats (e.g. Louette and De Meester 2005), which reflects good dispersal capacity. Dispersal of dormant stages can happen by wind, water, or animal vectors. The best documented way of dispersal is by birds (Figuerola et al. 2002; Frisch et al. 2007), but insects may also play a role (Van de Meutter et al. 2008). Dispersal in both space and time, i.e., hatching from a dormant egg bank, can be viewed as a risk spreading strategy, and bet-hedging strategies have indeed been documented with respect to hatching of dormant eggs (Cáceres and Tessier 2003).

In addition to the role they play in local survival, dispersal, and colonization of other habitats, dormant stages strongly influence local dynamics and community composition, as they contribute to a dormant egg bank that in most habitats acts as a very large reservoir of individuals. This influences local population dynamics, e.g., massive hatching in spring, but also increases species diversity (cf. storage effect, Cáceres 1997; Sears and Chesson 2007) and impacts rates of evolution (Hairston and DeStasio 1988; Hedrick 1995).

Dormant eggs and egg banks can also be used as tools in research and applications. Dormant egg banks provide an archive of past community and genetic composition. Using layered sediment cores, this archive has been probed to reconstruct changes in community composition (Mergeay et al. 2004), genetic structure (Weider et al. 1997; Mergeay et al. 2006, 2007), and to reconstruct microevolution through time using a resurrection ecology approach (Kerfoot et al. 1999). This latter approach involves hatching dormant eggs from different time intervals and quantifying changes in genotypic trait values through time. It has been successfully applied to document evolutionary responses to cyanobacteria occurrence associated with eutrophication (Hairston et al. 1999, 2001), evolutionary responses to changes in predator pressure (Cousyn et al. 2001), and co-evolutionary arms races between *Daphnia* and its microparasites (Decaestecker et al. 2007). Other studies have used dormant egg banks to assess local species diversity, either by looking at dormant stages or by hatching them and analyzing the resulting communities (Vandekerkhove et al. 2005). In terms of applications, increased reliability and hatching success of *Daphnia* dormant eggs would increase the ease with which dormant eggs can be used to obtain young individuals of known quality from stocks that can be kept on a shelf, greatly reducing the energy and time needed to maintain active cultures. This can be useful in aquaculture and ecotoxicology (Persoone et al. 1992) and would be most beneficial to standardize experimental material in research programs.

11.3 Characteristics of the Diapause State

The biochemical information available on the diapause state of *Daphnia* dormant eggs is very limited. Below we briefly discuss metabolic rate and energy reserves as well as protein and RNA levels and chaperones. Most information is derived from studies on *Artemia* rather than *Daphnia*. To our knowledge, the exploratory study of Pauwels et al. (2007) is the first to compare the biochemical characteristics of *Daphnia* dormant eggs with those of parthenogenetic eggs.

11.3.1 Metabolic Rate and Energy Reserves

During prolonged dormancy, it is essential that metabolic rates are strongly reduced so that energy reserves are not depleted. In addition, dormant eggs most often end up in microhabitats with very low oxygen concentrations, Clegg (1997) reports that metabolic rate in *Artemia* cysts is undetectably low, suggesting a complete arrest of metabolism. They also noticed that there were no changes in the energy reserves of the cysts. The levels of trehalose, glycogen, and glycerol remained unchanged during dormancy. No measurements of metabolic activity of *Daphnia* dormant eggs have been reported in the literature. Hand and Carpenter (1986) suggest that a low intracellular pH (pHi) may be a mechanism that inhibits metabolism in dormant cysts of *Artemia*. Trehalose is the most important energy source for early development of *Artemia* embryos. The conversion of trehalose into glucose is inhibited by low pHi during dormancy. This occurs because the hysteretic trehalase in *Artemia* embryos exists in two forms that differ in their polymerization state depending on the pHi. The unaggregated form of the enzyme is the physiologically active one. The acidic pHi that occurs during dormancy causes a conversion of unaggregated to polymerized trehalase and will thus inhibit the conversion of trehalose to glucose, leading to an arrest of trehalose metabolism during dormancy (Hand and Carpenter 1986).

11.3.2 Proteins and RNA

In order to maintain cellular integrity during metabolic standstill and as a protection against unfavorable environmental conditions, one expects molecular chaperones to be abundant in dormant stages. Clegg (2007) recently demonstrated for *Artemia* that protein denaturation only occurs to a mild extent during diapause and that the damage is reversible, probably because of the abundant presence of the small heat-shock protein p26. Specifically for *Daphnia*, Pauwels et al. (2007) observed higher levels of glycerol and stress protein *Hsp60* in dormant than in parthenogenetic eggs of *Daphnia magna*. p26 and artemin, another protein chaperone detected in *Artemia* (Warner et al. 2004; Villeneuve et al. 2006; Chen et al. 2007), were not detected in dormant eggs of *Daphnia* (Clegg and Campagna 2006).

The disaccharide trehalose, another important energy source for cellular metabolism, has chaperone activity in anhydrobiotic animals (Hontoria et al. 1998; Hengherr et al. 2008) and could play a role in maintaining structural integrity during diapause in *Daphnia*. To our knowledge, the presence of trehalose in dormant eggs of *Daphnia* has not yet been investigated.

Because of their hypometabolic state, one would expect the amount of mRNA to be very low in resting eggs, but some studies report otherwise. Hofmann and Hand (1992) found no significant differences in mRNA levels between dormant and active *Artemia* embryos, and no net degradation of mRNA. In addition, Van Breukelen et al. (2000) found that the half-life of mRNA is increased during dormancy. Maintenance of mRNA reserves during dormancy could be a mechanism to enable the quick resumption of development when conditions become favorable again.

11.4 Diapause Termination and Hatching

When dormant eggs are produced, they enter diapause, which is defined as an intrinsic state of dormancy in which the egg is not responsive to hatching stimuli (Stross 1987). When diapause is broken, the eggs go into quiescence and become sensitive to hatching cues. Hatching of dormant eggs in *Daphnia* thus involves two steps (Stross 1987). A cold shock, drought, low O_2/high CO_2 concentrations, or soaking the ephippia in a sodium hypochlorite solution are cues that break diapause (Pancella and Stross 1963; Stross 1971; Doma 1979). The effect of a cold shock and the optimal temperature to break diapause depend on, amongst other, geographic origin of the eggs (Schwartz and Hebert 1987). The mechanism by which exposure to sodium hypochlorite solution breaks diapause is open to speculation, but may involve a mechanical impact on the outer egg shell or a bleaching effect (Pancella and Stross 1963).

Once in a quiescent state, *Daphnia* eggs often show a hatching response upon exposure to higher temperatures (20°C), light in a long-day photoperiod, and refreshing of the medium (Pancella and Stross 1963; Stross 1966, 1969; Davison 1969; De Meester and De Jager 1993b; Vandekerkhove et al. 2005), while chemicals associated with the presence of active animals tend to suppress hatching rates (Lass et al. 2005). In general, the conditions that stimulate hatching mimic spring conditions in the field. Overall, pond species (e.g., *D. magna*; De Meester and De Jager 1993a, b) tend to show higher and less variable hatching than lake dwelling species (e.g., Carvalho and Wolf 1989a, b).

In addition to observations on environmental cues that may in nature be important to induce hatching, several studies have reported on experiments in which eggs were exposed to specific artificial stimuli in an effort to induce hatching of dormant eggs. Hagiwara et al. (1995) reported that prostaglandins (*E1*, *E2*, and *F2α*) and H_2O_2 induce hatching of *Brachionus plicatilis* eggs in the dark, and thus can shortcut the need of light exposure for dormant eggs to hatch. Dumont et al. (1992) showed that treatment of anostracan cysts with retinoic acid and calcium ionophore A23187 enhances and accelerates hatching in freshwater anostracans, and Busa and Crowe (1983) used ammonia to trigger an increase of the internal pH and break dormancy in *Artemia*. These stimuli and hatching responses provide some clues on the mechanisms of hatching. Based on these indications in other species, in the following paragraph, we provide plausible scenarios for the mechanisms of hatching in *Daphnia*.

11.4.1 Light and Photoreceptors

As mentioned above, dormant eggs of *Daphnia* need a light stimulus to initiate development. Blue and near UV-light (350–490 nm) activate *Daphnia* eggs most effectively (Davison 1969), but a pulse of blue light can subsequently suppress embryonic development initiated by white light (Davison and Stross 1986). This kind of light-induced reversal of activation is well known in plants (Shinomura 1997). Davison and Stross (1986) suggest the possibility that light-sensitive molecules analogous to plant phytochromes are present in *Daphnia* eggs. Phytochromes are common plant photoreceptor molecules, most of which are sensitive to red and far-red (600–700 nm) light (for various functions of phytochromes, see Bae and Choi 2008). Ahmad and Cashmore (1993) isolated a gene for a blue and ultraviolet (UV-A) photoreceptor protein in *Arabidopsis thaliana*. These proteins, called cryptochromes, are common in many plant species, bacteria, fungi, and animals. They greatly resemble and probably evolved from DNA photolyases, enzymes responsible for the blue light-dependent repair of DNA damage caused by UV-light (Gehring and Rosbash 2003; Lin and Todo 2005; Essen 2006). They appear to play an important role in the entrainment of the circadian clock in *Drosophila*, mice, and humans (Lin and Todo 2005; Hoang et al. 2008) and in the regulation, together with other photoreceptors, of several light-dependent growth and develoment responses in plants (Ahmad and Cashmore 1993; Li and Yang 2007). As yet, however, the expression of cryptochromes in *Daphnia* eggs has not been investigated.

Porphyrin and haem pigment have also been suggested as photoreceptors to explain blue light sensitivity of dormant stages in crustaceans. Van der Linden et al. (1986) observed that intact *Artemia* cysts are particularily sensitive to light intensities around 525 and 575 nm. Decapsulated cysts, however, show an extra peak at 420 nm, a wavelength normally absorbed by haematin, the main pigment in the cyst shell, which is sensitive for green–blue light of 400–500 nm. The authors suggest the involvement of haematin in the cyst shell combined with a haem protein photoreceptor in the gastrula to regulate hatching. Since extended exposure to light and water cause oxidation and bleaching of haematin in the cyst shell, they propose a hatching delaying role for the pigment in keeping the photoreceptor from being stimulated when the hydration conditions are not favorable. These researchers also looked for a potential role for gastrula carotenoids in light reception, but concluded that these pigments could not be involved because their absorption spectrum does not correspond with the action spectrum of the cysts.

Other photoreceptor candidates with homologues in the entire light spectrum are the opsins. Opsins belong to the family of G-protein coupled receptors (GPCRs) and are the common light receptors for visual perception. The opsin family is a very diverse group of receptors with functions in retinal and extra-retinal photoreception. The activation mechanism of these proteins is conserved, with only small variation throughout the animal kingdom. The occurrence of opsins is not limited to eyes. Santillo et al. (2006) found evidence of the expression of an opsin photoreceptor (rhodopsin) in the eyeless invertebrate *Hydra*. Although the presence of opsin photoreceptors in *Daphnia* dormant eggs has not been studied, genes for opsin proteins are present and diverse in *Daphnia*. Soetaert (2007) identified two opsin proteins in a microarray study on *Daphnia magna*, and several tens of opsin genes have been detected in the *Daphnia pulex* genome (Cáceres, Thomas and Oakley: information posted on website of the Daphnia Genomics Consortium (http://daphnia.cgb.indiana.edu/)).

11.4.2 Oxidation

Irrespective of the implication of light receptors in the hatching process, reactive oxygen species (ROS) generated in the water after illumination may be a potential trigger for hatching. As mentioned earlier, prostaglandins and hydrogen peroxide appear to have promising effects on the reactivation of dormant forms in several aquatic organisms. Van der Linden et al. (1991) observed that light activation of *Artemia* cysts is followed by a drop in cyclic adenosine monophosphate (cAMP) levels and that the same phenomenon occurred after treatment with H_2O_2. This led the authors to conclude that H_2O_2 mimics light activation. Hagiwara et al. (1995) reported hatching of dormant eggs of the rotifer *Brachionus plicatilis* in the dark after the addition of prostaglandins E_1, E_2, and $F_{2\alpha}$ or H_2O_2 to the hatching medium. The same effect was seen in a study on abalones by Morse et al. (1976). The authors' explanation for this phenomenon is that hydrogen peroxide induces ROS, which mimic the effect of short wavelength radiation on the medium. Short wavelength radiation indeed produces ROS by photolysis of organic material in the upper layers of the water. These radicals can induce the production of prostaglandins by oxidizing unsaturated fatty acids in the plasmamembrane of dormant eggs (Fig. 11.1). Prostaglandins bind with a prostanoid receptor, which is a member of the GPCR family, and trigger cell activation (Fig. 11.2). Recently, Heckmann

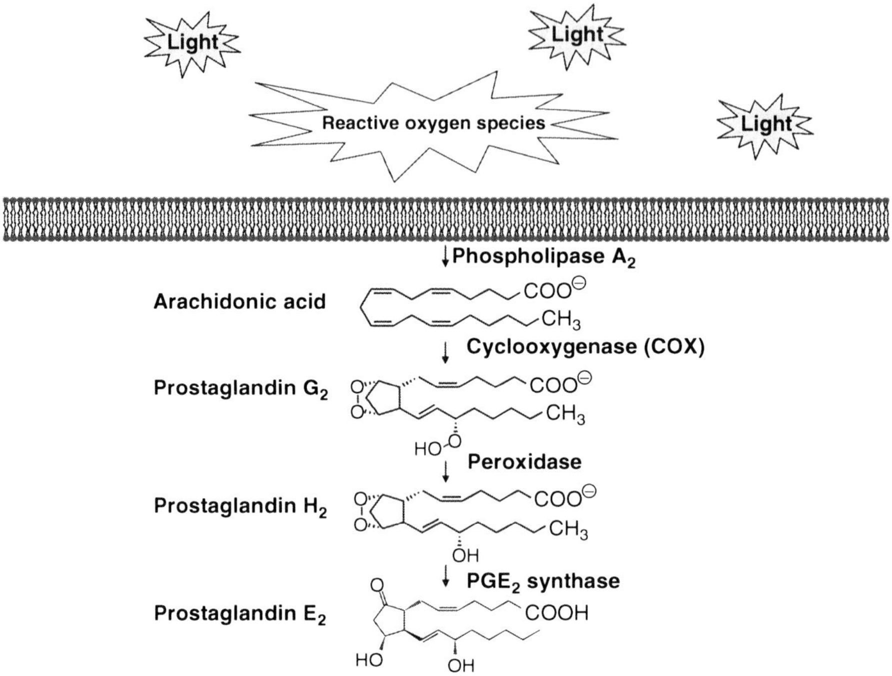

Fig. 11.1 Scheme of the formation of prostaglandin E_2, through several short-lived intermediates, following oxidation of plasma membrane phospholipids by reactive oxygen species (ROS) produced in the medium after irradiation (simplified after Heckmann et al. 2008)

et al. (2008) found evidence in the *D. pulex* genome for the existence of two possible prostanoid GPCRs.

11.4.3 Downstream Cellular Activation

Whether the activation mechanism involves light receptors or redox signaling, there are indications that downstream of this first stimulus the Ca^{2+}-calmodulin pathway is important for the reactivation of dormant embryos through activation of enzymes (Dumont et al. 1992) (Fig. 11.2). The classical Ca^{2+}-calmodulin pathway is triggered by stimulation of a GPCR. The GPCRs are a large family of membrane proteins with seven transmembrane regions that transduce extracellular signals to the interior of the cell. They are coupled with a heterotrimeric GTP-binding protein (G-protein) on the interior side of the membrane. The ligands binding to GPCRs are very diverse and include light-sensitive compounds, hormones, and neurotransmitters. Ligand binding on the receptor induces the exchange of GDP for GTP on the G-protein α subunit and the dissociation of the α subunit from the $\beta\gamma$-heterodimer. The GTP-α subunit complex mediates intracellular signaling by activating

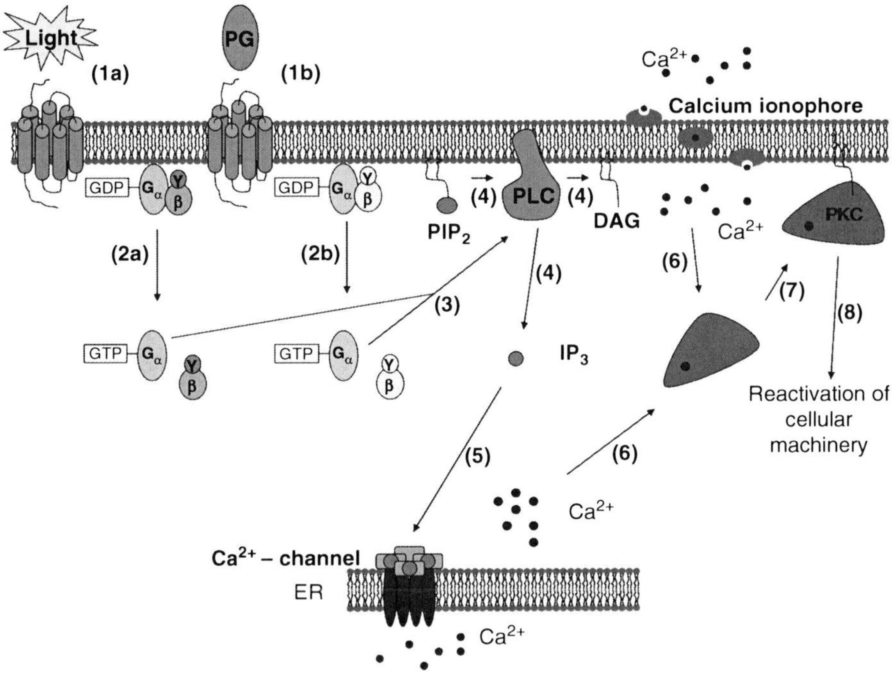

Fig. 11.2 Hypothetical hatching cascade in dormant eggs of *Daphnia* after activation by light or prostaglandins (PG). (**1a**) Light activates a GPCR [e.g., from the opsin family], which leads to (**2a**) the exchange of GDP for GTP on the G-protein α subunit and the dissociation of the α subunit from the $\beta\gamma$-heterodimer. (**1b**) Prostaglandins, generated after oxidation of the plasma membrane phospholipids, bind to a prostanoid GPCR and (**2b**) activate G-protein signal transduction as described above. (**3**) The GTP-α subunit activates phospholipase C (PLC). (**4**) PLC in turn converts phosphatidylinositol biphosphate (PIP_2) to inositol triphosphate (IP_3) and diacylglycerol (DAG). (**5**) IP_3 releases Ca^{2+} from the intracellular stores by activating Ca^{2+}-channels. (**6**) Free Ca^{2+} in the cytosol, originating from intracellular stores or from the extracellular environment and then transported through the membrane with the help of a Ca^{2+}-ionophore, binds to calmodulin. (**7**) The Ca^{2+}-calmodulin complex in turn activates protein kinase C, which (**8**) restarts the cell machinery by reactivating cellular enzymes

phospholipase C (PLC). PLC in turn converts phosphatidylinositol biphosphate (PIP_2) to inositol triphosphate (IP_3) and diacylglycerol (DAG). IP_3 releases Ca^{2+} from the intracellular stores by activating Ca^{2+}-channels. Free Ca^{2+} in the cytosol binds to calmodulin and the Ca^{2+}-calmodulin complex in turn activates protein kinase C, which restarts the cell machinery by reactivating cellular enzymes (Birnbaumer 2007). Dumont et al. (1992) showed that cyst hatching of the anostracans *Thamnocephalus platyurus* and *Streptocephalus dichotomus* increased after treatment with Calcium Ionophore A23187. Calcium Ionophore A23187 is a mobile ion carrier with a high sensitivity for divalent cations, such as Mg^{2+} and Ca^{2+}. This observation indicates that availability of transmembrane calcium pumps might be a limiting factor affecting cyst hatching. A similar mechanism may be important in

Daphnia, but so far no studies have examined this possibility. If the mechanism applies, treatment with a calcium ionophore could provide a short cut for enzyme activation by bypassing G-protein activation.

11.4.4 pH

Finally, we mention that in *Artemia salina* there is evidence for the importance of intracellular pH in termination of diapause. Busa and Crowe (1983) first demonstrated the role of intracellular pH (pHi) in the induction and termination of dormancy in *Artemia*. Although they never directly measured the pHi of the cysts, they showed that termination of dormancy could be induced by alkalinisation of the pHi through the addition of ammonia to the hatching medium, while induction of dormancy could be triggered by a decrease of the pHi through acidification of the medium with CO_2.

11.5 Research Perspectives

It is clear from the above that the current state of insight into the biochemical pathways that may be involved in the hatching of dormant eggs of the water flea *Daphnia* is very limited. Most of the research on the mechanisms of diapause termination has been carried out on anostracans, especially *Artemia* (see Chap. 10). Some exploratory research has also been carried out on rotifers (e.g., *Brachionus*: Hagiwara et al. 1995), but almost no work has been specifically directed at *Daphnia* dormant eggs (Pauwels et al. 2007). Perspectives for future work are, amongst others, the following:

(a) Testing of experimental treatments that impact hatching in other species, such as prostaglandins (e.g., rotifers: Hagiwara et al. 1995, abalones: Morse et al. 1976), hydrogen peroxide (e.g., abalones: Morse et al. 1976), Ca-ionophores (e.g., anostracans: Dumont et al. 1992), and manipulation of the pHi. The results of such exploratory analyzes may provide valuable indications on the mechanisms of diapause termination in *Daphnia*.
(b) Screening *Daphnia* dormant eggs for the presence of light receptors (e.g., opsins, haem pigments, cryptochromes) may help resolve a crucial step in the hatching sequence. Understanding changes in the amount and quality of light receptor and shielding pigments may be critical for our insight into important ecological phenomena such as the high interindividual variation in response to hatching cues and the development of bet-hedging strategies.
(c) There is a strong need for an in-depth comparative biochemical analysis of dormant eggs and parthenogenetic eggs. In addition to identifying mechanisms important to dormancy and hatching, this analysis would provide mechanistic

underpinning of the ecological differences between hatchlings from dormant and parthenogenetic eggs (Arbaciauskas and Lampert 2003). Comparative proteomics of eggs in various stages of dormancy and resumption of development may allow us to acquire a more complete picture of the hatching mechanisms of *Daphnia* dormant eggs, one that is less biased toward mechanisms that have been suggested based on work on other organisms.

References

Abatzopoulos TJ, Beardmore JA, Clegg JS, Sorgeloos P (eds) (2002) *Artemia*: basic and applied biology. Kluwer Academic Publishers, Dordrecht, The Netherlands, p 286

Ahmad M, Cashmore AR (1993) HY4 gene of *A. thaliana* encodes a protein with characteristics of a blue-light photoreceptor. Nature 366:162–166

Arbaciauskas K, Lampert W (2003) Seasonal adaptation of ex-ephippio and parthenogenetic offspring of *Daphnia magna*: differences in life history and physiology. Funct Ecol 17:431–437

Bae G, Choi G (2008) Decoding of light signals by plant phytochromes and their interacting proteins. Annu Rev Plant Biol 59:281–311

Birnbaumer L (2007) The discovery of signal transduction by G proteins. A personal account and an overview of the initial findings and contributions that led to our present understanding. Biochim Biophys Acta 1768:756–771

Brendonck L, De Meester L (2003) Egg banks in freshwater zooplankton: evolutionary and ecological archives in the sediment. Hydrobiologia 491:65–84

Busa WB, Crowe JH (1983) Intracellular pH regulates transitions between dormancy and development of brine shrimp (*Artemia salina*) embryos. Science 221:366–368

Cáceres CE (1997) Temporal variation, dormancy, and coexistence: a field test of the storage effect. Proc Natl Acad Sci USA 94(17):9171–9175

Cáceres CE, Tessier AJ (2003) How long to rest: the ecology of optimal dormancy and environmental constraint. Ecology 84(5):1189–1198

Carvalho GR, Wolf HG (1989a) Resting eggs of lake-*Daphnia* I. distribution, abundance and hatching of eggs collected from various depths in lake sediments. Freshw Biol 22(3):459–470

Carvalho GR, Wolf HG (1989b) Resting eggs of lake-*Daphnia* II. in situ observations on the hatching of eggs and their contribution to population and community structure. Freshw Biol 22 (3):471–478

Chen T, Villeneuve TS, Garant KA, Amons R, MacRae TH (2007) Functional characterization of artemin, a ferritin homolog synthesized in *Artemia* embryos during encystment and diapause. FEBS J 274:1093–1101

Clegg JS (1997) Embryos of *Artemia franciscana* survive four years of continous anoxia: the case for complete metabolic rate depression. J Exp Biol 200:467–475

Clegg JS (2007) Protein stability in *Artemia* embryos during prolonged anoxia. Biol Bull 212:74–81

Clegg JS, Campagna V (2006) Comparisons of stress proteins and soluble carbohydrate in encysted embryos of *Artemia franciscana* and two species of *Parartemia*. Comp Biochem Phys B 145:119–125

Cousyn C, De Meester L, Colbourne JK, BrendonckL VD, Volckaert F (2001) Rapid local adaptation of zooplankton behavior to changes in predation pressure in absence of neutral genetic changes. Proc Natl Acad Sci USA 98:6256–6260

Davison J (1969) Activation of ephippial eggs of *Daphnia pulex*. J Gen Physiol 53:565–575

Davison EA, Stross RG (1986) A blue light-reversible reaction in an animal system (*Daphnia pulex*). Experientia 42:620–622

De Coen WM, Janssen CR (1998) The use of biomarkers in *Daphnia magna* toxicity testing. Hydrobiologia 367:199–209

De Meester L, De Jager H (1993a) Hatching of *Daphnia* sexual eggs. I. Intraspecific differences in the hatching responses of D. magna eggs. Freshw Biol 30:219–226

De Meester L, De Jager H (1993b) Hatching of *Daphnia* sexual eggs. II. The effect of age and a second stimulus. Freshw Biol 30:227–233

Decaestecker E, Gaba S, Raeymaekers JAM, Stoks R, Van Kerckhoven L, Ebert D, De Meester L (2007) Host-parasite 'Red Queen' dynamics archived in pond sediment. Nature 450:870–873

Doma S (1979) Ephippia of *Daphnia magna* Straus – technique for their mass-production and quick revival. Hydrobiologia 67(2):183–188

Dos Santos MMM, Persoone G (1998) Hatching of *Daphnia magna* resting eggs: the effect of storage duration and storage temperature. Arch Hydrobiol Spec Issues Adv Limnol 52:253–262

Dumont H, Casier P, Munuswamy N, De Wasche C (1992) Cyst hatching in Anostraca accelerated by retinoic acid, amplified by Calcium Ionophore A23187, and inhibited by Calcium-channel blockers. Hydrobiologia 230:1–7

Essen L-O (2006) Photolyases and cryptochromes: common mechanisms of DNA repair and light-driven signaling? Curr Opin Struct Biol 16:51–59

Feder ME, Mitchell-Olds T (2003) Evolutionary and ecological functional genomics. Nat Rev Genet 4:651–657

Figuerola J, Green AJ, Santamaria L (2002) Comparative dispersal effectiveness of wigeongrass seeds by waterfowl wintering in south-west Spain: quantitative and qualitative aspects. J Ecol 90(6):989–1001

Frisch D, Green AJ, Figuerola J (2007) High dispersal capacity of a broad spectrum of aquatic invertebrates via waterbirds. Aquat Sci 69(4):568–574

Gehring W, Rosbash M (2003) The coevolution of blue-light photoreception and circadian rhytms. J Mol Evol 57:S286–S289

Hagiwara A, Hoshi N, Kawahara F, Tominaga K, Hirayama K (1995) Resting eggs of the marine rotifer *Brachionus plicatilis* Müller: development, and effect of irradiation on hatching. Hydrobiologia 313(314):223–229

Hairston NG, DeStasio BT (1988) Rate of evolution slowed by a dormant propagule pool. Nature 336:239–242

Hairston NG, Lampert W, Cáceres CE, Hotmeier CL, Weider LJ, Gaedke U, Fischer JM, Fox JA, Post DM (1999) Rapid evolution revealed by dormant eggs. Nature 401:446

Hairston NG, Hotmeier CL, Lampert W, Weider LJ, Fischer JM, Cáceres CE, Fox JA, Gaedke U (2001) Natural selection for grazer resistance to toxic cyanobacteria: evolution of phenotypic plasticity? Evolution 55(11):2203–2214

Hand SC, Carpenter JF (1986) pH-induced metabolic transitions in *Artemia* embryos mediated by a novel hysteretic trehalase. Science 232:1535–1537

Heckmann L-H, Sibly RM, Timmermans MJTN and Callaghan A (2008) Outlining eicosanoid biosynthesis in the crustacean *Daphnia*. Front Zool 5, article 11

Hedrick PW (1995) Genetic polymorphism in a temporally varying environment: effects of delayed germination or diapause. Heredity 75:164–170

Hengherr S, Heyer AG, Köhler HR, Schill RO (2008) Trehalose and anhydrobiosis in tardigrades – evidence for divergence in responses to dehydration. FEBS J 275(2):281–288

Hoang N, Schleicher E, Kacprzak S, Bouly J-P, Picot M, Wu W, Berndt A, Wolf E, Bittl R, Ahmad M (2008) Human and *Drosophila* cryptochromes are light activated by flavin photoreduction in living cells. PLOS Biol 6(7):1559–1569

Hofmann GE, Hand SC (1992) Comparison of messenger RNA pools in active and dormant *Artemia franciscana* embryos: evidence for translational control. J Exp Biol 164:103–116

Hontoria F, Crowe JH, Crowe LM, Amat F (1998) Trehalose prevents imbibitional damage in anhydrobiotic cysts of *Artemia* by depressing the phase transition temperature in lipids. Adv Limnol 52:451–462
Kerfoot WC, Robbins JA, Weider LJ (1999) A new approach to historical reconstruction: combining descriptive and experimental paleolimnology. Limnol Oceanogr 44(5):1232–1247
Lampert W (2006) *Daphnia*: model herbivore, predator and prey. Pol J Ecol 54(4):607–620
Lass S, Vos M, Wolinska J, Spaak P (2005) Hatching with the enemy: *Daphnia* diapausing eggs hatch in the presence of fish kairomones. Chemoecology 15:7–12
Li Q-H, Yang H-Q (2007) Cryptochrome signaling in plants. Photochem Photobiol 83:94–101
Lin C, Todo T (2005) The cryptochromes. Genome Biol 6, article 220
Louette G, De Meester L (2005) High dispersal capacity of cladoceran zooplankton in newly founded communities. Ecology 86(2):353–359
Mergeay J, Verschuren D, Van Kerckhoven L, De Meester L (2004) Two hundred years of a diverse *Daphnia* community in Lake Naivasha (Kenya): effects of natural and human-induced environmental changes. Freshw Biol 49(8):998–1013
Mergeay J, Declerck S, Verschuren D, De Meester L (2006) *Daphnia* community analysis in shallow Kenyan lakes and ponds using dormant eggs in surface sediments. Freshw Biol 51:399–411
Mergeay J, Vanoverbeke J, Verschuren D, De Meester L (2007) Extinction, recolonization, and dispersal through time in a planktonic crustacean. Ecology 88(12):3032–3043
Morse DE, Duncan H, Hooker N, Morse A (1976) Hydrogen peroxide induces spawning in mollusks with activation of prostaglandin endoperoxide synthetase. Science 196:298–300
Pancella JR, Stross RG (1963) Light induced hatching of *Daphnia* resting eggs. Chesapeake Sci 4:135–140
Pauwels K, Stoks R, Verbiest A, De Meester L (2007) Biochemical adaptation for dormancy in subitaneous and dormant eggs of *Daphnia magna*. Hydrobiologia 594:91–96
Persoone G, Blaise C, Snell T, Janssen C, Van Steertegem M (1992) Cyst-based toxicity tests: 2. Report on an international intercalibration exercise with three cost-effective Toxkits. Z Angew Zool 79(1):17–36
Philippi T, Seger J (1989) Hedging one's evolutionary bets, revisited. Trends Ecol Evol 4 (2):41–43
Santillo S, Orlando P, De Petrocellis L, Cristino L, Guglielmotti V, Musio C (2006) Evolving visual pigments: hints from the opsin-based proteins in a phylogenetically old "eyeless" invertebrate. Biosystems 86:3–17
Schwartz SS, Hebert PDN (1987) Methods for the activation of the resting eggs of *Daphnia*. Freshw Biol 17(2):373–379
Sears ALW, Chesson P (2007) New methods for quantifying the spatial storage effect: an illustration with desert annuals. Ecology 88:2240–2247
Shinomura T (1997) Phytochrome regulation of seed germination. J Plant Res 110:151–161
Soetaert A (2007) Development of cDNA microarrays in the freshwater flea *Daphnia magna* for toxicity characterisation of chemicals. Graduate thesis, Univerity of Antwerp, Belgium
Stross RG (1966) Light and temperature requirements for diapause development and release in *Daphnia*. Ecology 47(3):368–374
Stross RG (1969) Photoperiod control of diapause in *Daphnia*. III. Two-stimulus control of long-day, short-day induction. Biol Bull 137:359–374
Stross RG (1971) Photoperiod control of diapause in *Daphnia*. IV. Light and CO_2-sensitive phases within the cycle of activation. Biol Bull 140:137–155
Stross RG (1987) Photoperiodism and phased growth in *Daphnia* populations: coactions in perspective. In: Peters RH, De Bernardi R (eds) *Daphnia*, vol 45. Memorie dell'Istituto Italiano di Idrobiologia, Ornavasso, pp 415–437
Van Breukelen F, Maier R, Hand SC (2000) Depression of nuclear transcription and extension of mRNA half-life under anoxia in *Artemia franciscana* embryos. J Exp Biol 203:1123–1130

Van de Meutter F, Stoks R, De Meester L (2008) Size-selective dispersal of *Daphnia* resting eggs by backswimmers (*Notonecta maculata*). Ecol Lett 4(5):494–496

Van der Linden A, Vankerckhoven I, Caubergs R, Decleir W (1986) Action spectroscopy of light-induced hatching of *Artemia* cysts (Branchiopoda: Crustacea). Mar Biol 91:239–243

Van der Linden A, Gadeyne J, Van Onckelen H, Van Laere A, Decleir W (1991) Involvement of cyclic nucleotides in light-induced resumption of development of *Artemia* embryos. J Exp Biol 258:312–321

Vandekerkhove J, Declerck S, Brendonck L, Conde-Procuna JM, Jeppesen E, Sander Johansson LS, De Meester L (2005) Uncovering hidden species: hatching diapausing eggs for the analysis of cladoceran species richness. Limnol Oceanogr Methods 3:399–407

Villeneuve TS, Ma X, Sun Y, Oulton MM, Oliver AE, Mac Rae TH (2006) Inhibition of apoptosis by p26: implications for small heat shock protein function during *Artemia* development. Cell Stress Chaperones 11(1):71–80

Warner AH, Brunet RT, Mac Rae TH, Clegg JS (2004) Artemin is an RNA-binding protein with high thermal stability and potential RNA chaperone activity. Arch Biochem Biophys 424:189–200

Weider L, Lampert W, Wessels M, Colbourne JK, Limburg P (1997) Long-term genetic shifts in a microcrustacean egg bank associated with anthropogenic changes in the lake constance ecosystem. Proc Biol Sci 264(1388):1613–1618

Chapter 12
Metabolic Dormancy and Responses to Environmental Desiccation in Fish Embryos

Jason E. Podrabsky, Angèle Tingaud-Sequeira, and Joan Cerdà

Abstract Metabolic depression is relatively uncommon among the fishes with the greatest number of species exhibiting dormancy as embryos. Dormancy in fish embryos is largely associated with deposition of embryos into terrestrial habitats to avoid embryo predation or to survive intermittent drying of aquatic habitats. Killifish embryos in general, and especially the embryos of annual killifish, are highly adapted for life at the interface between land and water and thus have evolved a suite of characters that allows them to survive in an aerial environment. Here we review the available literature on embryonic dormancy and dehydration tolerance in killifish embryos.

12.1 Introduction

Examples of metabolic dormancy can be found in every Class of the Phylum Vertebrata including the most diverse group of vertebrates, the fish. Dormancy in fish is found in both adult and embryonic life stages, but at this point in time a single species has not been illustrated to enter metabolic dormancy in more than one life history stage.

Dormancy in adult fish is induced in response to unfavorable environmental conditions, especially changes in water and food availability, and seasonal changes

---

J.E. Podrabsky (✉)
Department of Biology, Portland State University, P.O. Box 751, 97207-0751, Portland, OR, USA
e-mail: podrabsj@pdx.edu

A. Tingaud-Sequeira,
Laboratory of Institut de Recerca i Tecnologia Agroalimentàries (IRTA)-Institut de Ciències del Mar, Consejo Superior de Investigaciones Científicas (CSIC), 08003 Barcelona, Spain
Génomique et Physiologie des Poissons, Université Bordeaux 1, UMR NuAGe 33405 Talence, France

J. Cerdà (✉)
Laboratory of Institut de Recerca i Tecnologia Agroalimentàries (IRTA)-Institut de Ciències del Mar, Consejo Superior de Investigaciones Científicas (CSIC), 08003 Barcelona, Spain
e-mail: joan.cerda@irta.es

E. Lubzens et al. (eds.), *Dormancy and Resistance in Harsh Environments*, Topics in Current Genetics 21,
DOI 10.1007/978-3-642-12422-8_12, © Springer-Verlag Berlin Heidelberg 2010

in temperature. Aestivation in response to a loss of water is found in a number of different fish lineages including the dipnoans (African and South American lungfish; Smith 1930; Fishman et al. 1992; Greenwood 1986; dos Santos Ferreira da Silva et al. 2008) and a number of teleost groups such as the Galaxids (Canterbury mudfish and the Australian Salamanderfish; Eldon 1979; Berra and Allen 1989; Pusey 1989), and Synbranchids (swamp eels; Eduardo et al. 1979; Chew et al. 2005). In addition, the Japanese Sandeel (sand lance, *Ammodytes personatus*) appears to aestivate in the sand of estuaries to avoid high summer temperatures (Tomiyama and Yanagibashi 2004). Cold torpor or hibernation is probably the most common form of dormancy in adult fish and has been reported in a diversity of species such as hagfish (Hansen and Sidell 1983; Lesser et al. 1996), lampreys (Pletcher 1963), and a number of teleost fish such as the crucian carp (Hyvarinen et al. 1985; Nilsson and Renshaw 2004). In many of these species, cold torpor is also likely accompanied by anoxia tolerance (Hansen and Sidell 1983; Hyvarinen et al. 1985; Nilsson and Renshaw 2004). In marine environments winter hibernation appears to be mostly an adaptation to low food availability and resembles in many aspects hibernation in small mammals (Campbell et al. 2008).

Embryonic dormancy is a more widespread form of dormancy in fish from a geographic, ecological, and evolutionary perspective. This increased diversity is likely driven by the variety of environments that fish inhabit. For example, hypoxia and anoxia commonly occur in freshwater habitats, and thus it is not surprising that a number of species produce embryos that can tolerate short or transient (1–2 days), and even prolonged environmental hypoxia and anoxia. In addition, many species of fish have developed life history strategies for either hiding embryos from predators or synchronizing development with favorable environmental conditions. In many cases, hiding of embryos means exposing them to a terrestrial existence. While this may increase survival due to reduced predation, it requires the ability to survive aerial exposures, and perhaps long bouts of severe dehydration. Thus, evolution of embryonic dormancy in fish is very likely tied to dehydration tolerance. There are three major types of dormancy that may occur during the development of fish embryos: delayed hatching, embryonic diapause, and anoxia-induced quiescence. Each of these types of dormancy will be addressed in this chapter.

12.2 Delayed Hatching

Delayed hatching is found in a number of different fish lineages including the Atherinids (Tewksbury and Conover 1987; Martin 1999), Fundulidae (Harrington 1959; Koenig and Livingston 1976; Taylor et al. 1977), and Galaxidae (McDowall and Charteris 2006; Charteris et al. 2003) and is typically associated with the deposition of eggs in a terrestrial environment (see Martin 1999 for an excellent review on delayed hatching). In some marine fish, such as the Halibut (*Hippoglossus hippoglossus*) hatching can be delayed by continued exposure to light (Helvik and Walther 1993). Based on the few studies available, metabolism and development are slowed, but not arrested during delayed hatching. For example, the oxygen consumption of California grunion (*Leuresthes tenuis*) embryos is not decreased during delayed hatching, but remains relatively constant (Darken et al. 1998). Although there is not a depression of metabolism per se during delayed hatching, the expected ontogenetic increase in metabolism associated with hatching and a larval existence is delayed. In a sense, this is a depression of metabolic requirements associated with larval life. In addition, because development does not cease during delayed hatching, a stable level of oxygen consumption in the face of a presumably increasing mass of metabolically active tissue may actually indicate a depression of metabolism.

An extensively studied model of delayed hatching among killifishes is *Fundulus heteroclitus*, a marine, nonmigratory teleost typically inhabiting coastal marshes and inland systems of North America (Bigelow and Schroeder 1953; Lee et al. 1980; Scott and Crossman 1998), Hawaii (Randall 1987), the Philippines (Seale 1910), and southern Europe (Gutierrez-Estrada et al. 1998). Natural populations of *F. heteroclitus* have a semilunar reproductive cycle that synchronizes reproductive maturity with the occurrence of spring tides (Taylor 1986). Eggs are laid in multiple clutches at the high water mark during high spring tides associated with new and full moons, usually in "protected" environments such as mussel shells or marsh chord grasses (Taylor 1999). The developing embryos protected by the egg envelope (chorion) are adapted to survive aerially, and eggs laid at one high tide are ready to hatch with the next high tide (Taylor 1999). Thus, hatching occurs only upon immersion in water. As in other teleosts, egg hatching is triggered by the release of the hatching enzyme from the hatching gland cells located in the buccal cavity of the opercular epithelium (Armstrong and Child 1965; Taylor 1999). Recent studies have shown that the choriolytic enzymatic system of *F. heteroclitus*, as well as that of Japanese medaka (*Oryzias latipes*), is composed in fact by two enzymes, the high choriolytic enzyme (HCE) and the low choriolytic enzyme (LCE), which act cooperatively for the complete digestion of the chorion (Kawaguchi et al. 2005).

Studies on the hatching mechanism of *F. heteroclitus* embryos indicate that hypoxia may be one of the main factors regulating hatching (DiMichele and Taylor 1980, 1981; DiMichele and Powers 1984a). Thus, the respiratory demands of developing embryos may be fulfilled by the oxygen supply from the air, whereas hatching during immersion is cued when the metabolic rate of the embryo exceeds a limit set by the diffusion of oxygen through the water column (DiMichele and Powers 1982, 1984a, b; Powers et al. 1991). This hypothesis is supported by the finding that populations of *F. heteroclitus* from North America adapted to different conditions of dissolved oxygen levels, temperature, photoperiod, and latitude show slightly different hatching times associated with genetic polymorphisms in heart-type lactate dehydrogenase, which may affect erythrocyte ATP concentrations and hence hemoglobin–oxygen affinity (DiMichele and Powers 1982, 1984a; Powers et al. 1991). These specific adaptations have likely arisen due to the restricted distribution of the nonmigratory *F. heteroclitus* populations, thus allowing for the development over multiple generations of reproductive adaptations to the temperature and climate of their immediate environment (Burnett et al. 2007). However, the molecular mechanisms regulating the release of hatching enzymes in response to variations in oxygen availability are still unknown.

12.2.1 *Advanced Hatching: The Case of* F. heteroclitus *Embryos*

As mentioned earlier, delayed hatching in nonannual killifish, such as *F. heteroclitus*, is not associated with an arrest of development. However, *F. heteroclitus* embryos are exposed to an aerial environment during low tides, and therefore physiological adaptations to regulate embryonic development in response to environmental dehydration are likely. The existence of these mechanisms has been recently demonstrated in a series of laboratory-controlled experiments showing that developmental rates of *F. heteroclitus* embryos increase in response to aerial incubation (Tingaud-Sequeira et al. 2009). In this study, developmental rates were stimulated in morula and blastula embryos exposed to air for up to 12 days under approximately 100% relative humidity, and hence the time of hatching was advanced with respect to embryos continuously immersed in seawater (Fig. 12.1). This ability of embryos to accelerate development and differentiation in response to water removal may be maternally inherited, since zygotic transcription is possibly not yet activated at these embryonic stages (O'Boyle et al. 2007).

The acceleration of development of *F. heteroclitus* embryos exposed to air may be related to elevated oxygen availability that can support increased metabolic rates. However, Tingaud-Sequeira et al. (2009) also observed large variations in hatching times between batches of embryos collected during the year that were

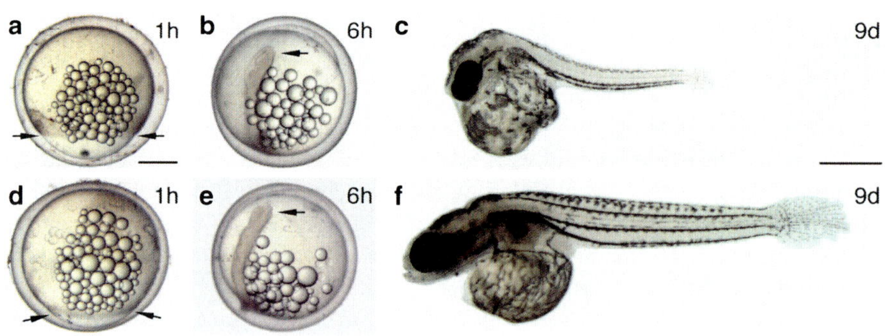

Fig. 12.1 Photomicrographs of *F. heteroclitus* embryos after 1 and 6 h, and approximately 9 days, of incubation in seawater (**a–c**) or in air (**d–f**) from the blastula stage. The *arrows* indicate the marginal zone of the blastoderm at late gastrulation (**a** and **d**), or the developing embryo at the neurula stage (**b** and **e**). Note that development of air-incubated embryos is accelerated with respect to embryos maintained in water, which is more evident when embryos are manually released from the egg envelope (**c** and **f**). *Scale bars*, 0.5 mm

incubated constantly in water. In addition, *F. heteroclitus* embryos can also accelerate development when incubated with reduced salinity (Cerdà, unpublished data). Together, these observations suggest that oxygen availability may not be the only environmental cue controlling developmental rate, and other stressful conditions may trigger the same response. In fact, embryos exposed to air for 6 h show an increased accumulation of mRNAs encoding proteins associated with oxidative stress such as glutathione S-transferase and peroxiredoxin 6 (ATS and JC unpublished data).

Although it is clear that the molecular mechanisms involved in the environmental control of embryonic development in *F. heteroclitus* need to be investigated further, the data available suggest that this species is tightly adapted to its specific reproductive strategy. During the reproductive season, gonad maturity and spawning readiness coincide with new and full moons, and spawning is thereby synchronized with the semilunar cycle of tides in the tide marsh habitat (Taylor 1999; Taylor et al. 1979). Accelerated development of embryos exposed to air may increase the ability of embryos to hatch during the next high tide (Taylor 1999). In the laboratory, the lunar and tidal spawning cycles of parent fish may be partially lost (Hsiao and Meier 1989; Hsiao et al. 1994, 1996; Taylor 1991), but the ability of embryos to accelerate development in response to aerial exposure is conserved. This strategy of *F. heteroclitus* embryos thus resembles that of tadpoles of desert-dwelling amphibians that are able to detect changes in the water level of the aquaria and accelerate development and metamorphosis, mimicking their behavior in the natural habitat (Denver 1997).

12.3 Embryonic Diapause

Embryonic diapause is found in a wide range of fish lineages including several families of teleost fish and a number of cartilaginous fish such as the guitarfish (Enajjar et al. 2008), several species of rays (White et al. 2002 and references therein), and sharks (Simpfendorfer 1992). Interestingly, most of these fish are found in tropical waters, and thus diapause appears to be important for synchronizing parturition with seasonal food or habitat availability. Diapause has also been reported for embryos of the autumn-spawning bitterling (*Acheilognathus rhombeus*), a small Asian cyprinid, as a mechanism for delaying development due to cold winter temperatures (Kawamura and Uehara 2005). Although very little has been done to investigate diapause in this species, it appears to be genetically determined and restricted to the autumn-spawning populations, which benefit from delayed development by emerging in the spring when food availability and temperatures are presumably more suitable for growth and survival of larvae. The highest diversity of species exhibiting embryonic diapause, and the most well studied group are the members of the Order Cyprinodontiformes, known as the annual killifish. Most of our knowledge about the physiological, biochemical, and molecular aspects of embryonic dormancy in fish comes from studies of this group, notably the genera *Nothobranchius* from Africa and *Austrofundulus* from South America.

12.3.1 Evolutionary History of Diapause in Annual Killifish

The annual killifish are classified into two Families, the Rivulidae (approximately 160 species) from South and Central America, and the Aplocheilidae (approximately 189 species) from Africa. A number of studies have used morphological and molecular markers to investigate the evolutionary history of diapause in the annual killifish. Murphy and Collier (1997) used molecular techniques to define both the Rivulidae and Aplocheilidae as monophyletic clades. However, their analysis was not able to determine if diapause evolved once in the common ancestor of these two clades, or independently. Hrbek and Larson (1999) used phylogenies based on mitochondrial DNA sequences to examine the evolution of diapause in the Rivulidae. Their analysis indicates two independent origins of diapause in South America. Thus, it appears likely that diapause may have arisen several times in the annual killifishes and independently on the continents of Africa and South America. A closer examination of the various groups, especially those that contain members that produce diapausing and nondiapausing embryos, might yield interesting insights into the evolution of this complex life history.

Annual killifish in Africa and South America share a surprising number of similarities that are collectively known as the "annual" life history. The similarities are found at all levels of organization including: habitat requirements, coloration, morphology, physiology, and most importantly embryological development (Wourms 1972a, b, c; Brosset 2003). In fact, the similarities in development observed in both the African and South American annual killifish are remarkable. They share three distinct developmental stages where arrest of development can occur in the form of diapause. In addition, all annual killifish, but not other closely related killifish, share a unique developmental process of dispersion and subsequent reaggregation of the embryonic blastomeres (Wourms 1972b). This sequence of events separates the cell migrations of epiboly (sheets of cells enveloping the yolk mass) from the formation of the embryonic axis. In nonannual killifish, these two events are coupled as they are in all other teleosts. While the evolutionary history of annual killifish in Africa and South America remains debatable, either option is incredibly interesting from a biological perspective; either it is an impressive case of convergence or a complex series of gains and losses of the annual life history in many species within this group (e.g., Murphy et al. 1999). Close examination of the molecular mechanisms that control entrance and exit from diapause in African and South American annual killifish should shed light on the evolution of the annual life history on these continents. From a practical perspective, the two groups are almost indistinguishable in terms of the developmental physiology of diapause, and thus the following discussion on diapause in annual killifish will draw on evidence collected in both African (*Nothobranchius* and *Aphyosemion*) and South American (*Austrofundulus*) species.

12.3.2 The Life History of Annual Killifish

Annual killifish are found in regions of the tropics and subtropics that experience pronounced dry and rainy seasons. They are typically found in ephemeral ponds in savanna and desert habitats, but they also exploit the margins of seasonally flooded forests. All annual killifish exhibit a life history pattern that includes the arrest of development in at least one of three possible stages of embryonic diapause (Wourms 1972c; Murphy and Collier 1997). These stages extend the total time of development through the dry season so that embryos are ready to hatch when water is available. Diapause was originally described by Peters (Peters 1963, 1965) in African species and by Wourms (Wourms 1964, 1967, 1972a, b, c) in South American and African species. Wourms also described the major events associated with development of annual killifish (Wourms 1972a) including the occurrence of dispersion and subsequent reaggregation of the embryonic blastomeres associated with the occurrence of diapause I (Wourms 1972b). In addition, he described three separate stages where diapause may occur in annual killifish and named them diapause I, II, and III (Wourms 1972c).

12.3.3 Diapause I

Diapause I occurs early in development prior to the formation of the embryonic axis and is associated with a stage of development unique to annual killifish called the dispersed cell phase. Diapause I appears to be facultative in most species and is induced by low temperatures and hypoxia (Wourms 1972c; Inglima et al. 1981). Very little mechanistic data is available for this stage of diapause, although Peters describes a substantial tolerance of anoxia in Diapause I embryos of *Nothobranchius* (Peters 1963). It appears that diapause I in *Nothobranchius* may be induced by chemical signals produced by adult fish. In *N. guentheri*, if embryos are incubated in the presence of adult fish without aeration, almost 100% of the embryos arrest development at diapause I despite conditions conducive for development (Inglima et al. 1981). This study found no affect of up to 8 ppm ammonia on the rate of embryonic development or the occurrence of diapause. Similar results were reported for embryos of *N. korthausae* (Denucé 1989). In both studies, the effect of the presence of adults was not species-specific. In *N. korthausae*, tissue extracts from ovary, muscle, and liver of adult fish all induced diapause I. The substance responsible for the arrest of development in embryos of *N. korthausae* was determined to be a polar hydrophilic substance that as of yet remains unidentified (Denucé 1989). Induction of diapause I by the presence of adults has not been observed in our laboratory stock of the South American annual killifish *Austrofundulus limnaeus* (JEP, unpublished data).

12.3.4 Diapause II

Diapause II is obligate in most species studied, although some report this diapause as facultative (see Discussion in Wourms 1972c and Murphy and Collier 1997). This diapause occurs after the formation of the embryonic axis in an embryo possessing 38–42 pairs of somites, the foundations of the central nervous system, optic cups, olfactory and lens placodes, otic vesicles, and a functional tubular heart (Wourms 1972a, c). Most of the physiological data on diapause has been focused on this stage due to its obligate nature, and the incredible robustness of the embryos to environmental insults.

Metabolism is depressed during diapause II by greater than 90% compared with actively developing embryos in both African and South American species (Levels et al. 1986; Podrabsky and Hand 1999). In addition, diapause II embryos of *A. limnaeus* appear to have a significant portion of their total metabolic heat dissipation due to anaerobic processes, even when incubated under aerobic conditions (Podrabsky and Hand 1999). Importantly, lactate is not accumulated during this time. Diapause II embryos maintain a high energetic status as assessed by ATP/ADP ratios near 60. However, despite high levels of ATP, AMP levels are elevated in diapause II embryos, and these levels are negatively correlated with the rate of metabolism (Podrabsky and Hand 1999). This information led to the speculation that activation of the AMP-activated protein kinase may play a role in suppressing biosynthetic reactions during diapause II (Podrabsky and Hand 1999). Rates of protein synthesis are depressed by over 90% during diapause II and can account for approximately 36% of the metabolic depression observed (Podrabsky and Hand 2000).

12.3.5 Diapause III

Diapause III occurs in the late prehatching stage embryo and appears to be obligate in most species (Wourms 1972a, c). Embryonic development is essentially complete in this stage of development and a significant portion of the yolk mass has been consumed. The metabolism of diapause III embryos is depressed compared with those just completing development that do not enter diapause, but are induced to hatch (Levels et al. 1986; Podrabsky and Hand 1999). In *A. limnaeus*, the metabolism of diapause III embryos slowly declines over a period of 4–6 weeks (Podrabsky and Hand 1999). While postdiapause II development is largely aerobic as indicated by combined calorimetry and respirometry, late diapause III embryos appear to have a significant anaerobic contribution to their total heat dissipation (Podrabsky and Hand 1999).

12.3.6 Environmental Control of Diapause II

Incubation temperature can have a marked affect on the occurrence of diapause in annual killifish. Markofsky and Matias (1977) found that in *N. guentheri* diapause I and II could be induced if embryos were collected and incubated at 19.4°C, while incubation temperatures of 22.7 or 26.5°C caused the embryos to develop directly to diapause III. *N. korthausae* embryos incubated at 18 and 25°C entered diapause II when kept in the dark or exposed to a short-day photoperiod, while incubation at 32°C or exposure to a 12:12 h photoperiod caused embryos to develop past diapause II and III (Levels and Denucé 1988). In embryos of *A. limnaeus*, incubation at 20 or 25°C induces embryos to arrest development at diapause II, while incubation at 30°C causes embryos to bypass diapause II (termed escape embryos by Wourms 1972c) and develop directly to diapause III.

The photoperiod experienced by the adult fish and by the embryos can affect the occurrence of diapause in annual killifish. Markofsky et al. (1979) report that short-day photoperiods experienced by adults can induce the production of embryos that enter diapause II. Levels and Denucé (1988) found that embryos were capable of responding directly to photoperiod cues with short-day photoperiods inducing diapause II and III, and long-day photoperiod causing embryos to skip these diapause stages. In embryos of *A. limnaeus* exposure to long- and short-day photoperiods causes embryos to break diapause II and develop to diapause III (Podrabsky and Hand 1999).

The above data on the effects of temperature and photoperiod on diapause are consistent with a general pattern of induction of diapause II in the late part of the rainy season and during the dry season (cooler temperatures and shorter day lengths) and a bypassing of diapause II during the rainy season (warmer temperatures and longer day lengths). This pattern is consistent with a bet hedging strategy for the production of embryos that may hatch during the current rainy season being produced early in the rainy season, and diapausing embryos that must endure long-term desiccation of the pond being produced late in the rainy season.

12.3.7 Alternate Developmental Pathways Associated with Escape Embryos

Embryos that do not enter diapause I or II but instead develop directly to hatching or diapause III are called escape embryos. These embryos develop quickly enough to hatch during a single rainy season and are likely very important in supporting multiple generations of fish in a single pond in the event of multiple inundations and dryings during a single rainy season. When escape embryos are compared with those entering diapause II, there is a clear difference in morphology and development. Escape embryos have a more advanced development of anterior structures relative to posterior structures when compared with embryos that are developing toward diapause II (Fig. 12.2; Podrabsky, Garrett and Kohl, unpublished observations). Thus, it appears that two alternate developmental pathways

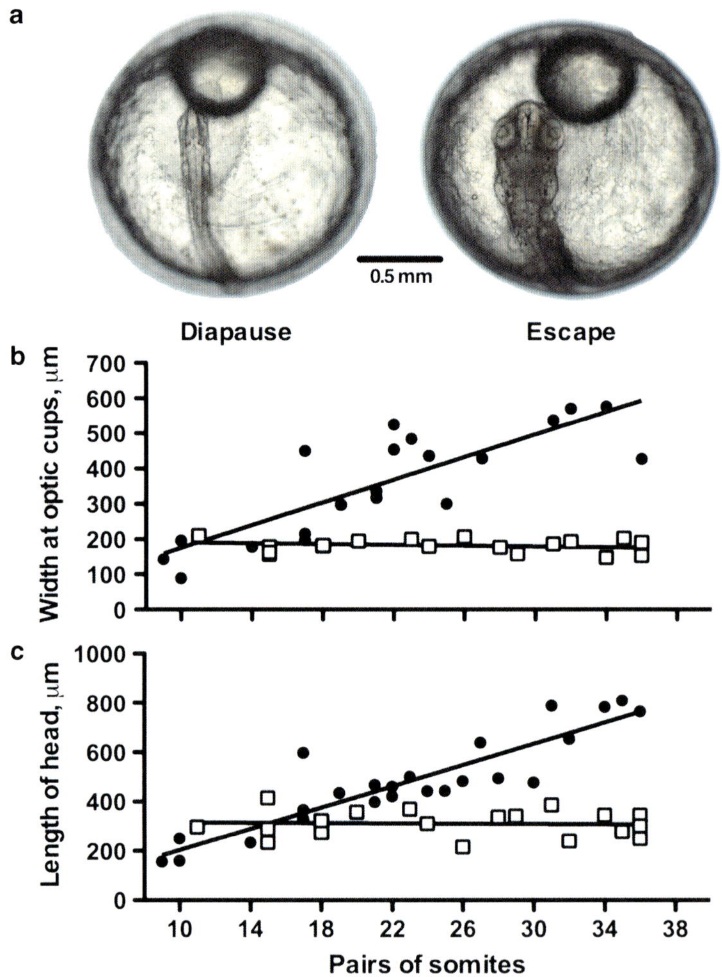

Fig. 12.2 Two alternate developmental pathways are observed in embryos of *A. limnaeus*. (**a**) Both images are of embryos possessing 24 pairs of somites. The embryo on the *left* is destined to enter diapause II while the embryo on the *right* is an escape embryo and will bypass this stage of diapause. (**b** and **c**) When embryos are staged according to the addition of pairs of somites, escape embryos (*closed circles*) have larger heads with more advanced morphology when compared with embryos destined to enter diapause II (*open squares*). Symbols represent measurements from a single embryo

exist in *A. limnaeus*, which differ in the timing of early developmental events. This finding adds another amazing dimension to the biology of annual killifish and represents one of the most extreme and perhaps the only case of alternate developmental pathways during early development of a vertebrate. Alternate developmental pathways may have profound effects on the physiology and ecology of embryos at hatching. Do these embryos differ metabolically or physiologically after passing though different developmental pathways? Do both pathways produce similar larvae? Are life history trade-offs associated with the different developmental trajectories? Will one pathway be favored under a specific set of environmental circumstances? The consequences of developing on either of these alternate developmental trajectories on larval and adult physiology and ecology are currently unknown but are under investigation.

12.4 Diapause and Tolerance to Environmental Stress

The natural environment of killifish in general and especially the annual killifishes is highly variable both spatially and temporally. Habitats may range from deserts and savannas to flooded forests and grasslands, as well as salt marshes and mangrove swamps. Each of these environments has a unique suite of environmental characters, but all are ephemeral in nature and will eventually if not regularly expose the fish to periods of dryness. The most extreme conditions survived by annual killifish are found in areas of tropical savannas and coastal deserts in both Africa and South America. In these areas, they inhabit small bodies of water that undergo large fluctuations, both daily and annually, in a number of biologically relevant environmental factors. Rainy season ponds inhabited by *A. limnaeus* in Venezuela (Podrabsky et al. 1998) may experience oxygen concentrations varying from hyperoxic during the day to extremely hypoxic at night. The pH can range over 3–4 units between ponds and several units in a single pond each day. Temperatures range from the low 20s to the low 40s on a daily basis. Ponds can be dilute with high quantities of organic acids, or highly turbid with as high as 1 ppt salinity (Podrabsky et al. 1998). The sediments where embryos are deposited are almost certainly extremely hypoxic or anoxic during periods of inundation (Podrabsky et al. 1998). The rainy season in this part of Venezuela lasts from May through August, but rain events can be highly variable and unpredictable during the rainy season. It is likely that many ponds inhabited by annual killifish experience several cycles of inundation and drying during a single rainy season. The dry season will typically last for 6–8 months, but inundation of specific ponds may be separated by several years depending on local weather patterns. Thus, embryos of *A. limnaeus* may be trapped in drying mud for months to years. The time frame for drying of their habitat has not yet been described, and thus the microenvironment experienced by embryos during the dry season is currently unknown. The habitats of various species of *Nothobranchius* from Africa have been described (Reichard et al. 2009; Watters 2009) and are very similar to those described here for *A. limnaeus*.

While all three stages of diapause may have increased tolerance to environmental stress, diapause II is definitely the most resistant stage. In fact, diapause II embryos may survive conditions that are far more extreme than the embryos are likely to experience in their natural habitat. This "excessive" resistance is a common theme in organisms that enter into states of profound metabolic dormancy (Jonsson 2003). Very little molecular evidence exists to explain this increased tolerance. Molecular and physiological evidence are accumulating and suggest that diapausing embryos prepare for environmental stress prior to entry into diapause as part of their natural developmental program. Thus, even though embryos in diapause may have a limited ability to respond to environmental stress through the production of large amounts of new proteins (Podrabsky and Hand 2000), they may already be protected because they develop a "stress tolerant" phenotype prior to entering diapause. For example, a heat inducible form of the molecular chaperone hsp70 is constitutively expressed in early embryos of *A. limnaeus* and the highest levels of this protein are observed during diapause II (Podrabsky and Somero 2007). Elevated levels of heat shock proteins would presumably protect the embryos from a variety of stresses during the dry season.

12.4.1 Temperature Tolerance

Temperatures may range from 10 to 42°C in the habitat of annual killifish from Africa and South America, with average temperatures falling in the high 20s and low 30s (Podrabsky et al. 1998; Reichard et al. 2009; Watters 2009). In general, temperatures of annual killifish habitat will range on the warmer end of the spectrum with very few days spent below 15°C. When given a choice, adult *A. limnaeus* choose temperatures around 26–27°C (Podrabsky et al. 2008). Temperatures may vary as much as 15°C on a daily basis during the rainy season (Podrabsky et al. 1998). However, no data are currently available for temperatures experienced by embryos during the dry season. Temperature tolerance in annual killifish embryos has been studied most extensively in the African annual killifish *N. guentheri*, and the data suggest that embryos are well suited to survive temperatures common in their habitat. Matias and Markofsky (1978) found that diapause II embryos could survive up to 5 h at −8°C and could survive long-term without abnormalities at temperatures ranging from 3.4 to 32.7°C. Diapause II embryos were able to survive for several days at 40°C. While no direct experiments to determine the temperature tolerance of *A. limnaeus* embryos have been conducted, it appears that their temperature tolerance is very similar (at least in the high range) to that reported for *N. guentheri* (JEP unpublished data). However, in a study of adult *A. limnaeus* exposed to a cycling temperature regime from 20 to 37°C on a daily basis, fertilization rates were found to decline significantly (Podrabsky et al. 2008). This study also determined the temperature preference of the adult fish, which in males appears to shift to cooler temperatures after exposure to temperature cycling. Taken together, these two pieces of evidence may suggest that spermatogenesis is compromised in annual killifish at high temperatures.

12.4.2 Salinity Tolerance

Many annual killifish habitats are found in areas that will receive significant loads of salts over time due to multiple inundation and drying cycles. In addition, the clay soils that appear to be critical for annual killifish survival in especially xeric environments (Watters 2009) may act to attract cations and cause them to accumulate over time. While ponds may begin as rather dilute solutions, there will inevitably be an increase in the salinity of a pond as it dries, and levels of calcium and other minerals can become very concentrated in the interstitial water of the soil just prior to drying which eventually leads to salt precipitation. Thus, embryos of annual killifish may have to endure rather high levels of salts in their natural environment. Embryos of *A. limnaeus* are quite tolerant of elevated salinities and defend their internal osmolality of around 300 mOsm/kg water even when exposed to 50 ppt salinity medium (Machado and Podrabsky 2007). Embryos in diapause II can survive for over a week in 70 ppt salinity and for several months in 35 and 50 ppt salinity (Machado and Podrabsky 2007). Survival appears to be due to an exceptionally low permeability of diapause II embryos to salts and to water. Surprisingly, diapause II embryos exhibit extremely slow rates of water exchange with their aqueous medium. When exposed to a 100% deuterated water environment the embryos will initially float in the "heavy water." It takes over 10 days on average for a diapause II embryo to exchange enough water for deuterated water to sink. By comparison, zebrafish embryos of a similar developmental stage will sink in deuterated water in a matter of minutes to hours, making diapause II embryos over 1,000 times less permeable to water than zebrafish embryos (Machado and Podrabsky 2007). Consistent with the overall low permeability of *A. limnaeus* embryos to water, no increase in embryonic osmolality is observed after 4 days of exposure to 50 ppt salinity medium, which indicates they are not accumulating organic osmolytes to prevent osmotic imbalance. In addition, $Na^{+}K^{+}$-ATPase activity is low in embryos during the duration of development, and actually decreases in response to high salinities, which appears to indicate a reduction in ion pumping in response to a salinity challenge (Machado and Podrabsky 2007). Interestingly, ions and water are exchanged rather quickly across the egg envelope under aqueous conditions and during the initial stages of dehydration (Podrabsky et al. 2001; Machado and Podrabsky 2007). These data suggest that the cell membranes of the enveloping cell layer, either alone or in conjunction with the chemicals found in the perivitelline fluid, are the major site of resistance to water and ion exchange with the environment. The unique properties of the enveloping layer cells of annual killifish may also be key to their ability to survive in highly desiccating environments (see below).

12.4.3 Anoxia Tolerance

Inundated soils typically become hypoxic or anoxic within a few days of being covered by standing water and this generality has been confirmed for the sediments of ponds inhabited by *A. limnaeus* (Podrabsky et al. 1998). Therefore, embryos of annual killifish are likely hypoxic or anoxic for the greater part of the rainy season when the ponds are inundated. Exposure to elevated oxygen is likely limited to the time period when most of the standing water has evaporated, but there is still enough moisture available to support development and gas exchange (see below). Thus, it is not surprising that embryos of annual killifish have evolved a high tolerance to anoxia. Embryos of *A. limnaeus* are the most anoxia-tolerant vertebrates described (Podrabsky et al. 2007). Diapause II embryos can survive well over 90 days ($LT_{50} = 65$ days) in the complete absence of oxygen at 25°C. Even more impressive, embryos that have broken diapause II and begin to actively develop retain this substantial anoxia tolerance for at least 4 days, despite aerobic metabolic rates that are over 80 times greater than those of diapause II embryos (Podrabsky et al. 2007). Metabolism is supported by the production of lactate and small amounts of alanine and succinate during prolonged bouts of anoxia (Podrabsky et al. 2007). Interestingly, early embryonic stages through diapause II that exhibit extreme tolerance of anoxia accumulate up to 10 mM gamma aminobutyric acid (GABA) in response to anoxia while normoxic levels of this compound are not detectable (Podrabsky et al. 2007). Heart rate declines and eventually stops within the first 24 h of anoxia in embryos that are tolerant of long-term anoxia (diapause II and 4 days postdiapause II). In contrast, later stage embryos that do not have a substantial tolerance of anoxia (including diapause III embryos) experience a decreased but sustained heart rate during anoxia until the death of the embryo, indicating that metabolic depression is likely key for long-term survival of cardiac tissue during oxygen deprivation (Fergusson-Kolmes and Podrabsky 2007).

The ability to survive prolonged anoxia is supported by a suite of characters in all animals that includes suppression of metabolism, a reduction in the rate of protein synthesis, and reduced ion conductance through membranes (channel arrest) that leads to reduced costs for ion homeostasis (Hand and Hardewig 1996; Hand 1998; Krumschnabel 2000; Hochachka and Somero 2002). Diapause II embryos of *A. limnaeus* appear to be preadapted for survival of anoxia by having an intrinsically low metabolic rate (Podrabsky and Hand 1999), severely reduced rates of protein synthesis (Podrabsky and Hand 2000), and a generally low permeability to ions which allows for a greatly reduced need for ATP consumption via ion motive ATPases (Machado and Podrabsky 2007).

12.4.4 Dehydration Tolerance

Regular drying of ponds appears to be a requirement for long-term survival of most populations of annual killifish. Embryos may have to endure several weeks to several years without the availability of free water. Loss of water from the drying sediments in which they are encased is likely to be relatively slow, taking several weeks once the standing water is gone. Slowing of water loss is facilitated by the particular kinds of clay soils that are commonly found in annual killifish habitats, and these clay minerals may even be required for long-term survival of African annual killifish in a given location (Watters 2009). Most of what is known about survival during water deprivation in annual killifish embryos comes from a single study on embryos of *A. limnaeus* (Podrabsky et al. 2001). When exposed to dehydrating conditions, embryos of *A. limnaeus* respond by reducing evaporative water loss. All developmental stages in *A. limnaeus* can survive for long periods of time in 85% relative humidity (RH) if they are initially dried on damp filter paper for 3–4 days (Podrabsky et al. 2001). However, only diapause II embryos can survive prolonged exposure to environments below 85% RH. Survival of 75.5% RH is over 100 days and possibly considerably longer, while 50% RH is tolerated for over a month. During the initial 3–4-day exposure to dehydrating conditions, and even while placed on damp filter paper, embryos lose about 50% of their water (Fig. 12.3). This water can be accounted for almost entirely by loss of water from the perivitelline space, while the embryonic compartment remains fully hydrated. In fact, differential scanning calorimetry of embryos exposed to 1 week of 75.5% RH indicates a large and single hydrated compartment in the embryos that acts almost exactly like bulk water (Podrabsky et al. 2001). The loss of the perivitelline fluid is easily observed in embryos exposed to desiccating conditions, while the embryo and yolk compartments retain their initial size for several months (Fig. 12.3, Podrabsky et al. 2001). The egg envelope of annual fish is thicker than in most other species of fish (Schoots et al. 1982). This thickness and a unique protein composition/structure may contribute to dehydration avoidance in annual killifish. This supposition is supported by changes in the secondary structure of dehydrated egg envelope proteins that suggest a higher amount of intermolecular interactions between adjacent protein fibers (Podrabsky et al. 2001). In *F. heteroclitus* embryos, the egg envelope may be the main biological barrier to retain water in the embryonic compartments during air exposure at high relative humidity, although this structure itself may suffer significant water loss (Tingaud-Sequeira et al. 2009) (Fig. 12.4). However, the presence of the egg envelope cannot explain the prolonged resistance to dehydration observed in diapause II embryos of *A. limnaeus* (Podrabsky et al. 2001). The increased resistance to dehydration observed in diapause II is postulated to be due to either the excretion of some hydrophobic substance into the perivitelline fluid, or to the vitrification of the perivitelline fluid due to the massive loss of water from this compartment during the initial stages of dehydration. Neither of these possibilities has yet been explored (Podrabsky et al. 2001). Interestingly, microarray analysis of air-incubated *F. heteroclitus* embryos has revealed the accumulation of certain sugar-binding lectin mRNAs that might play a role to increase the viscosity of the perivitelline fluid (Tingaud-Sequeira and Cerdà, unpublished data).

At high relative humidity (near air saturation), gas exchange is presumably enhanced in embryos exposed to an aerial environment. However, under more

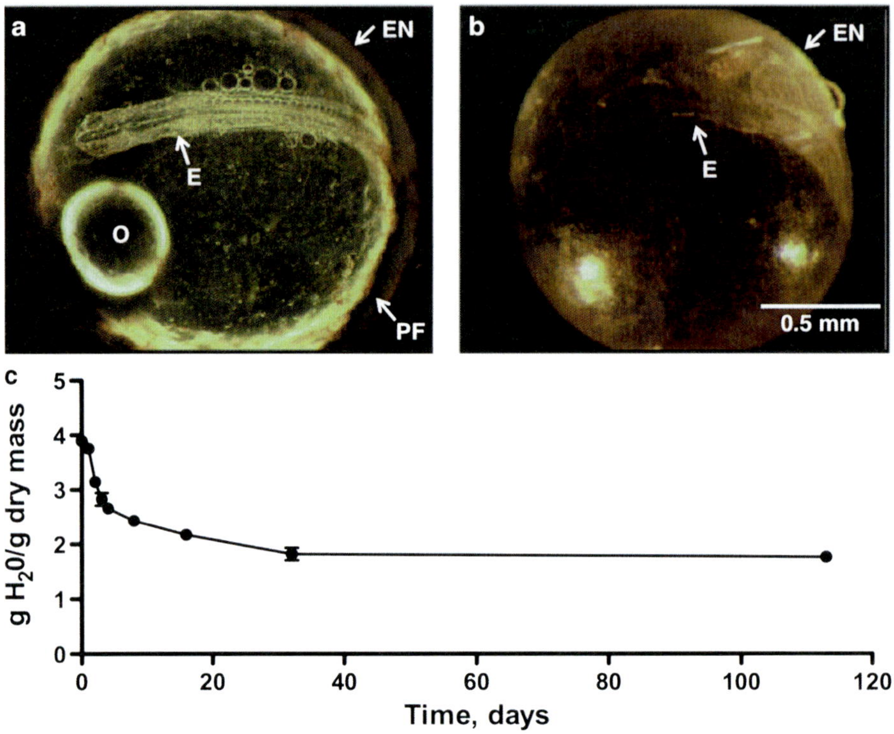

Fig. 12.3 Embryos of *A. limnaeus* survive environmental dehydration by reducing evaporative water loss. (**a**) An embryo under aqueous conditions illustrating the relationship between the egg envelope (*EN*), the perivitelline fluid (*PF*), and the embryo + yolk compartment (*E*). The large bright circle is an oil droplet (*O*) contained within the yolk. (**b**) A diapause II embryo exposed to a week of dehydration at 75.5% relative humidity. Note the lack of the perivitelline space. (**c**) Almost 50% of the total embryonic water is lost from the perivitelline fluid during the first 4 days of exposure to 75.5% relative humidity, but subsequently evaporative water loss from the embryonic compartment is very slow. Symbols are means ±SEM ($n = 3$). Data are from Podrabsky et al. (2001)

extreme dehydrating conditions, the induction of mechanisms that reduce evaporative water loss and thus promote long-term survival will inevitably limit gaseous exchange of oxygen and carbon dioxide as well. Thus, the extreme tolerance of dehydration observed in diapause II embryos of *A. limnaeus* may be related to their ability to survive a self-imposed anoxia during dehydrating conditions. It is highly likely that these two characters, anoxia tolerance and dehydration avoidance, evolved together. It is logical to predict that anoxia tolerance was acquired first, because this is a more common problem in aquatic environments, especially in inundated soils, and that dehydration tolerance followed. The interdependence of these two very different environmental stresses has not been addressed, but promises to be a very interesting avenue for future studies.

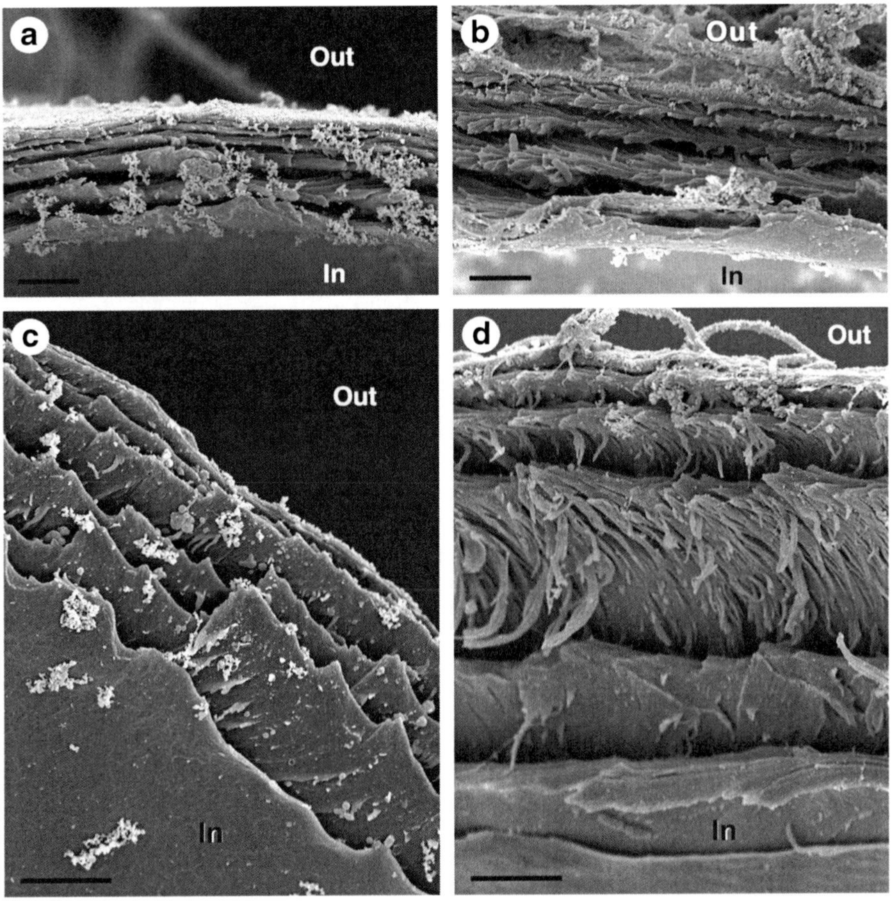

Fig. 12.4 Scanning electron micrographs of the egg envelope of 12 days old *F. heteroclitus* embryos incubated in water (**a** and **b**) or in air (**c** and **d**) from the blastula stage. Lateral (**a** and **c**) and internal (**b** and **d**) views of transversal sections of egg envelopes. In embryos incubated in air, the egg envelope layers are not as closely packed as in water-incubated embryos, resulting in a thicker chorion (*compare* **a** and **c**). In **d**, the *outer* layers of the egg envelope of air-incubated embryos show a fibrillar structure, possibly caused by evaporative water loss, while the *inner* layers had the same structural aspect as those of embryos continuously incubated in water. *Scale bars*, 5 μm

12.4.4.1 Role of Aquaporins During Dehydration Resistance

Aquaporins belong to a superfamily of integral membrane proteins widespread in all organisms that passively transport water and other noncharged small solutes across cell membranes (King et al. 2004). Aquaporins are implicated in various responses to osmotic challenge, including urine concentration in mammals (Nielsen et al. 1995), water shunting in sap-sucking insects (Beuron et al. 1995), tolerance of seawater consumption in birds (Müller et al. 2006), and intestinal water absorption in marine fish (Martinez et al. 2005; Raldúa et al. 2008). In many organisms, including plants, insects, and desert rodents, aquaporin transcription and/or protein concentration are differentially regulated during desiccation tolerance (Smith-Espinoza et al. 2003; Kikawada et al. 2008; Philip et al. 2008; Gallardo et al. 2005). These findings suggest that aquaporins might play a role during this process although their function has not been directly demonstrated yet.

In annual killifish, the potential role of aquaporins during dehydration tolerance, as well as during entrance or exit from diapause, has not been investigated. In *F. heteroclitus*, the ability of embryos to avoid dehydration and accelerate development has recently led to the investigation of the potential function of aquaporins during aerial exposure. *F. heteroclitus* aquaporin-0 (FhAqp0) and aquaporin-1 (FhAqp1), and the water and glycerol channel aquaporin-3 (FhAqp3), are all expressed during embryogenesis, but whereas the *fhaqp0* and *fhaqp1* mRNA levels are low until organogenesis starts, those of *fhaqp3* are highly and transiently expressed in blastomeres during gastrulation (Tingaud-Sequeira et al. 2009). Immunolocalization experiments revealed the presence of the FhAqp3 protein in the basolateral membrane of the cells of the enveloping layer (EVL) as well as in the membrane of ingressing and migrating blastomeres at the marginal region of the blastoderm. Remarkably, FhAqp3 transcript and protein levels seem to be down-regulated in air-exposed embryos with respect to embryos continuously immersed in seawater. However, in air-exposed embryos a reduction of FhAqp3 accumulation in the plasma membrane is only seen in the EVL, whereas the migrating blastomeres retain FhAqp3 in the membrane. The physiological significance of these findings is not known although it may be speculated that the regulation of FhAqp3 may be part of a coordinated set of adaptations to reduce evaporative water loss from the embryo through the EVL. FhAqp3 might also be controlled to regulate the migration of blastomeres (Papadopoulos et al. 2008), since gastrulation probably occurs at an increased rate in air-exposed embryos compared with water-incubated embryos, which would be consistent with the retention of FhAqp3 in the plasma membrane of the migrating blastomeres in these embryos. These hypotheses remain to be investigated, as well as the functional relationship of FhAqp3 with other aquaporins that may be expressed during early embryogenesis (as shown in mammalian embryos, Barcroft et al. 2003; Edashige et al. 2006), and play a role during dehydration resistance.

12.5 Future Prospects

The tolerance of annual killifish embryos to a variety of environmental stresses makes them an extremophile among the vertebrates. The possibility that embryonic diapause has evolved multiple times on at least two continents, and the remarkable similarity between these presumably convergent phenotypes suggests that diapause may be rather easily evolved in annual killifish. If this is in fact the case, then it may be feasible to identify the major molecular changes that underlie metabolic dormancy and increase tolerance of cell stress in vertebrate tissues by identifying a rather small number of genetic or epigenetic traits. In many situations, the conditions that these embryos endure are very similar to those that cause major damage in mammalian tissues as a consequence of injury or disease. For example, the long-term tolerance of *A. limnaeus* embryos to anoxia may provide a model for mediating the harmful effects of heart attack or stroke in humans. How is mitochondrial integrity maintained and how is apoptosis avoided after long-term oxygen deprivation? How is it possible to reduce metabolism to such low rates, and still maintain cell viability? The use of annual killifish embryos as a model for understanding how to engineer vertebrate cells to enter metabolic dormancy or survive environmental stress offers a new and promising route to developing novel insights and therapies for human diseases.

Acknowledgments The research conducted by the authors was financed by the U.S. National Science Foundation (IOS 0344578, JEP) and the American Heart Association (0335286 N, JEP), European Commission New and Emerging Science and Technologies (NEST) program (Contract No. 012674-2 Sleeping Beauty, JC), and by the Spanish Ministry of Science and Innovation (AGL2007-60262/ACU, JC).

References

Armstrong PB, Child JW (1965) Stages in the normal development of *Fundulus heteroclitus*. Biol Bull 128:143–168

Barcroft LC, Offenberg H, Thomsen P, Watson AJ (2003) Aquaporin proteins in murine trophectoderm mediate transepithelial water movements during cavitation. Dev Biol 256:342–354

Berra TM, Allen DR (1989) Burrowing, emergence, behavior, and functional morphology of the Australian salamanderfish, *Lepidogalaxias salamandroides*. Fisheries 14:2–10

Beuron F, Le Cahérec F, Guillam MT, Cavalier A, Garret A, Tassan JP, Delamarche C, Schultz P, Mallouh V, Rolland JP et al (1995) Structural analysis of a MIP family protein from the digestive tract of *Cicadella viridis*. J Biol Chem 270:17414–17422

Bigelow HB, Schroeder WC (1953) Fishes of the Gulf of Maine. US Fish & Wildlife Service. Fish Bull 53(74):1–577

Brosset A (2003) Convergent and divergent evolution in rain-forest populations and communities of cyprinodontiform fishes (*Aphyosemion* and *Rivulus*) in Africa and South America. Can J Zool 81:1484–1493

Burnett KG, Bain LJ, Baldwin WS, Callard GV, Cohen S, Di Giulio RT, Evans DH, Gómez-Chiarri M, Hahn ME, Hoover CA, Karchner SI, Katoh F, Maclatchy DL, Marshall WS, Meyer JN, Nacci DE, Oleksiak MF, Rees BB, Singer TD, Stegeman JJ, Towle DW, Van Veld PA, Vogelbein WK, Whitehead A, Winn RN, Crawford DL (2007) *Fundulus* as the premier teleost model in environmental biology: opportunities for new insights using genomics. Comp Biochem Physiol D 2:257–286

Campbell HA, Fraser KPP, Bishop CM, Peck LS, Egginton S (2008) Hibernation in an antarctic fish: on ice for winter. PLoS ONE 3:e1743. doi:10.1371/ journal.pone.0001743

Charteris SC, Allibone RM, Death RG (2003) Spawning site selection, egg development, and larval drift of *Galaxias postvectis* and *G. fasciatus* in a New Zealand stream. N Z J Mar Freshwater Res 37:493–505

Chew SF, Gan J, Ip YK (2005) Nitrogen metabolism and excretion in the swamp eel, *Monopterus albus*, during 6 or 40 days of estivation in mud. Physiol Biochem Zool 78:620–629

Darken RS, Martin KLM, Fisher MC (1998) Metabolism during delayed hatching in terrestrial eggs of a Marine fish, the Grunion *Leuresthes tenuis*. Physiol Zool 71:400–406

Denucé JM (1989) Developmental changes in annual fish embryos kept in the proximity of adult fishes. Acta Embryol Morphol Exper 10:87–94

Denver RJ (1997) Environmental stress as a developmental cue: corticotropin-releasing hormone is a proximate mediator of adaptive phenotypic plasticity in amphibian metamorphosis. Horm Behav 31:169–179

DiMichele L, Powers DA (1982) LDH-B genotype-specific hatching times of *Fundulus heteroclitus* embryos. Nature 296:563–564

DiMichele L, Powers DA (1984a) The relationship between oxygen consumption rate and hatching in *Fundulus heteroclitus*. Physiol Zool 57:46–51

DiMichele L, Powers DA (1984b) Developmental and oxygen consumption rate differences between lactate dehydrogenase B genotypes of *Fundulus heteroclitus* and their effect on hatching time. Physiol Zool 57:52–56

DiMichele L, Taylor MH (1980) The environmental control of hatching in *Fundulus heteroclitus*. J Exp Zool 214:181–187

DiMichele L, Taylor MH (1981) The mechanism of hatching in *Fundulus heteroclitus*: development and physiology. J Exp Zool 217:73–79

dos Santos Ferreira da Silva G, Giustia H, Sancheza AP, do Carmoa JM, Glass ML (2008) Aestivation in the South American lungfish, Lepidosiren paradoxa: effects on cardiovascular function, blood gases, osmolality and leptin levels. Respir Physiol Neurobiol 164:380–385

Edashige K, Tanaka M, Ichimaru N, Ota S, Yazawa K, Higashino Y, Sakamoto M, Yamaji Y, Kuwano T, Valdez DM Jr, Kleinhans FW, Kasai M (2006) Channel-dependent permeation of water and glycerol in mouse morulae. Biol Reprod 74:625–632

Eduardo J, Bicudo PW, Johansen K (1979) Respiratory gas exchange in the airbreathing fish, *Synbranchus marmoratus*. Environ Biol Fishes 4:55–64

Eldon GA (1979) Breeding, growth, and aestivation of the Canterbury mudfish, *Neochanna burrowsius* (Salmoniformes: Galaxiidae). N Z J Mar Freshwater Res 13:331–346

Enajjar S, Bradai MN, Bouain A (2008) New data on the reproductive biology of the common guitarfish of the Gulf of Gabès (southern Tunisia, central Mediterranean). J Mar Biol Assoc UK 88:1063–1068

Fergusson-Kolmes L, Podrabsky JE (2007) Differential effects of anoxia on heart rate in developmental stages of the annual killifish *Austrofundulus limnaeus* that differ in their tolerance of anoxia. J Exp Zool 307A:419–423

Fishman AP, Galante RJ, Winokur A, Pack AI (1992) Aestivation in the African lungfish. Proc Am Phil Soc 136:61–72

Gallardo PA, Cortes A, Bozinovic F (2005) Phenotypic flexibility at the molecular and organismal level allows desert-dwelling rodents to cope with seasonal water availability. Physiol Biochem Zool 78:145–152

Greenwood PH (1986) The natural history of African lungfish. J Morphol Suppl 1:163–179

Gutierrez-Estrada JC, Prenda J, Oliva F, Fernández-Delgado C (1998) Distribution and habitat preferences of the introduced mummichog *Fundulus heteroclitus* (Linneaus) in South-western Spain. Estuar Coast Shelf Sci 46:827–835

Hand SC (1998) Quiescence in *Artemia franciscana* embryos: reversible arrest of metabolism and gene expression at low oxygen levels. J Exp Biol 201:1233–1242

Hand SC, Hardewig I (1996) Downregulation of cellular metabolism during environmental stress: mechanisms and implications. Annu Rev Physiol 58:539–563

Hansen CA, Sidell BD (1983) Atlantic hagfish cardiac muscle: metabolic basis of tolerance to anoxia. Am J Physiol Regul Integr Comp Physiol 244:R356–R362

Harrington RW Jr (1959) Delayed hatching in stranded eggs of marsh killifish, *Fundulus confluentus*. Ecology 40:430–437

Helvik JV, Walther BT (1993) Development of hatchability in halibut (*Hippoglossus hippoglossus*) embryos. Int J Dev Biol 37:487–490

Hochachka PW, Somero GN (2002) Biochemical adaptation: mechanism and process in physiological evolution. Oxford University Press, New York

Hrbek T, Larson A (1999) The evolution of diapause in the killifish family Rivulidae (Atherinomorpha, Cyprinodontiformes): a molecular phylogenetic and biogeographic perspective. Evolution 53:1200–1216

Hsiao SM, Meier AH (1989) Comparison of semilunar cycles of spawning activity in *Fundulus grandis* and *F. heteroclitus* held under constant laboratory conditions. J Exp Zool 252:213–218

Hsiao SM, Greeley MS, Wallace RA (1994) Reproductive cycling in female *Fundulus heteroclitus*. Biol Bull 186:217–281

Hsiao SM, Limesand SW, Wallace RA (1996) Semilunar follicular cycle of an intertidal fish: the *Fundulus* model. Biol Reprod 54:809–818

Hyvarinen H, Holopainen IJ, Piironen J (1985) Anaerobic wintering of crucian carp (*Carassius carassius* L.). I: Annual dynamics of glycogen reserves in nature. Comp Biochem Physiol 82:797–803

Inglima K, Perlmutter A, Markofsky J (1981) Reversible stage-specific embryonic inhibition mediated by the presence of adults in the annual fish *Nothobranchius guentheri*. J Exp Zool 215:23–33

Jonsson KI (2003) Causes and consequences of excess resistance in cryptobiotic metazoans. Physiol Biochem Zool 76:429–435

Kawaguchi M, Yasumasu S, Shimizu A, Hiroi J, Yoshizaki N, Nagata K, Tanokura M, Iuchi I (2005) Purification and gene cloning of *Fundulus heteroclitus* hatching enzyme. A hatching enzyme system composed of high choriolytic enzyme and low choriolytic enzyme is conserved between two different teleosts, *Fundulus heteroclitus* and medaka *Oryzias latipes*. FEBS J 272:4315–4326

Kawamura K, Uehara K (2005) Effects of temperature on free-embryonic diapause in the autumn-spawning bitterling *Acheilognathus rhombeus* (Teleostei: Cyprinidae). J Fish Biol 67:684–695

Kikawada T, Saito A, Kanamori Y, Fujita M, Snigórska K, Watanabe M, Okuda T (2008) Dehydration-inducible changes in expression of two aquaporins in the sleeping chironomid, *Polypedilum vanderplanki*. Biochim Biophys Acta 1778:514–520

King LS, Kozono D, Agre P (2004) From structure to disease: the evolving tale of aquaporin biology. Nat Rev Mol Cell Biol 5:687–698

Koenig CC, Livingston RJ (1976) The embryological development of the diamond killifish (*Adinia xenica*). Copeia 1976:435–445

Krumschnabel G (2000) Cellular and molecular basis of anoxia-tolerance and -intolerance in vertebrates: comparative studies using hepatocytes from goldfish and trout. Recent Res Dev Comp Biochem Physiol 1:1–11

Lee DS, Gilbert CR, Hocutt VH, Jenkins RE, McAllister DC, Stauffer JR (1980) Atlas of North American freshwater fishes. North Carolina State Museum of Natural History, Publication of the North Carolina Biological Survey No. 1980-12

Lesser MP, Martini FH, Heiser JB (1996) Ecology of the hagfish. *Myxine glutinosa* L. in the Gulf of Maine I. Metabolic rates and energetics. J Exp Mar Biol Ecol 208:215–225

Levels PJ, Denucé JM (1988) Intrinsic variability in the frequency of embryonic diapauses of the annual fish *Nothobranchius korthausae*, regulated by light: dark cycle and temperature. Environ Biol Fishes 22:211–223

Levels PJ, Gubbels REMB, Denucé JM (1986) Oxygen consumption during embryonic development of the annual fish *Nothobranchius korthausae* with special reference to diapause. Comp Biochem Physiol 84A:767–770

Machado BE, Podrabsky JE (2007) Salinity tolerance in diapausing embryos of the annual killifish *Austrofundulus limnaeus* is supported by exceptionally low water and ion permeability. J Comp Physiol B 177:809–820

Markofsky J, Matias JR (1977) The effects of temperature and season of collection on the onset and duration of diapause in embryos of the annual fish *Nothobranchius guentheri*. J Exp Zool 202:49–56

Markofsky J, Matias JR, Inglima K, Vogelman JH, Orentreich N (1979) The variable effects of ambient and artificial light: dark cycles on embryonic diapause in a laboratory population of the annual fish *Nothobranchius guentheri*. J Exp Biol 83:203–215

Martin KLM (1999) Ready and waiting: delayed hatching and extended incubation of anamniotic vertebrate terrestrial eggs. Am Zool 39:279–288

Martinez AS, Cutler CP, Wilson GD, Phillips C, Hazon N, Cramb G (2005) Regulation of expression of two aquaporin homologs in the intestine of the European eel: effects of seawater acclimation and cortisol treatment. Am J Physiol Regul Integr Comp Physiol 288: R1733–R1743

Matias JR, Markofsky J (1978) The survival of embryos of the annual killifish *Nothobranchius guentheri* exposed to temperature extremes and the subsequent effects on embryonic diapause. J Exp Zool 204:219–228

McDowall RM, Charteris SC (2006) The possible adaptive advantages of terrestrial egg deposition in some fluvial diadromous galaxiid fishes (Teleostei: Galaxiidae). Fish Fish 7:153–164

Müller C, Sendler M, Hildebrandt JP (2006) Downregulation of aquaporins 1 and 5 in nasal gland by osmotic stress in ducklings, *Anas platyrhynchos*: implications for the production of hypertonic fluid. J Exp Biol 209:4067–4076

Murphy WJ, Collier GE (1997) A molecular phylogeny for Aplocheiloid fishes (Atherinomorpha, Cyprinodontiformes): the role of vicariance and the origins of annualism. Mol Biol Evol 14:790–799

Murphy WJ, Nguyen TN, Taylor EB, Collier GE (1999) Mitochondrial DNA phylogeny of West African Aplocheiloid killifishes (Cyprinodontiformes, Aplocheilidae). Mol Phylogenet Evol 11:343–350

Nielsen S, Chou CL, Marples D, Christensen EI, Kishore BK, Knepper MA (1995) Vasopressin increases water permeability of kidney collecting duct by inducing translocation of aquaporin-CD water channels to plasma membrane. Proc Natl Acad Sci USA 92:1013–1017

Nilsson GE, Renshaw GMC (2004) Hypoxic survival strategies in two fishes: extreme anoxia tolerance in the North European crucian carp and natural hypoxic preconditioning in a coral-reef shark. J Exp Biol 207:3131–3139

O'Boyle S, Bree RT, McLoughlin S, Grealy M, Byrnes L (2007) Identification of zygotic genes expressed at the midblastula transition in zebrafish. Biochem Biophys Res Commun 358:462–468

Papadopoulos MC, Saadoun S, Verkman AS (2008) Aquaporins and cell migration. Pflugers Arch 456:693–700

Peters N (1963) Embryonale anpassungen oviparer Zahnkarpfen aus periodisch austrocknenden Gewässern. Int Rev Ges Hydrobiol 48:257–313

Peters N (1965) Diapause und embryonale missbildung bei eierlegenden Zahnkarpfen. Wilhelm Roux Arch Entwickl Mech Org 156:75–87

Philip BN, Yi SX, Elnitsky MA, Lee RE Jr (2008) Aquaporins play a role in desiccation and freeze tolerance in larvae of the goldenrod gall fly, *Eurosta solidaginis*. J Exp Biol 211:1114–1119

Pletcher FT (1963) The life history and distribution of lampreys in the Salmon and certain other rivers in British Columbia, Canada. M.Sc. Thesis, The University of British Columbia, Vancouver, BC, p 195

Podrabsky JE, Hand SC (1999) The bioenergetics of embryonic diapause in an annual killifish, *Austrofundulus limnaeus*. J Exp Biol 202:2567–2580

Podrabsky JE, Hand SC (2000) Depression of protein synthesis during diapause in embryos of the annual killifish *Austrofundulus limnaeus*. Physiol Biochem Zool 73:799–808

Podrabsky JE, Somero GN (2007) An inducible 70 kDa-class heat shock protein is constitutively expressed during early development and diapause in the annual killifish *Austrofundulus limnaeus*. Cell Stress Chaperones 12:199–204

Podrabsky JE, Hrbek T, Hand SC (1998) Physical and chemical characteristics of ephemeral pond habitats in the Maracaibo and Llanos region of Venezuela. Hydrobiologia 362:67–78

Podrabsky JE, Carpenter JF, Hand SC (2001) Survival of water stress in annual fish embryos: dehydration avoidance and egg envelope amyloid fibers. Am J Physiol Integr Regul Comp Physiol 280:R123–R131

Podrabsky JE, Lopez JP, Fan TWM, Higashi R, Somero GN (2007) Extreme anoxia tolerance in embryos of the annual killifish *Austrofundulus limnaeus*: insights from a metabolomics analysis. J Exp Biol 210:2253–2266

Podrabsky JE, Clelen D, Crawshaw LI (2008) Temperature preference and reproductive fitness of the annual killifish *Austrofundulus limnaeus* exposed to constant and fluctuating temperatures. J Comp Physiol A 194:385–393

Powers DA, Lauerman T, Crawford D, DiMichele L (1991) Genetic mechanisms for adapting to a changing environment. Annu Rev Genet 25:629–659

Pusey BJ (1989) Aestivation in the teleost fish *Lepidogalaxias salamandroides* (Mees). Comp Biochem Physiol 92A:137–138

Raldúa D, Otero D, Fabra M, Cerdà J (2008) Differential localization and regulation of two aquaporin-1 homologs in the intestinal epithelia of the marine teleost *Sparus aurata*. Am J Physiol Regul Integr Comp Physiol 294:R993–R1003

Randall JE (1987) Introductions of marine fishes to the Hawaiian Islands. Bull Mar Sci 41:490–502

Reichard M, Polacik M, Sedlacek O (2009) Distribution, colour polymorphism and habitat use of the African killifish *Nothobranchius furzeri*, the vertebrate with the shortest life span. J Fish Biol 74:198–212

Schoots AFM, Stikkelbroeck JJM, Bekhuis JF, Denucé JM (1982) Hatching in Teleostean fishes: fine structural changes in the egg envelope during enzymatic breakdown in vivo and in vitro. J Ultrastruct Res 80:185–196

Scott WB, Crossman EJ (1998) Freshwater Fishes of Canada. Galt House Publications Ltd., Oakville, Canada

Seale A (1910) The successful transference of black bass into the Philippine Islands with notes on the transportation of live fish long distances. Philippine J Sci 5:153–159

Simpfendorfer CA (1992) Reproductive strategy of the Australian sharpnose shark, *Rhizoprionodon taylori* (Elasmobranchii: Carcharhinidae), from Cleveland Bay, northern Queensland. Aust J Mar Freshwater Res 43:67–75

Smith HW (1930) Metabolism of the lung-fish, *Protopterus aethiopicus*. J Biol Chem 88:97–130

Smith-Espinoza CJ, Richer A, Salami F, Bartels D (2003) Dissecting the response to dehydration and salt (NaCl) in the resurrection plant *Craterostigma plantagineum*. Plant Cell Environ 26:1307–1315

Taylor MH (1986) Environmental and endocrine influences on reproduction of *Fundulus heteroclitus*. Am Zool 26:159–171

Taylor MH (1991) Entrainment of the semilunar reproductive cycle of *Fundulus heteroclitus*. Proc 4th Int Symp Reprod Physiol Fish, Sheffield: Fish Symp 91:157–159

Taylor MH (1999) A suite of adaptations for intertidal spawning. Am Zool 39:313–320

Taylor MH, DiMichele L, Leach GJ (1977) Egg stranding in the life cycle of the Mummichog, *Fundulus heteroclitus*. Copeia 1977:397–399

Taylor MH, Leach GJ, DiMichele L, Levitan WM, Jacob WF (1979) Lunar spawning cycle in the mummichog, *Fundulus heteroclitus* (Pisces: Cyprinodontidae). Copeia 1979:291–297

Tewksbury HT, Conover DO (1987) Adaptive significance of intertidal egg deposition in the Atlantic Silverside *Menidia menidia*. Copeia 1987:76–83

Tingaud-Sequeira A, Zapater C, Chauvigné F, Otero D, Cerdà J (2009) Adaptive plasticity of killifish (*Fundulus heteroclitus*) embryos: dehydration-stimulated development and differential aquaporin-3 expression. Am J Physiol Regul Integr Comp Physiol 296:R1041–R1052

Tomiyama M, Yanagibashi S (2004) Effect of temperature, age class, and growth on induction of aestivation in Japanese sandeel (*Ammodytes personatus*) in Ise Bay, central Japan. Fish Oceanogr 13:81–90

Watters BR (2009) The ecology and distribution of *Nothobranchius* fishes. J Am Killifish Assoc 42:37–76
White WT, Hall NG, Potter LC (2002) Reproductive biology and growth during pre- and postnatal life of *Trygonoptera personata* and *T. mucosa* (Batoidea: Urolophidae). Mar Biol 140:699–712
Wourms JP (1964) Comparative observations on the early embryology of *Nothobranchius taeniopygus* (Hilgendorf) and *Aplochcheilichthys pumilis* (Boulenger) with special reference to the problem of naturally occurring embryonic diapause in teleost fishes. In: East African Freshwater Research Organization, Annual Report, Jinja, pp 68–73
Wourms JP (1967) Annual Fishes. In: Wilt FH and Wessels NK (eds). Methods in Developmental Biology. Thomas Y. Crowell Co., New York, pp 123–137
Wourms JP (1972a) Developmental biology of annual fishes I. Stages in the normal development of Austrofundulus myersi Dahl. J Exp Zool 182:143–168
Wourms JP (1972b) The developmental biology of annual fishes II. Naturally occurring dispersion and reaggregation of blasotmeres during the development of annual fish eggs. J Exp Zool 182:169–200
Wourms JP (1972c) The developmental biology of annual fishes III. Pre-embryonic and embryonic diapause of variable duration in the eggs of annual fishes. J Exp Zool 182:389–414

Chapter 13
Mammalian Hibernation: Physiology, Cell Signaling, and Gene Controls on Metabolic Rate Depression

Kenneth B. Storey, Gerhard Heldmaier, and Mark H. Rider

Abstract During the hibernating season, small mammals may suppress their metabolic rate during cyclic periods of deep torpor by as much as 99% as compared with normothermia. Endocrine regulation of metabolic depression is still poorly understood but recent studies suggest involvement of hormones including iodothyronamine, leptin, and ghrelin. At the intracellular level, suppression of many metabolic functions is achieved via reversible protein phosphorylation of metabolic enzymes, protein synthesis translation factors, and ion pumps. Potential roles for signaling enzymes such as the AMP-activated protein kinase in the coordination of metabolic suppression have been analyzed. Recent advances in the control of global gene expression have identified participating mechanisms including histone modifications that affect chromatin structure, SUMOylation to suppress transcription factor action, and differential regulation of mRNA transcripts by interaction with micro-RNA species. However, despite global transcriptional suppression, selected transcription factors are active during torpor bouts triggering the up-regulation of specific genes that serve the hibernation phenotype.

13.1 Metabolic Depression in Hibernation

Mammalian hibernation is an amazing phenomenon. Hibernating mammals have the remarkable ability to reduce metabolic rate to a fraction of basal and allow their body temperature (Tb) to fall to near ambient. They typically maintain this torpid

K.B. Storey (✉)
Institute of Biochemistry, Carleton University, 1125 Colonel By Drive, Ottawa, ON, Canada K1S 5B6
e-mail: kenneth_storey@carleton.ca

G. Heldmaier
Faculty of Biology, Marburg University, D-35032 Marburg, Germany

M.H. Rider
Université catholique de Louvain and de Duve Institute, Avenue Hippocrate 75, 1200 Brussels, Belgium

E. Lubzens et al. (eds.), *Dormancy and Resistance in Harsh Environments*, Topics in Current Genetics 21,
DOI 10.1007/978-3-642-12422-8_13,

state with depressed metabolism and low body temperature for 5–15 days and then spontaneously interrupt torpor by an arousal. During arousal animals raise their Tb by endogenous heat production so that Tb rises from the torpid level (often about 5°C) to normothermia at 36°C within a few hours. This aroused state is maintained for about 1 day before they enter deep torpor again. This succession of entry into torpor, deep torpor, arousal, and aroused state is called a hibernation bout. The entire hibernation season is a sequence of 10–20 such hibernation bouts, depending on the duration of the season and the length of the deep torpor periods. At the beginning of the hibernation season most species display short torpor bouts, which lengthen toward midwinter and then decrease again toward the end of the hibernation season. This is observed when squirrels and marmots are hibernating in their natural environment, as well as in the laboratory under constant conditions (Wang 1979; Nicol and Andersen 2000; Buck and Barnes 2000; Ortmann and Heldmaier 2000; Carey et al. 2003; Heldmaier et al. 2004). This indicates that hibernators actively control the timing of hibernation bouts as well as the depth and duration of deep torpor.

The extent of metabolic depression in hibernation can range up to 97–99% compared with metabolic rate in normothermia. For example, data for hibernating dormice, *Glis glis*, are shown in Table 13.1. This species may display rather long torpor bouts and the longest period spent in deep torpor lasted for 30 days. Obviously, the small amount of energy gained from oxidative metabolism, which is between 1 and 2% of resting energy requirements (3% basal metabolic rate) is enough to maintain the entire cellular and metabolic machinery intact and even allows the initiation of a spontaneous arousal. Similar low levels of metabolic rate, ventilatory, and cardiac activity in deep hibernation have also been observed in other hibernators including chipmunks, 13-lined ground squirrels, arctic ground squirrels, marmots, hedgehogs, bats, and marsupials (Lyman and O'Brien 1960; for review see: Geiser 1988, 2004; Heldmaier and Ruf 1992; Milsom et al. 2001; Heldmaier et al. 2004). In this deep torpor, hibernators are still able to respond to tactile or thermal stimuli and they also thermoregulate, controlling their metabolic heat production to prevent body temperature falling below a desired level, or to prevent freezing when ambient temperature falls below 0°C. When arousal is initiated, either spontaneously or by an external stimulus, they can raise their metabolic rate within minutes from this ignition level to the high normothermic

Table 13.1 Reduction of body temperature *Tb*, metabolic rate *MR*, ventilation rate *VR* and heart rate *HR* during deep torpor in hibernating dormice, *Glis glis*. Compiled from Wilz and Heldmaier (2000) and Elvert and Heldmaier (2005)

	Tb (°C)	MR (ml O_2 g^{-1} h^{-1})	Energy (mWatt g^{-1})	VR (breath min^{-1})	HR (beats min^{-1})
Normothermic at 7°C ambient	36.7	1.42	7.726	114	314.8
Deep torpor at 7°C ambient	7.2	0.017	0.092	0.41*	8.2
Reduction (%)		98.8%	98.8%	99.4%	97.4%

*At this temperature dormice are apneic for 20–25 min and then take 9–11 breaths within 2 min

level of heat production even though their Tb has not yet increased much above the hibernation level.

Despite convincing evidence for a major depression of physiological functions in hibernators, it is still not fully known how they adjust their metabolic organization at this low level. Basal metabolic rate at thermoneutrality (BMR) is generally considered as the minimum energy requirement of endotherms to maintain integrity of the living organisms. In dormice BMR generates 3.3 mW g^{-1} (0.6 ml O_2 g^{-1} h^{-1}) which is reduced to 0.092 mW g^{-1} in torpor and obviously this is enough to maintain basic functions of a torpid endotherm. In the torpid state, the rates of many cellular activities such as transmembrane ion movements are suppressed (Storey and Storey 2004) whereas other activities such as transcription, translation, and protein degradation are almost completely abolished (van Breukelen and Martin 2001, 2002; Berriel Diaz et al. 2004). Several physiological functions are also suspended, like the immune system as well as digestive functions (for review see Carey et al. 2003). All these functions are temporarily suppressed in deep torpor but can be rapidly restored during arousal.

Entrance into torpor is initiated by spontaneous metabolic depression as illustrated in Fig. 13.1. Ventilation, oxygen consumption, and heart rate are rapidly reduced whereas Tb declines gradually as a consequence of diminished heat production by

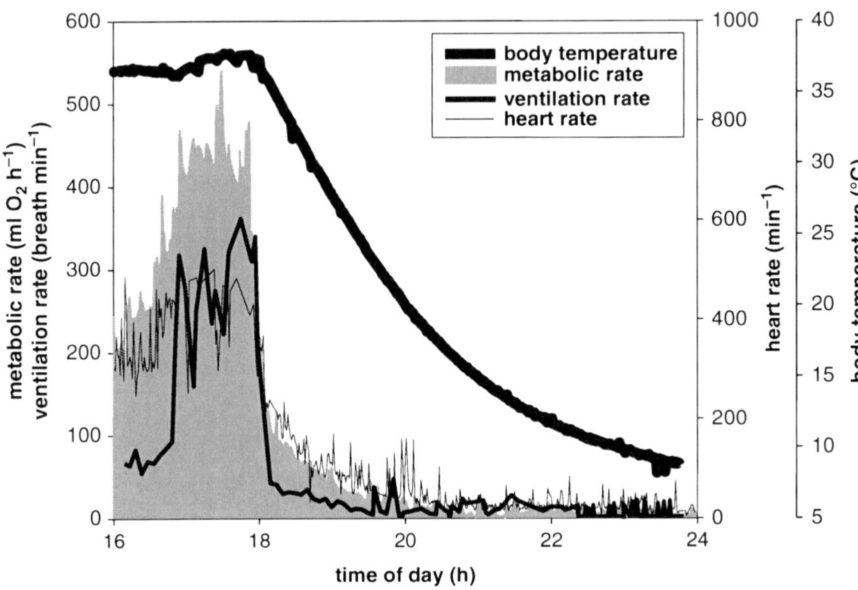

Fig. 13.1 Entrance into deep torpor in a hibernating dormouse *Glis glis* at an ambient temperature of 5°C. The dormouse was kept in an aviary and could enter hibernation spontaneously in its sleeping box. Tb and heart rate were recorded continuously by an intraperitoneally implanted transmitter (DSI). The sleeping box was designed as a whole body plethysmograph which allowed the simultaneous measurement of ventilation and metabolic rate. Modified from Elvert and Heldmaier (2005)

the body. This relationship is obvious from a comparison of the time required for 90% of the changes to occur. In *G. glis* suppression of ventilation frequency requires 82 min, metabolic rate 83 min, and heart rate 93 min, whereas Tb takes 343 min to reach the 90% threshold (Elvert and Heldmaier 2005). This delay of thermal transition as compared with metabolic and cardiac transition into hibernation has also been observed in other species including 13-lined ground squirrels, woodchucks, and marmots (Lyman 1958; Lyman and O'Brien 1960; Ortmann and Heldmaier 2000) as well as in species that exhibit daily torpor (Heldmaier and Ruf 1992; Heldmaier et al. 1999).

The initial rapid metabolic depression includes an inhibition of thermoregulatory heat production and a suppression of metabolic rate below the level of basal metabolic rate. The significance of the latter is also demonstrated by the fact that, for example, dormice may actually enter torpor at an ambient temperature of 28.6°C which is at the upper end of the thermoneutral zone in this species, and can even be considered to be a moderate heat exposure (Heldmaier and Elvert 2004). The ability to enter torpor at high ambient temperatures has also been demonstrated in species such as marsupials (Song et al. 1997), spiny mice (Ehrhardt et al. 2005; Gutman et al. 2006), and lemurs (Dausmann et al. 2009), indicating that this ability is widespread in mammals. Overall, then, it is concluded that metabolic depression during early entrance into torpor is not the consequence of low body temperature but that low body temperature is the consequence of reduced metabolic heat production.

Most hibernators develop substantial hypothermia while in deep hibernation and this will slow down all biochemical processes and contribute to the depression of metabolic rate. The above conclusion about an active metabolic depression in hibernation does not deny these thermodynamic constraints. There is a controversial discussion about the actual role of these temperature effects (Heldmaier and Ruf 1992; Guppy and Withers 1999; Geiser 2004). Hibernators are endotherms and like all endotherms they are able to actively control their metabolic rate. They have to cope with thermodynamic constraints to adjust their physiological and metabolic requirements in normothermia as well as in the torpid state. An interaction between active metabolic rate depression and temperature effects can be illustrated by several examples. The initial metabolic depression during entrance into hibernation does not immediately reach the minimum metabolic rate in hibernation. This final adjustment of minimum metabolic rate is only observed when Tb decreases below 20°C (Fig. 13.1), i.e., low temperature opens a thermodynamic window for maximum metabolic depression (Heldmaier and Elvert 2004). This is also supported by the observation of relatively high metabolic rates in hibernating fat-tailed lemurs on Madagascar that remain torpid in a relatively warm environment with ambient and body temperatures cycling between 13 and 33°C every day (Dausmann et al. 2005, 2009).

Another example that supports the active control of metabolic rate is the observation that at Tb values close to the freezing point, metabolic rate can be raised in deep torpor to generate more heat and prevent the freezing of body water. Thermoregulatory control of heat production is continued in deep torpor, which is also obvious from the maintenance of peripheral thermal sensitivity and hypothalamic thermoregulatory control in deep torpor (Florant and Heller 1977; Buck and Barnes 2000; Ortmann and Heldmaier 2000). The extent of metabolic rate reduction in deep torpor is more pronounced in small hibernators than in larger ones, which is also contradictory to simple temperature change as the explanation for metabolic depression. In addition, the extent of metabolic depression is often greater than that expected from known temperature effects on metabolic rate (Geiser 1988, 2004; Heldmaier and Ruf 1992) and is also variable for different organs and for different biochemical processes. For example, in deep hibernation the heart continues its pumping activity whereas digestive processes are suspended. Global rates of gene transcription and protein translation are strongly suppressed, yet selective expression of some genes and synthesis of some proteins continues. In summary, this suggests that hibernators are not simply victims of low temperature but exhibit controlled regulation of metabolism to maintain functional integrity at low Tb in the torpid state.

13.1.1 Endocrine Signaling

The endocrine signals that induce entrance into torpor as well as the metabolic switches operated by this signaling are still unknown. This is illustrated by the fact that entrance into hibernation still cannot be induced artificially. Indeed, to analyze the physiology of this transition, researchers still require patience and may have to observe their animals for weeks until they spontaneously enter hibernation or daily torpor. Recently, there have been several attempts to induce entrance into torpor using metabolic inhibitors. These experiments are based on the assumption that the first step of entrance may be associated with an inhibition of glycolysis or other major ATP-generating processes. One of the inhibitors that has been tested is 2-deoxyglucose which blocks glycolysis and forces a shift toward energy production via lipid utilization, which is one of the long-term pathway changes that occurs in natural torpidity. An injection of 2-deoxyglucose caused moderate hypothermia for a short period of time in Djungarian hamsters, but it remains an open question as to whether this really resembles natural and spontaneous torpor behavior with regard to duration, depth, incidence, and circadian timing of torpor bouts (Dark et al. 1996). Hypothermic responses were also observed in mice exposed to an H_2S atmosphere and in mice injected with 5′-AMP (Blackstone et al. 2005; Zhang et al. 2006; Swoap et al. 2007). The response of mice to 80 ppm H_2S looked rather similar to a spontaneous torpor bout (initial rapid metabolic inhibition, gradual development of hypothermia) but whether H_2S can trigger the same biochemical changes that occur during a spontaneous torpor bout remains to be demonstrated.

In mice a hypothermic response has also been observed following an injection of iodothyronamine, a newly discovered hormone that is derived by transamination from thyroid hormone (Scanlan et al. 2004). The hypothermic response to iodothyronamine was analyzed in Djungarian hamsters (*Phodopus sungorus*) and directly compared with the physiology of spontaneous daily torpor in this species.

An inhibitory activity was confirmed, but the extent of metabolic rate depression and the hypothermic response were less pronounced than in spontaneous torpor. Furthermore, iodothyronamine caused a rapid and long-lasting reduction in respiratory quotient (RQ), indicating a reduction in carbohydrate utilization, which was much more pronounced than the small change in RQ during spontaneous daily torpor (Braulke et al. 2008). Thyroid hormones are known to affect metabolic rate, which may include an inhibitory action of iodothyronamine. However, current results indicate that 3-iodothyronamine cannot be the primary candidate for the initiation of metabolic rate depression in torpor. Other endocrine regulators of energy homeostasis include leptin, an indicator for lipid storage, and ghrelin, an indicator for lack of food intake. Both hormones act on the hypothalamus and activate neuropeptide Y-dependent signaling of, for example, the arcuate nucleus. Injections of leptin reduced torpor incidence in several species of small mammals whereas injections of ghrelin prolonged torpor bouts in mice (Dark et al. 1996; Gluck et al. 2006). This indicates that the expression of torpor behavior is under control of the hypothalamic energy homeostasis network, but it is not known which endocrine or neural outputs of the hypothalamus are the signal(s) for control of torpor behavior and metabolic depression (for review see Drew et al. 2007). It is noteworthy here that modulation of AMP-activated protein kinase (AMPK – see below) activity in the hypothalamus by adipokines and gut-derived peptides plays a role in food intake. A decrease in hypothalamic AMPK activity is associated with decreased feeding, whereas AMPK activation increases food intake (Xue and Kahn 2006).

13.2 Metabolic Regulation by Reversible Phosphorylation

Although the extracellular signals that trigger metabolic depression are still unclear, considerable progress has been made in revealing the intracellular mechanisms that trigger, regulate, and coordinate subcellular responses to reversibly suppress intracellular metabolism. The next sections address issues including enzyme regulation, signaling cascades and the regulation of gene expression as they relate to metabolic suppression, and the reorganization of metabolism for long-term survival in the hypometabolic state.

The posttranslational modification of proteins by the addition or removal of covalently bound phosphate groups via the action of protein kinases or protein phosphatases is a widespread and extremely powerful method of metabolic control across the animal kingdom. The mechanism often provides complete on/off control of enzymes via changes in their kinetic properties, and more generally can affect the subcellular localization of proteins, protein–protein, protein–lipid, protein–DNA/RNA interactions, and protein stability. A large number of studies have shown that reversible phosphorylation is a critical mechanism of metabolic rate depression that is responsible for the coordinated, yet differential, suppression of many cellular processes when organisms enter hypometabolic or torpid states including during daily torpor, hibernation, aestivation, anaerobiosis, diapause, dauer state, and anhydrobiosis, to name a few (for review Storey and Storey 2004, 2007). With respect to the profound torpor bouts in hibernation, reversible phosphorylation is known to regulate enzymes of carbohydrate catabolism and other pathways of energy metabolism, ion motive ATPases, and protein synthesis (ribosomal initiation and elongation factors), among others. Reversible phosphorylation also controls many steps in signal transduction cascades (e.g., many protein kinases are themselves subject to on/off control by other kinases) as well as the activation of many transcription factors, so the mechanism undoubtedly participates in all of the cell signaling and gene expression events that occur over torpor-arousal cycles.

Multiple metabolic enzymes are regulated by reversible phosphorylation during hibernation (reviewed by Storey 1997) and recently, skeletal muscle creatine kinase and hexokinase (HK) from Richardson's ground squirrels (*Spermophilus richardsonii*) have been added to the list (Abnous and Storey 2007, 2008). HK offers an interesting example. This enzymatic reaction (glucose + ATP $\rightarrow$ G6P + ADP) gates the entry of glucose (taken up from blood) into glycolysis. As compared with the enzyme from euthermic animals, HK from muscle of animals in deep torpor showed 33% lower activity, reduced affinity for ATP (the K_m for ATP rose by 80%), and greater product inhibition by glucose-6-phosphate (the K_i for G6P decreased by 40%), all effects that would reduce HK action during torpor (Abnous and Storey 2008). Stimulation of AMPK in muscle extracts from squirrels in deep torpor restored HK activity and decreased the K_m for ATP, but did not affect euthermic HK (Fig. 13.2d). Consistent with this, ion exchange chromatography revealed two peaks of HK activity in torpid muscle, representing low and high phosphate forms, whereas only a single peak of phospho-HK was present in euthermic muscle (Fig. 13.2a). Incubations that stimulated AMPK or protein phosphatases shifted the elution profiles of these peaks appropriately (Fig. 13.2b) and stimulation of AMPK also enhanced ^{32}P-labeling of HK (Fig. 13.2c). Hence, the properties of HK appear to be modified when animals enter the torpid state in a manner that suppresses enzyme function by dephosphorylating the enzyme. The main HK isozyme (HK-II) in ground squirrel muscle is the form that is believed to target glucose toward anabolic fates including glycogen synthesis or the pentose phosphate cycle. Suppression of this HK action when animals enter torpor would therefore help to restrict the use of imported glucose for anabolic activities in the hypometabolic state. This makes sense because the main source of free glucose during torpor would be glucose released from glycogen stores in the liver and this glucose must be preferentially targeted to organs such as brain that have low internal glycogen reserves and little ability to use lipid fuels (except ketone bodies).

13.3 Metabolic Signaling in Hypometabolic States

Entry into torpor requires both global controls to slow all metabolic functions and differential regulation of many loci to reprioritize energy (ATP) use during torpor and to ensure that required cell preservation mechanisms are fully expressed.

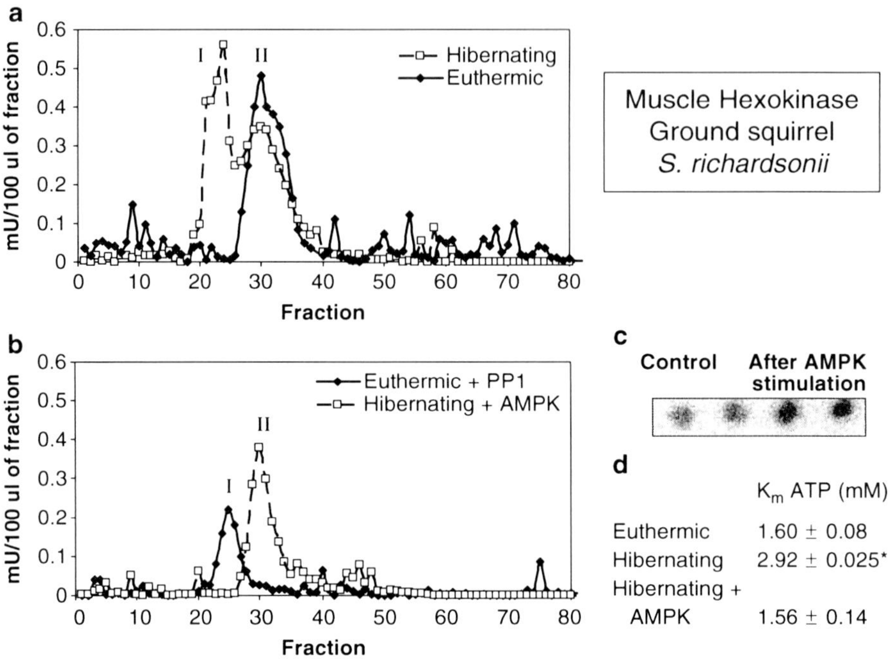

Fig. 13.2 Hexokinase from skeletal muscle of Richardson's ground squirrels, *Spermophilus richardsonii*. (**a**) Elution profiles of HK on DEAE Sephadex ion exchange columns (developed using a 0–400 mM KCl gradient in buffer) show a *single peak* in euthermia and a *double peak* in muscle from squirrels in deep torpor. (**b**) Preincubation of extracts from muscle of torpid animals under conditions that stimulated AMPK shifted all HK into the high phosphate form (*peak II*) whereas incubation of euthermic extracts under conditions that stimulated protein phosphatase 1 action converted the enzyme to the low phosphate (*peak I*) form. (**c**) Extracts from muscle of torpid squirrels were incubated with γ-^{32}P-ATP under control conditions (no additions) vs. conditions that stimulated AMPK. After immunoprecipitation to isolate HK, detection of ^{32}P-incorporation using a phosphorimager showed much stronger radiolabeling of HK after AMPK stimulation. (**d**) HK substrate affinity for ATP changed significantly (* – P <0.05) in muscle from torpid squirrels but AMPK treatment reversed this effect. Data compiled from Abnous and Storey (2008)

Much of this is regulated via signal transduction cascades responding to environmental, extracellular, or intracellular stimuli. Changes in the phosphorylation states of enzymes and functional proteins as well as changes in gene expression attest to the fact that differential regulation of signaling cascades must occur during entry into and exit from torpor (for review Carey et al. 2003; Storey and Storey 2004, 2007). Indeed, multiple examples of differential control of signal transduction enzymes are known to date. For example, cyclic AMP-dependent protein kinase (PKA) that is involved in intermediary energy metabolism and in brown fat activation is differentially regulated in the organs of hibernator species, Akt (also called PKB) which has a major role in insulin-mediated growth responses is suppressed in most tissues during torpor, and mitogen-activated protein kinases (MAPKs) show organ-specific responses with activation of the p38 MAPK signaling cascade prominent in skeletal muscle during torpor (MacDonald and Storey 1998, 2005; Eddy and Storey 2003, 2007; Abnous et al. 2008). Studies with Akt also show multiple levels of control. Not only was the amount of active phosphorylated Akt (Ser 473) reduced by 40% in *S. richardsonii* muscle during deep torpor but substrate affinities for peptide and ATP differed significantly between euthermic and torpid states and also between assay at high and low temperature (Abnous et al. 2008).

The importance of energy conservation during hibernation suggests that signaling cascades that respond to energy charge could also have important roles to play. The AMPK system is one such kinase cascade and is described below.

13.3.1 The AMP-Activated Protein Kinase (AMPK) System

AMPK is a stress kinase that acts as a sensor of cellular energy status in eukaryotes and plays a critical role in systemic energy balance (Hardie et al. 1998; Kahn et al. 2005; Hardie 2007). AMPK is a heterotrimer consisting of a catalytic α subunit and two regulatory subunits, β and γ. Multiple isoforms exist (α1, α2, β1, β2, γ1, γ2, γ3) giving 12 possible combinations of holoenzyme with different tissue distribution and subcellular localization. AMPK can be activated by changes in the intracellular AMP:ATP ratio, as occurs for example as a result of anoxia or other stresses (Hardie et al. 1998), or via an increase in intracellular Ca^{2+} (Kahn et al. 2005; Witters et al. 2006). AMP binds to the cystathionine β-synthase domains of the γ-subunits. The β-subunit, which contains a conserved glycogen binding domain, forms an interface between the α- and γ-subunits. AMPK is controlled by upstream kinases; for example, LKB1 (the Peutz-Jeghers protein) and calcium/calmodulin-dependent protein kinase kinase-β (CaMKKβ) activate AMPK by phosphorylating Thr172 in the activation loop of the catalytic α-subunits (Kahn et al. 2005; Witters et al. 2006). AMP is not only an allosteric stimulator of AMPK, but also activates AMPK by protecting Thr172 against dephosphorylation by protein phosphatases (Sanders et al. 2007). Since LKB1 is constitutively active, a rise in AMP will lead to AMPK activation by increasing the level of Thr172 phosphorylation.

AMPK activation leads to suppression of ATP-consuming processes and stimulates ATP production (Hardie et al. 1998; Hardie 2004, 2007). A summary of targets comprising metabolic enzymes, transcription factors, translation factors, and signaling proteins for which there is now good evidence for control by AMPK is given in Hue and Rider (2007) and Rider (2008). AMPK inhibits fatty acid synthesis by phosphorylating and inactivating acetyl-CoA carboxylase (ACC). As a consequence, malonyl-CoA concentrations fall and this not only inhibits fatty acid synthesis, but also stimulates fatty acid oxidation to help restore ATP levels. AMPK activation also inhibits glycogen synthesis, cholesterol/isoprenoid biosynthesis, lipolysis, and expression of gluconeogenic and lipogenic enzymes via the phosphorylation of glycogen synthase (GS), hydroxymethylglutaryl CoA reductase (HMGR), hormone-sensitive lipase (HSL), and transcription factors such as the CREB-regulated transcription coactivator 2 (TORC2).

Protein synthesis is a prominent consumer of ATP in cells (Wieser and Krumschnabel 2001) and is rigorously controlled at both the initiation and elongation steps (Proud 2007). Activation of the mammalian target of rapamycin complex 1 (mTORC1) leads to phosphorylation of the 40 S ribosomal protein S6 (rpS6), via phosphorylation and activation of p70 ribosomal protein S6 kinase (p70S6K), and this seems to be required for cell growth. mTORC1 activation also enhances translation initiation of capped mRNAs by phosphorylation of eukaryotic initiation factor-4E-binding protein-1 (4E-BP1). AMPK activation inhibits mTORC1 signaling (Proud 2007) at several levels by phosphorylating tuberous sclerosis complex-2 (Inoki et al. 2003), mTOR kinase (Cheng et al. 2004), and the mTOR binding partner raptor (Gwinn et al. 2008). Translation elongation consumes at least four equivalents of ATP for each peptide bond synthesized and is inhibited via the phosphorylation of eukaryotic elongation factor-2 (eEF2) at Thr56 by a dedicated Ca^{2+}/calmodulin-dependent kinase called eEF2 kinase (eEF2K) (Proud 2007). AMPK activation inhibits protein synthesis as a result of Thr56 eEF2 phosphorylation via the phosphorylation-induced activation of eEF2K (Horman et al. 2002, 2003; Proud 2007).

13.3.2 AMPK in Hypometabolic States

It is now becoming clear that AMPK plays a role in maintaining energy homeostasis in a number of hypometabolic states (Rider 2008). AMPK has been implicated in metabolic rate suppression during freezing in wood frogs (Rider et al. 2006), and in response to anoxia in frogs (Bartrons et al. 2004; Rider et al. 2006), freshwater turtles (Rider et al. 2009), carp (Stenslokken et al. 2008), and goldfish (Jibb and Richards 2008). In wood frogs, AMPK was activated in liver and skeletal muscle of frozen animals (Rider et al. 2006) and in liver, AMPK activation during freezing was associated with a rise in eEF2 Thr56 phosphorylation, which could inhibit protein synthesis. AMPK activation and eEF2 phosphorylation were also observed in *Rana perezi* liver in response to hypothermia and hypoxia (Bartrons et al. 2004). AMPK activation in liver during freezing could facilitate glycogen mobilization for glucose release by inhibiting glycogen synthase, whereas in skeletal muscle (and likely other organs as well), AMPK activation could stimulate glucose uptake for cryoprotection. In the red-eared slider turtle, *Trachemys scripta* elegans, subjected to 20 h of anoxic submergence at 4°C, AMPK was activated 2-fold in white skeletal muscle but decreased by 40% in heart and was unchanged in red skeletal muscle and liver. Moreover, AMPK activation in white muscle was associated with an increase in eEF2 Thr56 phosphorylation (Rider et al. 2009).

13.3.3 AMPK in Mammalian Hibernation

AMPK involvement in hibernation was assessed in 13-lined ground squirrels (*Spermophilus tridecemlineatus*). The enzyme was activated in white adipose tissue (WAT) during deep torpor but not in liver, skeletal muscle, brown adipose tissue (BAT), or brain (Horman et al. 2005). The increase in AMPK activity in WAT could result from a rise in AMP due to an increase in triacylglycerol/fatty acid substrate cycling that occurs during fat mobilization (Fig. 13.3). Indeed, it was recently shown that the fatty acyl-CoA synthase inhibitor, Triacsin C, blunted both AMPK activation and the rise in AMP:ATP ratio when adipocytes were treated to stimulate lipolysis (Gauthier et al. 2008). An increase in eEF2 Thr56 phosphorylation was seen in WAT, liver, and brain where it could be responsible for inhibiting protein synthesis during hibernation. Surprisingly, ACC phosphorylation at the inactivating Ser79 site increased dramatically in BAT from ground squirrels in deep torpor although AMPK activation was not seen, and no change in ACC phosphorylation occurred in any other tissues. Also, no change in phosphorylation state of glycogen synthase at the inactivating Ser7 AMPK site was observed in tissues from torpid animals as compared with euthermic controls.

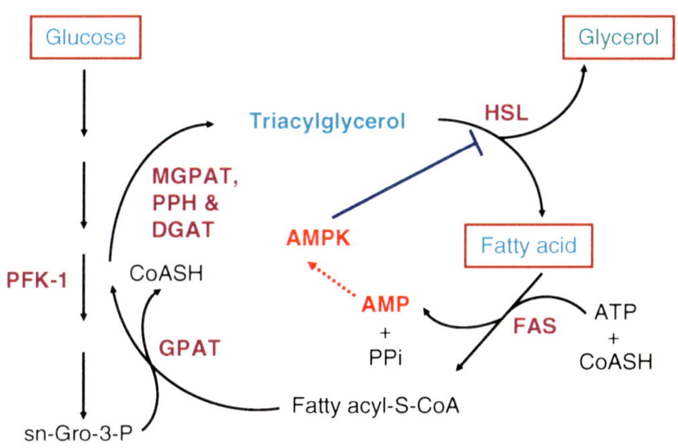

Fig. 13.3 AMPK activation via the triacylglycerol/fatty acid cycle in white adipose tissue. During triglyceride hydrolysis in incubated WAT or adipocytes, the concentration of fatty acids found in the medium is less than 3 equivalents expected from the amount of glycerol produced; this indicated that a substantial proportion of fatty acids released during lipolysis is recycled to triacylglycerol. Resynthesis involves fatty acid activation via esterification to CoA, catalyzed by long chain fatty acyl-CoA synthetase *FAS* isoenzymes, a process that produces AMP. The dominant AMPK α1 isoform in WAT is activated by β-agonists (Daval et al. 2005) and then phosphorylates and inactivates hormone-sensitive lipase *HSL*. In adipocytes treated to stimulate lipolysis, the FAS inhibitor Triacsin C blunted AMPK activation (Gauthier et al. 2008). Other abbreviations are: *PFK-1* 6-phosphofructo-1-kinase; *MGPAT* monoacylglycerol-3-phosphate *o*-acyltransferase; *PPH* phosphatidate phosphohydrolase; *DGAT* diacylglycerol *o*-acyltransferase; *GPAT* sn-glycerol-3-phosphate *o*-acyltransferase

Lastly, phosphorylation levels of the anti-lipolytic Ser565 AMPK site of hormone-sensitive lipase were unchanged in adipose tissues during the torpor phase.

Overall, the above data seem to indicate that AMPK has little role in global metabolic rate depression during hibernation. One reason for this might be that although total adenine nucleotide levels drop when animals are in deep torpor, there is no change in tissue "energy charge," i.e., the relative proportion of AMP in the total adenyline nucleotide pool does not change (MacDonald and Storey 1999). However, it cannot be excluded that AMPK would be activated by other means during entry into torpor or on arousal when high rates of fatty acid oxidation are needed for thermogenesis. Interestingly, AMPK was activated in WAT of transgenic mice overexpressing the mitochondrial uncoupling protein UCP1 (Matejkova et al. 2004). Also, chronic cold exposure in mice activated AMPK in both WAT and BAT (Mulligan et al. 2007) and in isolated white adipocytes, α1-AMPK (the main catalytic subunit isoform in these cells), was activated in response to β-agonist stimulation (Daval et al. 2005). It was recently suggested that AMP might play a role in torpor, since AMP levels were elevated in plasma of fasted mice and peripheral administration of the nucleotide induced a hypothermic state (Zhang et al. 2006). Therefore, an intriguing possibility would be that elevated levels of AMP during fasting would activate AMPK, ultimately resulting in a fall in body temperature. However, administration of AMP was found not to induce a state of true torpor in mice (Swoap et al. 2007). It would be interesting to investigate whether the shallow bouts of torpor induced by food deprivation in mice (Swoap 2008) would be affected in AMPK α1 or α2 knockout mice. Lastly, recent studies have implicated hypothalamic AMPK in the regulation of food intake and body weight and suggest that AMPK is a mediator of adipocyte-derived and gut-derived hormone effects on fatty acid oxidation and glucose uptake in peripheral tissues (Xue and Kahn 2006). Recall that roles for both leptin and ghrelin are implicated in torpor regulation, as discussed in Sect. 13.1. Hence, modulation of AMPK activity in the hypothalamus could play a prominent role in the suppression of food intake during the hibernating season and in regulating the prehibernation fattening that must occur before seasonal torpor bouts can begin.

13.4 Transcriptional Silencing and Epigenetic Mechanisms

One significant part of overall metabolic rate depression in hibernating species is global transcriptional silencing. Whereas selected genes are specifically up-regulated during entry into torpor or throughout the torpor period, gene transcription as a whole is strongly suppressed. Previous data gathered from nuclear run-on assays in ground squirrels or from rates of ^{3}H-uridine incorporation into RNA in hamsters and squirrels indicated that the overall rate of transcription was strongly reduced in organs of torpid animals (Bocharova et al. 1992; van Breukelen and Martin 2002; Osborne et al. 2004). Recent studies have documented a number of mechanisms that can contribute to this global gene silencing. Epigenetic control over DNA and chromatin is one of these and plays a key role in a host of processes including development, disease states (especially tumorigenesis), aging, and senescence as well as a significant role in inheritance (reviewed by Fraga and Esteller 2007).

The best known epigenetic mechanisms are DNA methylation and a variety of posttranscriptional modifications of histones that alter chromatin structure to regulate access of the transcriptional machinery to DNA. Hypermethylation of CpG islands in the promoter region of genes typically hinders their transcription by blocking transcription factor binding (Tate and Bird 1993). Methylation of CpG islands can also recruit histone modifying enzymes including histone deacetylases (HDACs) and histone methyltransferases (HMTs) that alter the local chromatin environment and further enhance or hinder transcription (Sarraf and Stancheva 2004). Indeed, hypermethylation of the promoter regions of key tumor suppressor genes is frequently linked with uncontrolled tumor growth and, for this reason, one of the hot new areas of cancer treatment is the use of DNA-demethylating agents to trigger the renewed expression of tumor suppressor genes (Mack 2006). The role of this mechanism in regulating deep torpor in hibernating species is not yet known but a recent study did show that DNA methylation patterns are responsible for the organ-specific expression of the HP-27 gene in chipmunks (Fujii et al. 2006). HP-27 codes for a plasma protein that is produced in the liver of euthermic animals but is conspicuously absent during torpor. CpG islands in the HP-27 promoter were hypomethylated in liver of euthermic squirrels but hypermethylated in other organs so that the gene was expressed only in liver. However, methylation status did not change between euthermic and hibernating states in either liver or kidney. Nonetheless, this methylation mechanism should be investigated more widely for it has great potential as a means for differential gene expression (up- or down-regulation) over cycles of torpor and arousal.

A recent study of histone modification confirms that this epigenetic mechanism clearly contributes to global gene silencing during torpor bouts in hibernating ground squirrels. Histones are subject to a variety of modifications including methylation, acetylation, phosphorylation, ubiquitylation, SUMOylation, citrullination, and ADP-ribosylation (Bhaumik et al. 2007). Methylation of lysine residues on histones leads to a closed chromatin structure whereas both acetylation and phosphorylation open up the DNA-protein structure to give transcription factors and the transcriptional machinery access to gene promoters. When skeletal muscle from ground squirrels (*S. tridecemlineatus*) was analyzed, the content of acetylated histone H3 (Lysine 23) was 25% lower in muscle from torpid animals compared with euthermic controls (Table 13.2) (Morin and Storey 2006). Animals in deep torpor also showed a 40% decrease in the content of phosphorylated histone H3 (Serine 10). Both of these changes are consistent with transcriptional suppression. Furthermore, the mechanism of change in histone acetylation was linked to an 82% increase in histone deacetylase (HDAC) activity in skeletal muscle from torpid squirrels and immunoblotting also showed higher HDAC1 and HDAC4 protein levels (Table 13.2).

Another parameter of transcriptional capacity was also suppressed during torpor in ground squirrels: RNA polymerase II (Pol II) activity (Table 13.2) (Morin and

Table 13.2 Effect of entry into the torpid state on indices of transcriptional control in skeletal muscle of hibernating 13-lined ground squirrels

	Ratio torpor: euthermic
Histone 3	
Total H3 protein	0.99 ± 0.16
Phospho-H3 (Ser 10) protein	0.61 ± 0.09*
Acetyl-H3 (Lys 23) protein	0.75 ± 0.04*
Histone deacetylase	
Activity	1.82 ± 0.20*
HDAC1 protein	1.21 ± 0.05*
HDAC4 protein	1.48 ± 0.09*
RNA polymerase II	
Activity	0.57 ± 0.04*
Total Pol II protein	1.05 ± 0.19
Phospho-Pol II (Ser 5) protein	1.79 ± 0.21*

Proteins were detected by immunoblotting with antibodies that detected total protein or the phosphorylation or acetylation of specific residues. For RNA polymerase II, the antibody detected phospho-Ser 5 in the heptapeptide repeats YSPTSPS of the C-terminal domain. Significance testing used the Student's t-test to compare torpor and euthermic values (* – significantly different, $P < 0.05$) and then ratios, torpor:euthermic, were calculated. Data are means ± SEM, $n = 3$–4 separate animals. Modified from Morin and Storey (2006)

Storey 2006). The maximum activity of Pol II in muscle from animals in deep torpor decreased to just 58% of the activity in euthermic muscle (measured at the same temperature) although total Pol II protein levels remained constant (assessed by immunoblotting). The content of phosphorylated (Ser 5) Pol II rose by 79% in muscle during torpor. Although the effect that phosphorylation at this site has on transcriptonal activity is still debatable (Kim et al. 1997), these data do show that Pol II is subject to differential phosphorylation between euthermic and torpid states as are many other enzymes and functional proteins in hibernating species (Storey and Storey 2004, 2007).

Lee et al. (2007) demonstrated the importance of yet another posttranscriptional modification to gene control in hibernation. This is SUMOylation. The amount of SUMO (small ubiquitin-related modifier) conjugated protein rose dramatically in brain, liver, and kidney of ground squirrels in deep torpor with a concomitant loss of free SUMO. Levels fell quickly again when torpor ended. In addition, expression levels of Ubc9, the E2-conjugating enzyme in the SUMO pathway, were closely correlated with SUMO conjugate levels in both brain and kidney. Transcription factors are primary targets of SUMOylation and this reversible posttranslational modification mainly has negative effects on gene expression (Girdwood et al. 2004). Significantly, microscopy and staining showed that SUMO conjugates were preferentially concentrated in the nuclei of neuronal cells during torpor whereas in active animals SUMO staining was dispersed throughout the cell (Lee et al. 2007). Hence, high levels of SUMO conjugates, as well as their nuclear localization, argue strongly for global suppression of transcription factor action during torpor thereby suppressing transcription of the genes under their control. As described later in this article, a number of transcription factors are specifically activated during torpor bouts and trigger the expression of selected genes; one would predict that these particular transcription factors would not be SUMOylated but no study has yet explored this idea.

13.5 Regulation of mRNA Transcripts: The New Frontier of microRNA

Control of gene expression is most often discussed in terms of the regulation of transcription and translation. Both gene level (dealt with in other parts of this review) and ribosome level controls are well known in hibernating mammals (reviewed by Storey and Storey 2004, 2007). With respect to the latter, a variety of studies have confirmed the inhibition of translation initiation and elongation factors via reversible phosphorylation during deep torpor well as a general dissociation of polysomes (Frerichs et al. 1998; Knight et al. 2000; Hittel and Storey 2002b; van Breukelen et al. 2004; Horman et al. 2005). However, between transcription and translation there are multiple opportunities for additional regulation by mechanisms that control mRNA processing, splicing, longevity, and availability to the ribosomes. Several studies have shown that total mRNA transcript levels do not change significantly over the torpor/arousal cycle (Williams et al. 2005; Crawford et al. 2007). This argues that there must be ways to stabilize/preserve transcripts in an inhibited form over torpor bouts that can range from days to weeks and yet still have these transcripts immediately available again for translation when the animal arouses. One way to do this is to store transcripts in physical association with monosomes or with ribonucleoprotein particles. Analysis of the distribution of mRNA transcripts in tissues from euthermic vs. torpid ground squirrels indicated that a high percentage of mRNA species are sequestered with monosomes in a translationally silent fraction during torpor (Frerichs et al. 1998; Knight et al. 2000; Hittel and Storey 2002a). Only transcripts that are preferentially translated while animals are in torpor, such as *fabp* (fatty acid binding protein), remain associated with polysomes (Hittel and Storey 2002b).

Another possibility for mRNA regulation is one of the "hot new topics" in molecular biology – microRNA (miRNA). miRNAs are small noncoding transcripts (19–25 nucleotides in length) that bind to target mRNAs; a perfect match between a miRNA species and its target typically directs the mRNA into degradation pathways, whereas an imperfect match leads to translational inhibition via storage in P-bodies (Bartel 2004; Gammell 2007). Many studies have now shown that specific temporal and spatial microRNA expression is required for normal cellular development and differentiation whereas many human diseases are associated with aberrant microRNA expression patterns; indeed, a new database summarizes the relationships between 299 human microRNAs and 94 human diseases (Gammell 2007; Flynt and Lai 2008; Jiang et al. 2008).

In recent work, reverse transcription PCR was used to quantify expression levels of selected miRNAs in four organs of 13-lined squirrels, comparing euthermic and deep torpor states. Torpor-specific differences in the levels of some miRNAs were seen; for example, in kidney, two miRNA species, mir-1 and mir-21, increased significantly by 2.0- and 1.3-fold, respectively, when animals were in torpor as compared with euthermic controls (Morin et al. 2008a). In mammals, mir-1 suppresses expression of a variety of mRNAs (Lim et al. 2005); in muscle, for example, mir-1 targets mRNAs that are involved in muscle cell proliferation and the expression of HDAC4, a histone deacetylase that leads to gene repression in muscle (Chen et al. 2006). By contrast, mir-21 is linked to anti-apoptotic properties (Chan et al. 2005). Both heart and skeletal muscle of ground squirrels showed significant reductions of 30–50% in mir-24 levels during deep torpor. This miRNA is linked to cell growth processes (Cheng et al. 2005) so it is possible that adjustments of miRNA levels, both of mir-24 and other species, could be an integral part of growth and proliferation repression during torpor. Immunoblot analysis also showed strong changes in Dicer protein levels in ground squirrels organs between euthermic and torpid states (Morin et al. 2008a). Dicer is the enzyme that controls the final step in miRNA processing, acting in the cytoplasm to cleave the larger miRNAs that are exported from the nucleus into mature miRNA molecules (Ding et al. 2009). Although much remains to be explored, there is clearly much potential for translational control via organ-specific and transcript-specific differential regulation of mRNAs under the control of miRNAs during seasonal hibernation.

13.6 Transcription Factors and Hibernation-Responsive Gene Expression

Although global suppression of transcription and translation is an integral part of transitions to/from deep torpor during seasonal hibernation, a variety of studies have shown that selected genes are up-regulated in torpid animals as evidenced by increased levels of specific mRNA transcripts or protein products (reviewed in Storey and Storey 2004). Some of these genes have been found by targeted analysis of specific gene/protein systems but many other candidate genes/proteins have been discovered by gene screening (e.g., cDNA library screening, DNA array screening) or proteomics approaches (e.g., 2D gel electrophoresis to visualize differential protein expression coupled with mass spectrometry for peptide mapping) (Gorham et al. 1998; Andrews et al. 1998; Fahlman et al. 2000; Hittel and Storey 2002a, b; Storey 2004; Epperson et al. 2004; Eddy et al. 2005; Williams et al. 2005; Brauch et al. 2005; Eddy and Storey 2008; Chen et al. 2008; Martin et al. 2008; Lee et al. 2008). Since gene expression is under the control of transcription factors (Tfs) that each regulate a group of genes that are dedicated to a particular cell function, another strategy for gene hunting is to identify the Tfs that are activated and then focus on the genes/proteins that are downstream of these. To be active, Tfs must be in the nucleus and in the correct conformation to bind to the specific response element for that Tf in the promoter region of genes. The active conformation is often produced either by phosphorylation of the Tf or by binding to a co-activator or specific ligand. Therefore, another way to search for genes that are critical to hibernation success expression is to look at which Tfs are active in the deep torpor phase (and/or at different stages over the torpor-arousal cycle) and use these as signposts to guide a targeted examination of the genes under their control. This approach is facilitated by the recent development of effective commercial methods for high throughput assessment of the levels of active Tfs. All of these methods begin with the preparation of nuclear extracts from tissues and then binding Tfs to the consensus binding site oligonucleotide for the Tfs of interest but are followed by different mechanisms for the separation of unbound material (Tfs or oligonucleotides) or detecting/quantifying binding among the methods from Active Motif Inc. (Carlsbad, CA) (http://www.activemotif.com/), Marligen Biosciences Inc. (Ijamsville, MD) (http://www.marligen.com/), or Panomics Inc. (Fremont, CA) (http://www.panomics.com/). All have proven effective in analyzing Tf involvement in animal freeze tolerance (Storey 2008) and hibernation (unpublished results from the Storey lab).

From studies using this approach or others, a variety of Tfs are now known to be important in regulating different aspects of hibernation-responsive gene expression. Prominent among these are the hypoxia inducible factor 1 (HIF-1) (Morin and Storey 2005), activating transcription factor 4 (ATF4) (Mamady and Storey 2008), peroxisome proliferator-activated receptor gamma isoform (PPARγ) and its co-activator PGC-1α (Eddy et al. 2005), and the NF-E2-related factor-2 (Nrf2) (Morin et al. 2008b). HIF-1 regulates a variety of genes that help organisms to deal with low oxygen stress by improving oxygen delivery (e.g., by stimulating capillary growth or red blood cell proliferation) or enhancing the capacity for oxygen-independent metabolism (e.g., by increasing the expression of glucose transporters and glycolytic enzymes) (Semenza 2003). The specific targets that are up-regulated by HIF-1 during torpor/arousal cycles in hibernators are not yet known but it is notable that HIF-1α protein levels (the subunit that is oxygen sensitive) and HIF-1 DNA-binding capacity were prominently elevated in the two thermogenic organs, brown adipose and skeletal muscle. This may suggest a link with maintaining oxygen-sensing and the capacity for oxygen delivery to these organs in the hypometabolic state to support the huge increase in oxygen consumption that is needed to fuel thermogenesis during arousal. ATF4 is involved with the unfolded protein response (UPR) which is a multifaceted response to cell stresses/conditions that elevate the load of unfolded proteins in the endoplasmic reticulum (ER). The UPR is a 3-pronged response to correct this and includes global protein synthesis inhibition to reduce the incoming load of unfolded proteins (via phosphorylation-induced inactivation of eukaryotic initiation factor 2α), selected synthesis of more ER chaperones to increase folding capacity and, if needed, enhanced degradation of unfolded proteins (Schroder and Kaufman 2005). At least the first two components of the UPR are evident during deep torpor in ground squirrels – translation inhibition and increased levels of ER chaperones, the glucose-regulated proteins (GRPs) (Mamady and Storey 2006, 2008). Indeed, elevated levels of ATF4 and the active form of its co-factor, CREB-1, correlated with strong increases in both the transcript and protein levels of the most important ER chaperone, GRP78, in both brain

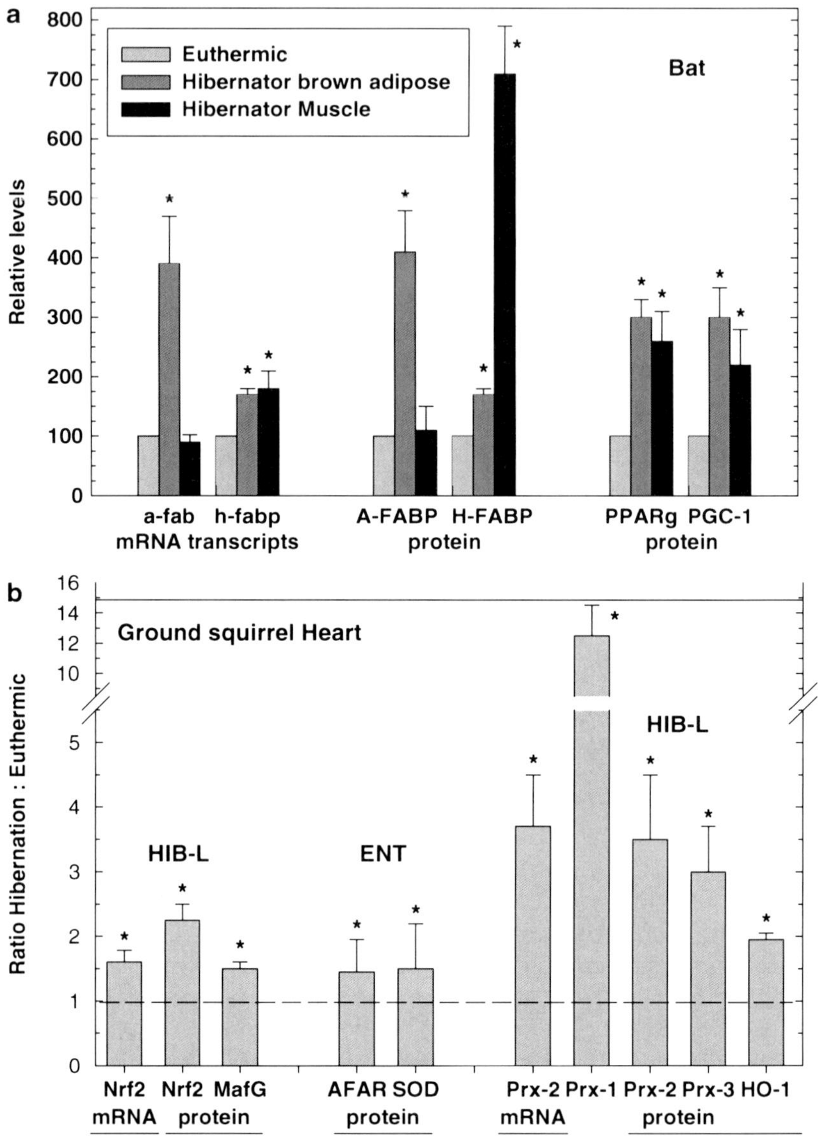

Fig. 13.4 Transcription factor responses in hibernating mammals and control of downstream gene expression. (**a**) Levels of PPARγ and its co-activator, PGC-1, are 2–3 fold higher in skeletal muscle and brown adipose tissue of torpid, vs. euthermic, bats (*Myotis lucifugus*). PPARγ controls the expression of fatty acid binding proteins and organ-specific up-regulation of the adipose *A* or heart *H* isoforms is confirmed by increased *fabp* mRNA transcript and FABP protein levels. (**b**) Nrf2 triggers the expression of various antioxidant genes. Both *nrf2* mRNA and Nrf2 protein levels increased during hibernation in heart of hibernating ground squirrels (*Spermophilus tridecemlineatus*), along with the Nrf2 binding partner, MafG. Protein levels of genes under Nrf2 control were elevated during entrance into torpor (ENT) or in long-term torpor (L-HIB; 3–5 days

and brown adipose of 13-lined ground squirrels during torpor. Elevated GRP78 may be an important mechanism that supports selective synthesis and/or correct refolding of protein to go ahead during cold torpor.

Reprioritization of fuel use is a requirement for successful hibernation with animals switching to a primary dependence on lipid fuels. PPARγ is a Tf that regulates a variety of genes that participate in fatty acid oxidation and therefore has a role in optimizing capacities for fatty acid transport/delivery and optimizing lipid-based energy metabolism in hibernator organs. PPARγ activation when animals enter torpor has been clearly linked with the expression of both the adipose (A) and heart (H) isoforms of fatty acid binding proteins (FABPs), the intracellular carrier proteins that move fatty acids through the cytosol, in a hibernating bat species (Eddy and Storey 2003, 2004). Figure 13.4a shows that 2–3 fold increases in the levels of PPARγ and its co-activator PGC-1α were correlated with ~4-fold increases in the mRNA and protein levels of A-FABP in brown adipose tissue and with a 7-fold increase in H-FABP content in heart. Finally, Nrf2 is a major Tf involved in antioxidant defense, a function that is important for hibernators for two reasons: (a) to ensure long-term protection from reactive oxygen species throughout an extended torpor bout, and (b) to address high rates of ROS generation associated with the massive increase in oxygen consumption and thermogenesis that supports arousal. An analysis of Nrf2 expression in ground squirrel heart found a 1.5-fold increase in *nrf2* mRNA transcripts in torpid animals and a similar rise in Nrf2 protein along with hibernation-responsive increases in the levels of multiple downstream antioxidant proteins including Cu/Zn superoxide dismutase, heme oxygenase, aflatoxin aldehyde reductase, and peroxiredoxins (Fig. 13.4b) (Morin and Storey 2007; Morin et al. 2008b). Based on the magnitude of the responses, it seems possible that peroxiredoxins are primary target genes for Nrf2 action during torpor; these are also well known to be stress-responsive in other systems (Ishii and Yanagawa 2007) and Prx1 was also up-regulated during torpor in bat heart (Eddy et al. 2005). Peroxiredoxins are intracellular antioxidant enzymes that reduce and detoxify a range of hydroperoxides with the use of thioredoxin as the electron donor. They are major protein components of mammalian cells typically constituting 0.1–0.8% of total soluble protein (Chae et al. 1999). Strong expression of peroxiredoxins during torpor may be a preparatory move that ensures that a high intracellular antioxidant potential is present to deal high rates of reactive oxygen species generation associated with the arousal process. Similarly, ascorbate, a major extracellular antioxidant rises by 3–5 fold in plasma during hibernation but is rapidly depleted when oxygen consumption rises sharply during arousal (Drew et al. 2002).

←

Fig. 13.4 (continued) a stable Tb of ~5–7°C) including aflatoxin aldehyde reductase *AFAR*, Cu/Zn superoxide dismutase *SOD*, peroxiredoxin *Prx* isoforms 1-3, and heme oxygenase 1 *HO-1*. Prx2 mRNA levels were also measured and found up-regulated. Data are means ± SEM, $n = 3–7$; * – significantly higher than corresponding euthermic values, $P < 0.05$. Compiled from Eddy and Storey (2003, 2004); Morin and Storey (2007) and Morin et al. (2008b)

13.7 Concluding Remarks

Hibernator species may look like uncomplicated "Sleeping Beauties," blissfully sleeping through the cold winter months. But we now know that hibernation is a multifaceted process involving not only metabolic rate depression but also a range of readjustments to cellular metabolism including altered patterns of fuel use and up-grading cell preservation mechanisms to sustain viability during cold torpor. Much progress has been made in recent years in understanding the events and the regulatory mechanisms operating to control cycles of torpor and arousal but there is still a huge amount to be learned. The endocrine trigger or triggers that set off metabolic suppression are still elusive as is the signal for arousal. The intracellular signal transduction cascades that bring about the massive suppression of oxygen consumption and inhibition of thermogenesis are similarly elusive. Well-known metabolic suppression responses to stresses such as starvation, hypoxia/ischemia, hypothermia, etc. that can be components of the torpid state are undoubtedly involved but their magnitude is not sufficient – there must be additional signals or a master coordinating switch. The timing mechanisms that determine both torpor bout duration and overall length of the hibernating season are also still unknown and particularly mysterious for species that spend the winter in completely dark burrows with no chance of receiving a photoperiod cue. Apart from understanding the regulation of hibernation itself, studies of hibernating mammals will help us better understand the evolution of endothermy and will hopefully lead to therapies that can be used to induce torpor in humans as a medical intervention strategy.

Acknowledgments Thanks to J.M. Storey for editorial review of the manuscript. Research in the Storey lab was supported by a discovery grant from the Natural Sciences and Engineering Research Council of Canada; KBS holds the Canada Research Chair in Molecular Physiology. Research in the Rider laboratory was supported by the Interuniversity Attraction Poles Program – Belgian Science Policy (P5/05 and P6/28), the Directorate General Higher Education and Scientific Research, French Community of Belgium, the Fund for Medical Scientific Research (Belgium), and the EXGENESIS Integrated Project (LSHM-CT-2004-005272) from the European Commission. Research in the Heldmaier lab was supported by the Deutsche Forschungsgemeinschaft (DFG HE 990/9 and 10).

References

Abnous K, Storey KB (2007) Regulation of skeletal muscle creatine kinase in a hibernating mammal. Arch Biochem Biophys 467:10–19

Abnous K, Storey KB (2008) Skeletal muscle hexokinase: regulation in mammalian hibernation. Mol Cell Biochem 319:41–50

Abnous K, Dieni CA, Storey KB (2008) Regulation of Akt during hibernation in Richardson's ground squirrels. Biochim Biophys Acta 1780:185–193

Andrews MT, Squire TL, Bowen CM, Rollins MB (1998) Low-temperature carbon utilization is regulated by novel gene activity in the heart of a hibernating mammal. Proc Natl Acad Sci USA 95:8392–8397

Bartel DP (2004) MicroRNAs: genomics, biogenesis, mechanism and function. Cell 116:281–297
Bartrons M, Ortega E, Obach M, Calve MN, Navarro-Sabate A, Bartrons R (2004) Activation of AMP-dependent protein kinase by hypoxia and hypothermia in the liver of frog *Rana perezi*. Cryobiology 49:190–194
Berriel Diaz M, Lange M, Heldmaier G, Klingenspor M (2004) Depression of transcription and translation during daily torpor in the Djungarian hamster (*Phodopus sungorus*). J Comp Physiol B 174:495–502
Bhaumik SR, Smith E, Shilatifard A (2007) Covalent modifications of histones during development and disease pathogenesis. Nat Struct Mol Biol 14:1008–1016
Blackstone E, Morrison M, Roth MB (2005) H_2S induces a suspended animation-like state in mice. Science 308:518
Bocharova LS, Gordon RY, Arkhipov VI (1992) Uridine uptake and RNA synthesis in the brain of torpid and awakened ground squirrels. Comp Biochem Physiol B 101:189–192
Brauch KM, Dhruv ND, Hanse EA, Andrews MT (2005) Digital transcriptome analysis indicates adaptive mechanisms in the heart of a hibernating mammal. Physiol Genomics 23:227–234
Braulke LJ, Klingenspor M, DeBarber A, Tobias SC, Grandy DK, Scanlan TS, Heldmaier G (2008) 3-Iodothyronamine: a novel hormone controlling the balance between glucose and lipid utilisation. J Comp Physiol B 178:167–177
Buck CL, Barnes BM (2000) Effects of ambient temperature on metabolic rate, respiratory quotient, and torpor in an Arctic hibernator. Am J Physiol 279:R255–R262
Carey HV, Andrews MT, Martin SL (2003) Mammalian hibernation: cellular and molecular responses to depressed metabolism and low temperature. Physiol Rev 83:1153–1181
Chae HZ, Kim HJ, Kang SW, Rhee SG (1999) Characterization of three isoforms of mammalian peroxiredoxin that reduce peroxides in the presence of thioredoxin. Diabetes Res Clin Pract 45:101–112
Chan JA, Krichevsky AM, Kosik KS (2005) MicroRNA-21 is an antiapoptotic factor in human glioblastoma cells. Cancer Res 65:6029–6033
Chen JF, Mandel EM, Thomson JM, Wu Q, Callis TE, Hammond SM, Conlon FL, Wang DZ (2006) The role of microRNA-1 and microRNA-133 in skeletal muscle proliferation and differentiation. Nat Genet 38:228–233
Chen J, Yuan L, Sun M, Zhang L, Zhang S (2008) Screening of hibernation-related genes in the brain of *Rhinolophus ferrumequinum* during hibernation. Comp Biochem Physiol B Biochem Mol Biol 149:388–393
Cheng SW, Fryer LG, Carling D, Shepherd PR (2004) Thr2446 is a novel mammalian target of rapamycin (mTOR) phosphorylation site regulated by nutrient status. J Biol Chem 279: 15719–15722
Cheng AM, Byrom MW, Shelton J, Ford LP (2005) Antisense inhibition of human miRNAs and indications for an involvement of miRNA in cell growth and apoptosis. Nucleic Acids Res 33:1290–1297
Crawford FI, Hodgkinson CL, Ivanova EA, Logunova LB, Evans GJ, Steinlechner S, Loudon AS (2007) The influence of torpor on cardiac expression of genes involved in the circadian clock and protein turnover in the Siberian hamster (*Phodopus sungorus*). Physiol Genom 31:521–530
Dark J, Miller DJ, Licht P, Zucker I (1996) Glucoprivation counteracts effects of testosterone on daily torpor in Djungarian hamsters. Am J Physiol 270:R398–R403
Dausmann KH, Glos J, Ganzhorn JU, Heldmaier G (2005) Hibernation in the tropics: lessons from a primate. J Comp Physiol B 175:147–155
Dausmann KH, Glos J, Heldmaier G (2009) Energetics of tropical hibernation. J Comp Physiol B 179:345–357
Daval M, Diot-Dupuy F, Bazin R, Hainault I, Viollet B, Vaulont S, Hajduch E, Ferre P, Foufelle F (2005) Anti-lipolytic action of AMP-activated protein kinase in rodent adipocytes. J Biol Chem 280:25250–25257
Ding XC, Weiler J, Großhans H (2009) Regulating the regulators: mechanisms controlling the maturation of microRNAs. Trends Biotechnol 27:27–36

Drew KL, Tøien Ø, Rivera PM, Smith MA, Perry G, Rice ME (2002) Role of the antioxidant ascorbate in hibernation and warming from hibernation. Comp Biochem Physiol C 133: 483–492
Drew KL, Buck LC, Barnes BM, Christian SL, Rasley BT, Harris MB (2007) Central nervous system regulation of mammalian hibernation: implications for metabolic suppression and ischemia tolerance. J Neurochem 102:1713–1726
Eddy SF, Storey KB (2003) Differential expression of Akt, PPARγ, and PGC-1 during hibernation in bats. Biochem Cell Biol 81:269–274
Eddy SF, Storey KB (2004) Up-regulation of fatty acid-binding proteins during hibernation in the little brown bat, *Myotis lucifugus*. Biochim Biophys Acta 1676:63–70
Eddy SF, Storey KB (2007) p38^{MAPK} regulation of transcription factor targets in muscle and heart of hibernating bats, *Myotis lucifugus*. Cell Biochem Funct 25:759–765
Eddy SF, Storey KB (2008) Comparative molecular physiological genomics. Heterologous probing of cDNA arrays. Methods Mol Biol 410:81–110
Eddy SF, McNally JD, Storey KB (2005) Up-regulation of a thioredoxin peroxidase-like protein, proliferation-associated gene, in hibernating bats. Arch Biochem Biophys 435:103–111
Ehrhardt N, Heldmaier G, Exner C (2005) Adaptive mechanisms during food restriction in *Acomys russatus*: the use of torpor for desert survival. J Comp Physiol B 175:193–200
Elvert R, Heldmaier G (2005) Cardiorespiratory and metabolic reactions during entrance into torpor in dormice, *Glis glis*. J Exp Biol 208:1373–1383
Epperson LE, Dahl TA, Martin SL (2004) Quantitative analysis of liver protein expression during hibernation in the golden-mantled ground squirrel. Mol Cell Proteomics 3:920–933
Fahlman A, Storey JM, Storey KB (2000) Gene up-regulation in heart during mammalian hibernation. Cryobiology 40:332–342
Florant GL, Heller HC (1977) CNS regulation of body temperature in euthermic and hibernating marmots (*Marmota flaviventris*). Am J Physiol 232:R203–R208
Flynt AS, Lai EC (2008) Biological principles of microRNA-mediated regulation: shared themes amid diversity. Nat Rev Genet 9:831–842
Fraga MF, Esteller M (2007) Epigenetics and aging: the targets and the marks. Trends Genet 23:413–418
Frerichs KU, Smith CB, Brenner M, DeGracia DJ, Krause GS, Marrone L, Dever TE, Hallenbeck JM (1998) Suppression of protein synthesis in brain during hibernation involves inhibition of protein initiation and elongation. Proc Natl Acad Sci USA 95:14511–14516
Fujii G, Nakamura Y, Tsukamoto D, Ito M, Shiba T, Takamatsu N (2006) CpG methylation at the USF-binding site is important for the liver-specific transcription of the chipmunk HP-27 gene. Biochem J 395:203–209
Gammell P (2007) MicroRNAs: recently discovered key regulators of proliferation and apoptosis in animal cells. Cytotechnology 53:55–63
Gauthier MS, Miyoshi H, Souza SC, Cacicedo JM, Saha AK, Greenberg AS, Ruderman NB (2008) AMP-activated protein kinase is activated as a consequence of lipolysis in the adipocyte: potential mechanism and physiological relevance. J Biol Chem 283:16514–16524
Geiser F (1988) Reduction of metabolism during hibernation and daily torpor in mammals and birds: temperature effect or physiological inhibition? J Comp Physiol B 158:25–37
Geiser F (2004) Metabolic rate and body temperature reduction during hibernation and daily torpor. Annu Rev Physiol 66:239–274
Girdwood DW, Tatham MH, Hay RT (2004) SUMO and transcriptional regulation. Semin Cell Dev Biol 15:201–210
Gluck EF, Stephens N, Swoap SJ (2006) Peripheral ghrelin deepens torpor bouts in mice through the arcuate nucleus neuropeptide Y signaling pathway. Am J Physiol 291:R1303–R1309
Gorham DA, Bretscher A, Carey HV (1998) Hibernation induces expression of moesin in intestinal epithelial cells. Cryobiology 37:146–154
Guppy M, Withers P (1999) Metabolic depression in animals: physiological perspectives and biochemical generalizations. Biol Rev 74:1–40

Gutman R, Choshniak I, Kronfeld-Schor N (2006) Defending body mass during food restriction in *Acomys russatus*: a desert rodent that does not store food. Am J Physiol 290:R881–R891
Gwinn DM, Shackelford DB, Egan DF, Mihaylova MM, Mery A, Vasquez DS, Turk BE, Shaw RJ (2008) AMPK phosphorylation of raptor mediates a metabolic checkpoint. Mol Cell 30:214–226
Hardie DG (2004) The AMP-activated protein kinase pathway – new players upstream and downstream. J Cell Sci 117:5479–5487
Hardie DG (2007) AMP-activated/SNF1 protein kinases: conserved guardians of cellular energy. Nat Rev Mol Cell Biol 8:774–785
Hardie DG, Carling D, Carlson M (1998) The AMP-activated/SNF1 protein kinase subfamily: metabolic sensors of the eukaryotic cell? Annu Rev Biochem 67:821–855
Heldmaier G, Elvert R (2004) How to enter torpor: thermodynamic and physiological mechanisms of metabolic depression. In: Barnes BM, Carey HV (eds) Life in the cold: evolution, mechanisms, adaptation, and application. Biological Papers of the University of Alaska, #27, Fairbanks, pp183–198
Heldmaier G, Ruf T (1992) Body temperature and metabolic rate during natural hypothermia in endotherms. J Comp Physiol B 162:696–706
Heldmaier G, Klingenspor M, Werneyer M, Lampi BJ, Brooks SP, Storey KB (1999) Metabolic adjustments during daily torpor in the Djungarian hamster. Am J Physiol 276:E896–E906
Heldmaier G, Ortmann S, Elvert R (2004) Natural hypometabolism during hibernation and daily torpor in mammals. Respir Physiol Neurobiol 141:317–329
Hittel D, Storey KB (2002a) The translation state of differentially expressed mRNAs in the hibernating thirteen-lined ground squirrel (*Spermophilus tridecemlineatus*). Arch Biochem Biophys 401:244–254
Hittel DS, Storey KB (2002b) Differential expression of mitochondria-encoded genes in a hibernating mammal. J Exp Biol 205:1625–1631
Horman S, Browne G, Krause U, Patel J, Vertommen D, Bertrand L, Lavoinne A, Hue L, Proud C, Rider M (2002) Activation of AMP-activated protein kinase leads to the phosphorylation of elongation factor 2 and an inhibition of protein synthesis. Curr Biol 12:1419–1423
Horman S, Beauloye C, Vertommen D, Vanoverschelde JL, Hue L, Rider MH (2003) Myocardial ischemia and increased heart work modulate the phosphorylation state of eukaryotic elongation factor-2. J Biol Chem 278:41970–41976
Horman S, Hussain N, Dilworth SM, Storey KB, Rider MH (2005) Evaluation of the role of AMP-activated protein kinase and its downstream targets in mammalian hibernation. Comp Biochem Physiol B Biochem Mol Biol 142:374–382
Hue L, Rider MH (2007) The AMP-activated protein kinase: more than an energy sensor. Essays Biochem 43:121–137
Inoki K, Zhu T, Guan KL (2003) TSC2 mediates cellular energy response to control cell growth and survival. Cell 115:577–590
Ishii T, Yanagawa T (2007) Stress-induced peroxiredoxins. Subcell Biochem 44:375–384
Jiang Q, Wang Y, Hao Y, Juan L, Teng M, Zhang X, Li M, Wang G, Liu Y (2008) miR2 Disease: a manually curated database for microRNA deregulation in human disease. Nucleic Acids Res 37:D98–D104. doi:10.1093/nar/gkn714
Jibb LA, Richards JG (2008) AMP-activated protein kinase activity during metabolic rate depression in the hypoxic goldfish, *Carassius auratus*. J Exp Biol 211:3111–3122
Kahn BB, Alquier T, Carling D, Hardie DG (2005) AMP-activated protein kinase: ancient energy gauge provides clues to modern understanding of metabolism. Cell Metab 1:15–25
Kim E, Du L, Bregman DB, Warren SL (1997) Splicing factors associate with hyperphosphorylated RNA polymerase II in the absence of pre-mRNA. J Cell Biol 136:19–28
Knight JE, Narus EN, Martin SL, Jacobson A, Barnes BM, Boyer BB (2000) mRNA stability and polysome loss in hibernating Arctic ground squirrels (*Spermophilus parryii*). Mol Cell Biol 20:6374–6379
Lee YJ, Miyake S, Wakita H, McMullen DC, Azuma Y, Auh S, Hallenbeck JM (2007) Protein SUMOylation is massively increased in hibernation torpor and is critical for the cytoprotection

provided by ischemic preconditioning and hypothermia in SHSY5Y cells. J Cereb Blood Flow Metabol 27:950–962
Lee K, Park JY, Yoo W, Gwag T, Lee JW, Byun MW, Choi I (2008) Overcoming muscle atrophy in a hibernating mammal despite prolonged disuse in dormancy: proteomic and molecular assessment. J Cell Biochem 104:642–656
Lim LP, Lau NC, Garrett-Engele P, Grimson A, Schelter JM, Castle J, Bartel DP, Linsley PS, Johnson JM (2005) Microarray analysis shows that some microRNAs downregulate large numbers of target mRNAs. Nature 433:769–773
Lyman CP (1958) Oxygen consumption, body temperature and heart rate of woodchucks entering hibernation. Am J Physiol 194:83–91
Lyman CP, O'Brien RC (1960) Circulatory changes in the thirteen-lined ground squirrel during the hibernation cycle. Bull Mus Comp Zool 124:353–372
MacDonald JA, Storey KB (1998) cAMP-dependent protein kinase from brown adipose tissue: temperature effects on kinetic properties and enzyme role in hibernating ground squirrels. J Comp Physiol B 168:513–525
MacDonald JA, Storey KB (1999) Regulation of ground squirrel Na^+K^+-ATPase activity by reversible phosphorylation during hibernation. Biochem Biophys Res Commun 254: 424–429
MacDonald JA, Storey KB (2005) Mitogen-activated protein kinases and selected downstream targets display organ-specific responses in the hibernating ground squirrel. Int J Biochem Cell Biol 37:679–691
Mack GS (2006) Epigenetic cancer therapy makes headway. J Natl Cancer Inst 98:1443–1444
Mamady H, Storey KB (2006) Up-regulation of the endoplasmic reticulum molecular chaperone GRP78 during hibernation in thirteen-lined ground squirrels. Mol Cell Biochem 292:89–98
Mamady H, Storey KB (2008) Coping with stress: expression of ATF4, ATF6 and downstream targets in organs of hibernating ground squirrels. Arch Biochem Biophys 477:77–85
Martin SL, Epperson LE, Rose JC, Kurtz CC, Ané C, Carey HV (2008) Proteomic analysis of the winter-protected phenotype of hibernating ground squirrel intestine. Am J Physiol 295: R316–R328
Matejkova O, Mustard KJ, Sponarova J, Flachs P, Rossmeisl M, Miksik I, Thomason-Hughes M, Hardie GD, Kopecky J (2004) Possible involvement of AMP-activated protein kinase in obesity resistance induced by respiratory uncoupling in white fat. FEBS Lett 569:245–248
Milsom WK, Zimmer MB, Harris MB (2001) Vagal control of cardiorespiratory function in hibernation. Exp Physiol 86:791–796
Morin P, Storey KB (2005) Cloning and expression of hypoxia-inducible factor 1α from the hibernating ground squirrel, *Spermophilus tridecemlineatus*. Biochim Biophys Acta 1729: 32–40
Morin P, Storey KB (2006) Evidence for a reduced transcriptional state during hibernation in ground squirrels. Cryobiology 53:310–318
Morin P, Storey KB (2007) Antioxidant defense in hibernation: cloning and expression of peroxiredoxins from hibernating ground squirrels, *Spermophilus tridecemlineatus*. Arch Biochem Biophys 461:59–65
Morin P, Dubuc A, Storey KB (2008a) Differential expression of microRNA species in organs of hibernating ground squirrels: a role in translational suppression during torpor. Biochim Biophys Acta 1779:628–633
Morin P, Ni Z, McMullen DC, Storey KB (2008b) Expression of Nrf2 and its downstream gene targets in hibernating thirteen-lined ground squirrels, *Spermophilus tridecemlineatus*. Mol Cell Biochem 312:121–129
Mulligan JD, Gonzalez AA, Stewart AM, Carey HV, Saupe KW (2007) Upregulation of AMPK during cold exposure occurs via distinct mechanisms in brown and white adipose tissue of the mouse. J Physiol 580:677–684
Nicol SC, Andersen NA (2000) Patterns of hibernation of echidnas in Tasmania. In: Heldmaier G, Klingenspor M (eds) Life in the Cold. Springer, Heidelberg, pp 21–28

Ortmann S, Heldmaier G (2000) Regulation of body temperature and energy requirements of hibernating alpine marmots (*Marmota marmota*). Am J Physiol 278:R698–R704
Osborne PG, Gao B, Hashimoto M (2004) Determination in vivo of newly synthesized gene expression in hamsters during phases of the hibernation cycle. Jpn J Physiol 54:295–305
Proud CG (2007) Signalling to translation: how signal transduction pathways control the protein synthetic machinery. Biochem J 403:217–234
Rider MH (2008) Dealing with energy stress in hypometabolic states: role of the AMP-activated protein kinase. In: Lovegrove BG, McKechnie AE (eds) Hypometabolism in animals: torpor, hibernation and cryobiology. University of KwaZulu-Natal, Pietermaritzburg, pp 75–82
Rider MH, Hussain N, Horman S, Dilworth SM, Storey KB (2006) Stress-induced activation of the AMP-activated protein kinase in the freeze-tolerant frog *Rana sylvatica*. Cryobiology 53:297–309
Rider MH, Hussain N, Dilworth SM, Storey KB (2009) Phosphorylation of translation factors in response to anoxia in turtles, *Trachemys scripta elegans*: role of the AMP-activated protein kinase and target of rapamycin signalling pathways. Mol Cell Biochem 332:207–213
Sanders MJ, Grondin PO, Hegarty BD, Snowden MA, Carling D (2007) Investigating the mechanism for AMP activation of the AMP-activated protein kinase cascade. Biochem J 403:139–148
Sarraf SA, Stancheva I (2004) Methyl-CpG binding protein MBD1 couples histone H3 methylation at lysine 9 by SETDB1 to DNA replication and chromatin assembly. Mol Cell 15:595–605
Scanlan TS, Suchland KL, Hart ME, Chiellini G, Huang Y, Kruzich PJ, Frascarelli S, Crossley DA, Bunzow JR, Ronca-Testoni S, Lin ET, Hatton D, Zucchi R, Grandy DK (2004) 3-Iodothyronamine is an endogenous and rapid-acting derivative of thyroid hormone. Nat Med 10:638–642
Schroder M, Kaufman RJ (2005) The mammalian unfolded protein response. Annu Rev Biochem 74:739–789
Semenza GL (2003) Targeting HIF-1 for cancer therapy. Nat Rev Cancer 3:721–732
Song X, Körtner G, Geiser F (1997) Thermal relations of metabolic rate reduction in a hibernating marsupial. Am J Physiol 273:R2097–R2104
Stenslokken KO, Ellefsen S, Stecyk JA, Dahl MB, Nilsson GE, Vaage JI (2008) Differential regulation of AMP-activated kinase and AKT kinase in response to oxygen availability in crucian carp (*Carassius carassius*). Am J Physiol 295:R1803–R1814
Storey KB (1997) Metabolic regulation in mammalian hibernation: enzyme and protein adaptations. Comp Biochem Physiol 118:1115–1124
Storey KB (2004) Cold, ischemic organ preservation: lessons from natural systems. J Investig Med 52:315–322
Storey KB (2008) Beyond gene chips: transcription factor profiling in freeze tolerance. In: Lovegrove BG, McKechnie AE (eds) Hypometabolism in animals: hibernation, torpor and cryobiology. University of KwaZulu-Natal, Pietermaritzburg, pp 101–108
Storey KB, Storey JM (2004) Metabolic rate depression in animals: transcriptional and translational controls. Biol Rev Camb Philos Soc 79:207–233
Storey KB, Storey JM (2007) Putting life on 'pause' – molecular regulation of hypometabolism. J Exp Biol 210:1700–1714
Swoap SJ (2008) The pharmacology and molecular mechanisms underlying temperature regulation and torpor. Biochem Pharmacol 76:817–824
Swoap SJ, Rathvon M, Gutilla M (2007) AMP does not induce torpor. Am J Physiol 293: R468–R473
Tate PH, Bird AP (1993) Effects of DNA methylation on DNA-binding proteins and gene expression. Curr Opin Genet Dev 3:226–231
Van Breukelen F, Martin SL (2001) Translational initiation is uncoupled from elongation at 18°C during mammalian hibernation. Am J Physiol 281:R1374–R1379

Van Breukelen F, Martin SL (2002) Reversible depression of transcription during hibernation. J Comp Physiol B 172:355–361

Van Breukelen F, Sonenberg N, Martin SL (2004) Seasonal and state-dependent changes of eIF4E and 4E-BP1 during mammalian hibernation: implications for the control of translation during torpor. Am J Physiol 287:R349–R353

Wang LCH (1979) Time patterns and metabolic rates of natural torpor in the Richardson's ground squirrel. Can J Zool 57:149–155

Wieser W, Krumschnabel G (2001) Hierarchies of ATP-consuming processes: direct compared with indirect measurements, and comparative aspects. Biochem J 355:389–395

Williams DR, Epperson LE, Li W, Hughes MA, Taylor R, Rogers J, Martin SL, Cossins AR, Gracey AY (2005) Seasonally hibernating phenotype assessed through transcript screening. Physiol Genomics 24:13–22

Wilz M, Heldmaier G (2000) Comparison of hibernation, estivation and daily torpor in the edible dormouse, *Glis glis*. J Comp Physiol B 170:511–521

Witters LA, Kemp BE, Means AR (2006) Chutes and Ladders: the search for protein kinases that act on AMPK. Trends Biochem Sci 31:13–16

Xue B, Kahn BB (2006) AMPK integrates nutrient and hormonal signals to regulate food intake and energy balance through effects in the hypothalamus and peripheral tissues. J Physiol 574:73–83

Zhang J, Kaasik K, Blackburn MR, Lee CC (2006) Constant darkness is a circadian metabolic signal in mammals. Nature 439:340–343

Chapter 14
Lessons from Natural Cold-Induced Dormancy to Organ Preservation in Medicine and Biotechnology: From the "Backwoods to the Bedside"

Barry Fuller, Edgardo Guibert, and Joaquín Rodríguez

Abstract Hypothermia is a powerful modulator of all life processes, and this has been harnessed over the past 50 years in clinical sciences where tissues or organs for transplantation need to be stored outside the body for periods of time. However for human organs (as an obligate homoeothermic), cooling alone cannot provide sufficient time for the clinical logistics of transplantation, and a series of interventions to further control metabolism have been developed empirically. In retrospect, it can be seen that these approaches mimic to some degree the ways in which cold tolerance in the natural world has developed in evolutionary terms. This chapter reviews the history and the current state of the art of applied hypothermic preservation, and compares and contrasts what is known about natural cold tolerance, highlighting areas for further research and development to meet the challenges for organ and tissue preservation in the next few years.

14.1 Introduction

One of the most significant advances in medicine in the latter half of the twentieth century has been the development of organ and tissue transplantation, to enhance quality of life or, in many cases, reverse otherwise fatal derangements in body function (Starzl et al. 2000). This has a truly multi-disciplinary triumph, requiring major advances in surgery, immunology, virology and microbiology to name only a few of the essential disciplines, but underpinning the whole enterprise has been the necessity to access and preserve the required donor organs. However, one problem

---

B. Fuller (✉), *E. Guibert, and *J. Rodríguez
Division of Surgery & Interventional Sciences, Royal Free & University College Medical School, UCL, London, NW3 2QG, UK
*Fac. de Ciencias Bioquímicas y Farmacéuticas, Universidad Nacional de Rosario, S2002LRK, Rosario, Argentina
e-mail: bfuller@medsch.ucl.ac.uk; edgardo.guibert@gmail.com; jrodrig@fbioyf.unr.edu.ar

E. Lubzens et al. (eds.), *Dormancy and Resistance in Harsh Environments*, Topics in Current Genetics 21,
DOI 10.1007/978-3-642-12422-8_14,

which became clear from the very beginning was that of organ injury by hypoxia, because by definition, transplantation requires removing an organ from a donor patient, and physically moving it to the recipient patient and performing complex surgery to re-connect the blood supply (Calne et al. 1963). At a minimum, this can take 1–2 h, but if the donor and recipient patients are on different hospital sites, perhaps in different cities, hypoxic time can extend beyond 8 h. Successful cold preservation was therefore quickly identified as a key factor in being able to facilitate organ or tissue sharing between different locations within one country, or even between different countries (Calne et al. 1963), in order to best match the most needy patient to the available organ, or the best immune-matched donor/recipient pairing. As was stated by Calne and colleagues in 1963 in relation to their early work on kidney transplantation, "Ischaemic renal damage is extremely important to those interested in cadaveric kidney transplantation and has probably been responsible for more failures than immunological reactions." The whole of what eventually became a major clinical focus for use of hypothermia to preserve biological function was driven by these considerations, and this is what will be discussed in this chapter. Without the development of modern cold preservation protocols, organ transplantation would have remained an esoteric and infrequent operation of minor clinical significance, and the study of low temperature biology would have remained the domain of environmental scientists. In a logistical sense for transplantation, "preservation" equals "time", time to organise staff and facilities, transport organs, and perform essential laboratory testing when required. The ability to slow biological deterioration in organs removed from their normal physiological environment has moved transplantation into the realms of the routine treatment of choice for many chronic life-threatening diseases. For example, in the UK currently, about 3,500 organ transplants of various types are performed each year (NHS Blood and Transplant 2010). In most of these, there is an essential step when the organ is removed from the donor patient and before it can be transplanted, when the organ must be protected from hypoxic injury. Even if the hypoxic period is short (such in live-donor kidney operations), the organ physiology must be protected from injury and by appropriate preservation technologies. In addition, the growing shortage of organs for transplantation has meant that donor organs, which already have sub-optimal function because of donor age or pre-existing pathologies, have to be considered, and these organs (so-called marginal donor organs) withstand preservation less successfully. Therefore, the study of organ preservation, both as a basic and applied science, is deeply embedded in transplant services. Within this chapter, we will review the challenges which arise when organs are accessed and stored for transplantation – the problems of hypoxia (Sect. 2) and of applying cold (Sect. 3) to human organs. Next we discuss the history of organ preservation from the earliest concepts developed since the middle of the last century up to today (Sect. 4), and what knowledge became available in parallel on functional responses to cold tolerance from Nature (Sect. 5). Finally, we outline the benefits and limitations of the two main, but differing, approaches to organ preservation currently in use (controlled hypometabolism at hypothermia – Sect. 6, and aerobic metabolism in cold perfusion – Sect. 7) and highlight areas where ongoing research is being focused to improve the outcomes (Sect. 8).

14.2 Organ Injury from Hypoxia

To be able to ask the question "what is the best low temperature preservation protocol?", we first need to understand the various events associated with hypoxic injury to organs, which are complex and multifactorial. A large number of studies have been performed over the past two decades to define the cellular metabolic, molecular and ultrastructural changes (Lemaster and Thurman 1993; Pinsky et al. 1995; Jassem and Heaton 2004; Rauen and de Groot 2004; Kupiec-Weglinski and Busuttil 2005; Casillas-Ramírez et al. 2006) and these are beyond the remit of the current discussion. Central to the problem is the deprivation of oxygen supply to sustain normal mitochondrial function, leading to a host of changes in cellular functioning and homeostasis. In most mammalian systems, these changes become significant beyond about 30 min, and move towards irreversibility by 2 h. Mitochondria are key but by no means sole players in these cascades, with injured organelles initiating "death signalling", and contributing to further injury when the organ is re-supplied with oxygen during the transplantation procedure. The secondary and additive injury has been largely linked to production of oxygen free radicals by aberrant respiration via the injured mitochondria and a generalised inflammatory oxidative stress, known collectively as ischaemia/reperfusion (I/R) injury (Fuller et al. 1988; Jassem et al. 2002; Jassem and Heaton 2004). No one single pharmacological strategy can possibly control the ischaemic cascade, and thus intuitively, a global regulator of metabolism was sought. This central issue in organ preservation is the loss of homeostatic control of vital cellular processes, which impacts to varying degrees and with different times of onset whichever molecular processes are being studied. It is a complex and cumulative "pathological syndrome", which has been the focus of many previous studies (Fuller 1999; Piper et al. 2003; Teoh and Farrell 2003), and in many (but not all) ways resembles the development of irreversible injury in hypoxia. The control of ion distribution, pH, membrane charge, solute content and associated osmotic potential within cells all depend on a high level of basal metabolic activity, particularly for fuelling energy-requiring membrane pumps such as the Na^+/K^+ ATPase. Such significant energy requirements in mammalian systems can only be met by oxidative energy production via coupled mitochondrial electron transfer, and thus it is unsurprising that transplantation (requiring removal, cooling and manipulation of organs) has such a propensity to induce hypoxic damage. Energy levels (as reflected by ATP content) fall dramatically within minutes of the onset of hypoxia. Even though most cells possess various "emergency" pathways for energy generation, such as anaerobic glycolysis or other substrate inter-conversions (Churchill et al. 1994), these can only provide a small fraction of the required energy supply and can only function for limited periods. As pointed out above, the problems are compounded because this cascade itself predisposes the cells to additional damage when oxygen is supplied during organ reperfusion and is manifested as I/R injury.

In parallel to this, the effects of cooling in slowing biological activity have been observed by natural scientists for centuries, and therefore it was an intuitive step to turn towards hypothermia to delay the deterioration of the hypoxic organ, by a type of "enforced metabolic depression". However, in a non-adapted species such as man, enforced cooling equally presents problems, and these have remained the core focus of organ preservation research, reviewed in the next section.

14.3 Effects of Cooling and Hypothermic Preservation on Mammalian Cells

The physiological basis for the effects of cooling on metabolism was established over the past century by the pioneers such as Arrhenius (1915) and Heilbrunn (1937). Reaction rates (V) are given by the Arrhenius relationship, where the temperature (T) is given in Degrees Absolute, R is the gas constant, Ea is the activation energy of the particular process and A is a constant.

$$V = A^{\exp}(-Ea/RT)$$

Measurements of various metabolic reactions during cooling, when plotted as Logn V against $1/T$, provide a quasi-linear relationship with a slope of $-Ea/R$. This can be used to derive the activation energy of the specific reaction (assuming a single, rate limiting step). Rate depression by cooling can also be quantified as the Q_{10} relationship (Heilbrunn 1937), which reflects the fall in reaction rate for every 10°C temperature reduction. For many biochemical reactions, the Q_{10} between 40 and 20°C is about 2, whilst between 20 and 0°C, the value increases to between 3 and 5 (Fuhrman 1956; Zimmermann et al. 1982). These observations have been interpreted as the increasing and additive effects of cooling on cellular ultrastructure which impact on biochemical processes. For example, many important biological reactions are catalysed by enzymes embedded in the membrane lipid bilayers, or require substrate exchange across such bilayers. In essence, lipids generally are known to undergo a phase transition to a gel state on cooling, and in the complex mixtures of lipid classes in biological membranes, cooling induces a thermotropic separation of the different lipid classes as the respective melting point of a particular lipid is passed. This leads to de-mixing and lateral phase separation into gel and liquid crystalline phases in the plane of the membrane, and also an exclusion of integral proteins from the gel-phase areas (Feltkamp and van der Waerden 1982). Such gel formation acts to decrease global membrane fluidity (Crowe et al. 1989) and may also lead to increased "solute leakiness" through membranes where such packing faults have occurred. Cooling has been reported to increase trans-membrane diffusion of many solutes from small ions to larger molecules, such as disaccharide sugars (Fuller and So 2001). The change in fluidity affects the kinetics of many membrane-bound enzymes, such as the Na^+/K^+ ATPase, since it was shown that cooling of the enzyme when bound in the membrane had a greater inhibitory effect than when the enzyme was cooled as a free protein in solution (Willis et al. 1978).

Hypothermia may also impact on the cytoskeleton, where microtubules start de-polymerize and F-actin from microfilaments to precipitate, during storage of hepatocytes at 4°C (Stefanovich et al. 1996). In kidney cells, a cold-induced disruption of microtubules was also observed (Breton and Brown 1998) and was associated with a redistribution of integral membrane proteins. Mitochondria are also sensitive to reduced temperatures during organ preservation, and swelling or vacuolization has been demonstrated by electron microscopy in experiments on kidney (Fuller et al. 1985) and liver (Ellermann et al. 1992) storage. Collectively, these processes may be relatively slow in kinetic terms, and take some hours to be developed, but within practical organ preservation logistics (storage between 12–24 h), they may be significant.

Many of these changes are inescapable whether oxygen is supplied during the cooling process or not, but the magnitude of events and rapidity of their onset is much greater if hypoxia is a major contributing factor. The combined effects then lead to ion and solute redistributions, a linked, osmotically driven water influx, and activation of various degenerative processes. The destabilisation of homeostasis when hypoxia is coupled with cooling leads to a predictably complex cascade of biochemical events, which eventually become irreversible (see Fig. 14.1). Progressive dephosphorylation of ATP results in accumulation of breakdown products such as purines and nucleoside bases (e.g. inosine, hypoxanthine and xanthine), which can diffuse from the cell. The internal pH falls towards acid conditions as the switch to anaerobic glycolysis fuels accumulation of lactate. Failure of energy-linked transmembrane pumps results in an inward flux of Na^+ (enhanced by resident anions on intracellular proteins) and Cl^-, leading to associated water influx and cell swelling. Phosphate levels increase due to the dephosphorylation events, and control of other actively transported ions (such as calcium and magnesium) is lost. The high free calcium levels can cause mitochondrial damage and activate degradative enzymes (such as phospholipases and non-lysozomal proteases) which destabilise membranes, accelerating the abnormal leakage of solutes. Taken together with the lowered pH which also favours the activation of lysozomal digestive enzymes by destabilising lysozomal membranes, an "auto-digestive" state is enhanced. Depending on the time and conditions experienced during cold hypoxia, progressive cell death by both necrosis and apoptosis has been observed.

For cell survival, two strategies are of paramount importance: (1) inhibition of the many and various contra-homeostatic alterations wherever possible, and (2) rapid reversal of any changes in a coherent manner on rewarming to allow cell physiological and functional recovery. On balance, the lack of oxygen supply in the cold flush-stored organ has a greater and earlier impact than cooling per se. If oxygen is provided, even in non-adapted mammals, mitochondrial electron transport and ATP production is surprisingly robust after cooling. It has been known for some time (Pegg 1978) that in kidneys subjected to oxygenated hypothermic perfusion for up to 48 h, both the total content of adenine nucleotides and relative ratios of ATP to lower phosphorylated-state adenine nucleotides are unchanged compared with those measured in kidneys at normal body temperatures. In addition, in organs previously depleted of ATP by cold hypoxic storage, a

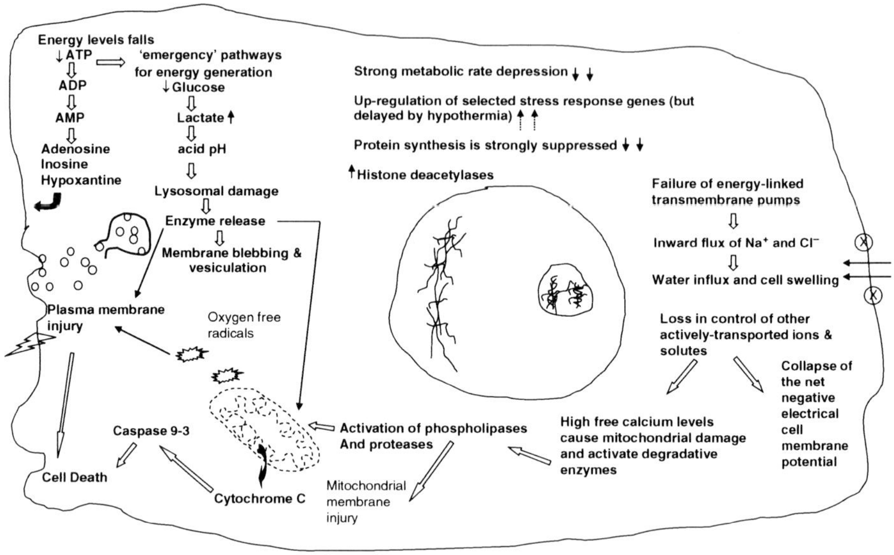

Fig. 14.1 Schematic of a cell showing different process during cold hypoxia: A continuum of complex processes that involve multiple cellular and molecular pathways come into play during cold hypoxia. Strong metabolic rate depression is achieved by a coordinated suppression of the rates of energetically costly cell functions such as transmembrane ion pumping and protein synthesis. Whereas cooling has a benefit in sharply reducing metabolic rate and hence prolonging the time that an organ can survive without or with low oxygen supply, metabolism of most mammals is optimised for function within a very narrow temperature window. When cooled below this, multiple injuries can arise due to problems such as temperature-dependent changes in enzyme affinities for substrates and regulators, changes in subunit association of oligomeric proteins and changes in membrane permeability due to temperature effects on lipid fluidity. ATP synthesis is greatly reduced and ATP breakdowns accumulate and can diffuse across cell membrane, forcing the cell to obtain energy from alternative sources, e.g. anaerobic glycolysis, producing lactate and a diminution in intracellular pH, activating lysosomes. Membrane ion pumps fail because of a lack of energy and inhibition from cold-induced alterations in local membrane viscosity, leading to a influx of sodium (and chloride), loss of potassium and magnesium, and an increase in calcium, producing a collapse of the electrical potential difference across the cell membrane because energy-driven membrane ion pumps suffer greater low temperature impediment than do the opposite passive diffusions of ions through ion channels. Cell and mitochondrial swelling gradually increase. There is a release of Cytochrome C from mitochondria, activating cell death via by apoptosis. Beyond the deficit in energy for synthesis, there are changes in gene regulation and protein translation, with an altered assembly of chromatin, mainly influenced by a higher activity of enzymes involved in chromatin condensation

progressive and quite rapid restoration of ATP can be measured when a brief period of oxygenated perfusion was instituted (Fuller et al. 1990). Of course, energy balance reflects both production and consumption, and hypothermia imposes selective inhibition of some metabolic processes such as protein synthesis, which is almost undetectable below 10°C (Fuller 2003) compared with transmembrane ion pumping which is reduced but still significant, so comparisons made solely on energy levels between normo- and hypothermia are not strictly valid.

Finally, it should be recognised that all of the changes during hypothermia described above impact on the ability of the cells to recover when rewarming takes place, where additional injury may be produced within the now well-recognised syndrome termed ischaemia/reperfusion (I/R) injury. As discussed above, a great deal of information on the biochemical and molecular changes during organ I/R has been described which are linked to oxidative stress, cell death signalling via apoptosis and associated local and systemic liberation of pro-inflammatory agents such as cytokines, which compound and exacerbate the original hypothermic injury (Jassem et al. 2002; Fuller et al. 1988; Fuller 1999; Lopez-Neblina et al. 2005; Daemen et al. 2003). Any strategies aimed at organ preservation have needed to be developed with this additional I/R insult in mind.

14.4 History of Organ Preservation for Transplantation

In retrospect, the background concepts for organ preservation arose from work on organ physiology and the identification of factors important to keep organs alive outside the body. In the early part of the twentieth century, studies of the essential support provided to organs by the circulatory system produced attempts to perfuse organs in vitro, such as the kidney and liver by simple syringe methods, and the later use of sophisticated perfusion circuits using artificial pumps allowed experiments on detailed physiology of organ function (Brodie 1903). In conjunction with this, investigators began to formulate synthetic perfusates – solutions of electrolytes, solutes, vitamins, etc. capable of replacing blood in some respects and avoiding thrombus formation, which frequently followed from using whole blood in an isolated circuit (Locke and Rosenheim 1907; Carrel and Lindbergh 1935, 1938). Throughout the same period, the strong modifying effect of hypothermia on organ function was also starting to be appreciated (Bickford and Winton 1937).

Although the microsurgical methods necessary to permit organ transplantation were developed from the early 1900s, it was not until the 1950s that consistent operative success was achieved, first in renal transplantation (Hume et al. 1955), whilst ischaemic injury became a recognised major problem. Over the same time, there were co-incidental reports about the protection provided by hypothermia and afforded to kidneys in vivo during ischaemia (Schloerb et al. 1959). One of the first systematic investigations on use of hypothermia in organ preservation for transplantation was performed by Calne et al. in the kidney. These authors investigated the relative merits of cooling kidneys by simple surface cooling of the organ or by perfusion of the renal artery with cooled heparinised blood (Calne et al. 1963) and concluded that vascular flush-perfusion was more efficient method of cooling the organ mass. However, use of chilled heparinised or diluted blood still led to many problems with vascular stasis on re-implantation of the graft, and thus the search for better preservation solutions, preferably synthetic solutions which could be reliably manufactured and sterilised, became the focus of organ preservation research.

Some of the first attempts to use acellular "synthetic" preservation solutions were made using infusions such as ice-chilled lactated Ringer's, or a glucose-salts mixture into the renal artery of donor kidney grafts immediately after their removal from the body cavity (Stickel et al. 1968). The solution described by Collins and his colleagues (Collins et al. 1969), which was designed to mimic, in a simple fashion, the intracellular electrolyte balance, was the first notable attempt to advance organ preservation based on an understanding of the ongoing changes in mammalian cells during hypothermic exposure, and this solution was used for almost two decades. Using this, renal allograft preservation was feasible for 1–2 days, long enough to allow tissue matching and sharing of organs over a wide geographic area. Alongside these approaches for flush cooling, proponents of continuous hypothermic perfusion continued to develop their methods for oxygenated low temperature perfusion (Humphries et al. 1964a, b, 1968a, b). However, the logistics and reliability of the perfusion equipment available at that time meant that gradually, flush cooling and ice-storage became the most widely used preservation method. Even today, intravascular flush cooling is the first step in the preservation of all whole-organ grafts, via variations on in situ flush cooling with cold electrolyte solutions infused into accessible sites such as the distal aorta for abdominal organs or thoracic aorta for heart and lungs. Expansion of transplantation as a clinical therapy necessitated methods of multiple organ procurement from the same donor, by which the kidneys, liver, heart and lungs or various combinations of these organs could be removed without jeopardising any of the individual organs. This has led to the adoption of so-called "Flexible techniques" (Starzl et al. 1984) in which all organs to be transplanted are cooled in situ by vascular flush with chilled acellular solutions, rapidly removed in a bloodless field, and dissected on a back table.

From the above discussions, it will have become apparent that there is a fundamental choice to be made in applied organ preservation, whether to supply oxygen to sustain low-level oxidative metabolism and energy-consuming homeostatic control of the intracellular milieu, or whether to allow hypoxia to develop and attempt to modulate or even capitalise on, the inevitable biochemical changes such that organ viability is maintained for a practically useful period. Hypothermia is central to both approaches, but thereafter they differ in philosophy, complexity, portability and cost. In some ways, each approach mirrors specific scenarios in nature where survivals against cold and hypoxia have developed on an evolutionary basis.

14.5 Cold Survival Strategies: The Links Between Medicine and Nature

The effects of cold on the natural world have been part of our collective knowledge-base since the establishment of scientific method over the past four centuries, and there have been many pioneering studies. To name just some, in the seventeenth century Boyle, in his treatise (Boyle 1683), collected many reports of experiments on low temperatures, whilst Hunter (Palmer 1825), a century later, compared hibernating animals such as hedgehogs and bats with normal mammals in relation to their physiological activities and body temperatures at different seasons of the year, and classified these as "perfect" or "imperfect" species depending on how they regulated their body temperatures in seasonally cold environments.

In the natural world, it has become clear over the past 20 years that a reversible metabolic rate depression is an additional central strategy (in addition to body cooling) for surviving periods of stress (cold, hypoxia, starvation) across all kingdoms of life. In animals, cold-associated hypometabolism can be linked to seasonal climatic changes (as experienced by some lower vertebrates such as frogs or turtles which over-winter in cold, oxygen-depleted environments at the bottom of ponds or buried under the forest floor, or by small mammals entering periods of hibernation in sub-artic regions of the world). A great deal of information has been gained on the physiological and metabolic consequences of adapted cold survival (Storey and Storey 1990, 2004a, b, 2007; Storey 2002). Storey and Storey (2007) and (see Storey et al. 2010) have identified several key criteria which enable cold, oxygen-limited survival, and amongst these, the major ones which link conceptually with organ preservation for transplantation are: (1) a global metabolic rate suppression; (2) metabolic pathways capable of delivering energy (ATP supply) and phenotypic alterations to prioritise ATP use; and (3) enhancement of defence mechanisms (at the molecular and cellular levels) to allow cohesive return to normal metabolism during arousal. Co-ordinated net suppression of metabolism is frequently around 90% in the hypometabolic state.

It is not surprising that in natural cold tolerance, maintenance of intracellular homeostasis a key priority of the residual metabolic activity. Correct ion distribution with maintenance of cell membrane potential is crucial to survival, all-be-it via much reduced ATP turnover because there is an associated reduction in ion leakage following specific molecular mechanisms – termed "channel arrest" (Hochachka 1986; Hochachka et al. 1996). These in turn are co-ordinated through transcriptional and translational signalling events such as the HIF-1a cascade and either GMP or AMP-dependent-protein kinase activity (Brooks and Storey 1994; Holden and Storey 1997) leading to reversible protein phosphorylation (Bickler and Buck 2007) as one example. An example of the prioritised changes in metabolism was given by Hochachka's group (Hochachka et al. 1996) studying turtle hepatocytes, where membrane Na^{+}/K^{+} ATPase activity consumed 60% of available ATP production in the hypoxic state, whereas it comprised only about 30% of ATP consumption under normal conditions. Likewise, DNA micoarray technology has shown that conditions such as mammalian hibernation require exquisite "fine tuning" of transcriptional activity – there is not the "blanket" reduction in gene expression that might be expected from low temperatures, but a highly selective increase in targeted gene products alongside a down-regulation of other genes. Indeed, a surprising number of genes may be upregulated in hibernation, many of which are as yet unidentified, but others can be identified which fall into two main categories – those which preserve metabolism and cellular molecular architecture and those that address specific cell or organism needs during stress (such as antioxidant defences or fatty acid oxidation as an energy supply).

Beyond the many regulated phenotypic changes which are essential to survive under conditions such as hibernation, there is one other inescapable factor – the need for low (but continuous) body perfusion via blood circulation derived from an intermittent cardiac activity. It has been known for more than 30 years that even in deep hibernation, rhythmical heart beat can be detected (Lyman 1982; Lyman et al. 1982). A slow, uniform heart rate accompanies deep hibernation (Johansson 1996). Essential organ perfusion is maintained, and autoregulation of coronary flow can be demonstrated. Lyman reported in hibernating 13-lined ground squirrels that at a body temperature of about 10°C, a heart rate of 2–3 beats per minute and a systolic pressure around 60 mm Hg could be measured (Lyman 1982). The circulatory events are accompanied by infrequent breathing activity, interspersed with longer periods of apnoea (Milsom et al. 1999). Nevertheless, the continued body perfusion and episodic breathing permit some exchange to supply oxygen and remove carbon dioxide, although the relationship between oxygen delivery and demand in tissues of hibernators has not to our knowledge been clearly defined. This will also be complicated by the numerous energy-sparing adaptations which have already been discussed (Storey and Storey 2004a, b). But, at a prosaic level, successful hibernation requires tissue perfusion as an essential prerequisite.

There is yet another essential factor in hibernation, which is worth noting in relation to organ preservation, and this is that successful cold tolerance is strictly time limited. Even where evolution has combined to successfully integrate the molecular changes, hibernators have to arouse at frequent intervals during a long overall seasonal cold period. For example, in a 13-lined ground squirrel hibernating between November and March, around 20 periods of brief arousal and re-entry were recorded (Andrews 2007). Even at its best, successful mammalian cold tolerance is very finely balanced on a kinetic basis.

14.6 Application of Cooling and Additional Hypometabolism by Manipulation of Preservation Solutions

Given the successful strategy for manipulating basal metabolism in natural cold tolerance described above, it is not surprising that applied cryobiologists attempted to harness this approach in organ preservation. By the early 1970s, several pieces of evidence accumulated in organ preservation studies which pointed to the fact that transmembrane movements of ions and water were still active at low temperatures. Keeler et al. demonstrated that ion diffusion and oedema in renal tissues could be identified during cold preservation (Keeler et al. 1966) and could be altered by changing the composition of the solution surrounding the cells. This led to the development of the notion of "intracellular" solutions by Collins and his colleagues (Keeler et al. 1966; Collins et al. 1969), which extended safe preservation time for kidneys up to 48 h. These solutions minimised the loss of intracellular potassium as a main strategy, by raising the external potassium concentration to that found inside the cell, and concomitantly reducing sodium levels (see Table 14.1), and by doing so, also reduced to some extent futile cycling of energy through the membrane sodium–potassium ATPase pump which would consume ATP in trying to maintain intracellular sodium/potassium balance. This imposed hypometabolic state, by preventing complete exhaustion of cellular adenine nucleotide reserves, sustained cell survival by multiple pathways, which even now are not fully understood (Pegg 1978; Salahudeen et al. 2000). For example, it has been suggested that by maintaining low, residual adenine nucleotide levels in the cell, mitochondrial stability was enhanced during cold storage.

A second important factor that was established by the studies from Collins and other contemporaries (Keeler et al. 1966; Collins et al. 1969) was the importance of the selection of osmotically active components, particularly sugars and large

Table 14.1 Compositions of preservation solutions

	EuroCollins	UW (Viaspan®)	Celsior®	Custodiol®
NaCl	10.0	–	100.0	15.0
KCl	108.0	–	15.0	9.0
Potassium hydrogen 2-ketoglutarate	–	–	–	1.0
Potassium Lactobionate	–	100.0	80.0	–
$MgCl_2.6H_2O$	–	–	13.0	4.0
$NaCO_3H$	10.0	–	–	–
NaOH	–	25	–	–
$MgSO_4$	–	5.0	–	5.0
Glutamic acid	–	–	20.0	–
Glutathione,	–	3.0	3.0	–
Glucose	180.0	–	–	–
Histidine. HCl. H_2O	–	–	30.0	18.0
Histidine	–	–	–	180.0
Adenosine,	–	5.0	–	–
Allopurinol	–	1.0	–	–
Tryptophan	–	–	–	2
Mannitol	–	–	60.0	30.0
Raffinose	–	30.0	–	–
$CaCl_2$. $2H_2O$	–	–	0.25	0.015
Pentafraction (HES) g/L	–	50	–	–
K_2HPO_4	60.0	25.0	–	–
Insulin U/L	–	40.0	–	–
Dexamethasone, mg/L	–	16.0	–	–
Penicilin G UI/L	–	200.000	–	–
Na^+	10	25	100	15
K^+	115	125	15	10
pH	7.30 (0°C)	7.40 (25°C)		7.02–7.20 (25°C)
Osmolality: (mosmol/kg H_2O)	340	320		310

The constituents of four commonly used and commercially available preservation solutions. Values are shown as mmol/L unless otherwise indicated

molecular weight anions, in solution development. Even with the altered sodium/potassium balance in the synthetic solutions, control of other ionic and solute redistributions was not possible, and cell swelling was noted during cold preservation. The addition of agents such as glucose (in Collins' solution) or mannitol (in Marshall's Hypertonic Citrate solution, another solution developed at the time) in high milimolar concentrations was found to be efficacious (Fuller 1999). The replacement of chloride in the solution with either phosphate (Collins' solution) or citrate (Marshall's solution) also bolstered control of osmotic balance, since these anions only diffuse across cell membranes very slowly at hypothermia. Again there is a parallel with the natural world, where seasonal accumulation of sugars is a common over-wintering strategy, although this may have more to do with freeze avoidance.

Over the next decade, various minor modifications of preservation solutions were reported (such as supplementation of Collins' solution with sucrose rather than glucose), but the next major improvement came with the development of the University of Wisconsin Solution by Southard and Belzer (1995). Whilst maintaining the high potassium concentration, a new set of constituents (lactobionate as the high molecular weight anion, and raffinose as the sugar osmolyte) were included (see Table 14.1). In addition, agents to combat oxidative stress during the hypothermic period (Rauen and De Groot 2002), and leading into the early reperfusion phase, were added. These included reduced glutathione as a thiol anti-oxidant, and allopurinol, which has both antioxidant effects and inhibits one of the enzymatic sources of free radical production (xanthine oxidase). Cytoprotection was also sought by addition of insulin and dexamethasone, although prospective evidence for addition of such agents took much later to establish (see McAnulty et al. 2004; Salahudeen et al. 2000, and discussion below). The UW solution was initially developed for preservation of pancreas, and a further addition was that of hydroxyl-ethyl starch as colloid, which was thought to improve interstitial oedema in that organ. At the same time, UW solution was quickly applied to preservation of liver and kidneys (Southard and Belzer 1995) with notable successes. Much effort was invested by various groups to understand the importance of the various components in the UW solution, for example, the value of adding the colloid was unclear. The inclusion of antioxidants does seem to be generally beneficial during cold storage (Salahudeen et al. 2000) although the optimal antioxidant selections from the range of available agents have yet to be fully determined and may be enhanced by addition of other trophic factors (McAnulty et al. 2004).

The lactobionate/raffinose formulation first used in UW solution has remained a central plank of hypometabolic flush solutions over the past decade. It does seem that the control of oedema, plus other attributed effects, such as possible metal ion chelation by the lactobionate, may hold greater importance than the "intracellular" balance of potassium or sodium, since other recently developed solutions (still based on lactobionate/raffinose) such as Celsior (see Table 14.1) have a more normal, plasma-like, high sodium balance and yet are claimed to be equally effective to UW solution (Faenza et al. 2001). Also of note has been the development of solutions employing the cytoprotective effects of amino acids, particularly for small bowel preservation (Salehi et al. 2008). However, it does appear that we may have reached close to the maximum achievable benefit with current static hypometabolic flush solutions in organ preservation, unless new molecular targets are identified in the future. Some of these new ideas will be discussed in Sect. 8 below.

14.7 Hypothermic Machine Perfusion Preservation

The other approach to organ preservation, which was used in the early days of transplantation, but is now showing a resurgence of interest, is hypothermic machine perfusion preservation (HMP). In the early part of the twentieth century, a great deal of work was carried out on isolated perfusion of body organs, where the focus was to achieve perfusion at normal body temperatures. By the 1940s, a considerable degree of sophistication was achieved in these technologies as reported by Carrel and Lindbergh (1935, 1938). There was only a minor interest in organ perfusion at sub-physiological temperatures, which was used to perturb normal physiology by temperature reduction (Bickford and Winton 1937). However, the arrival of clinical transplantation as a therapeutic reality in the 1960s quickly demanded a reappraisal of isolated perfusion for organ preservation, initially for the kidney. There was already the conceptual basis for application of hypothermic perfusion in renal transplantation, and the group led by Humphries made several important contributions to the acceptance of renal hypothermic perfusion preservation in transplantation (Humphries et al. 1964a, b, 1968a, b). Another major pioneer was the late F.O. Belzer who developed one of the first reliable transportable HMP machines for kidneys. Interest in HMP for other organs was also active at around the same time (Belzer et al. 1970; Hobbs et al. 1968; Ruiz and Lillihie 1974), but in reality these were never developed at the time into widespread clinical application.

The aim of continuous HMP is to stimulate basal but essential levels of aerobic metabolism. The organ is perfused via the vascular bed with continuous delivery of oxygen and nutrients from the perfusate, whilst waste products are continuously removed. The essential requirements are a circuit built from sterile non-thrombogenic and non-allergenic delivery tubing, a sterile organ chamber, a pump and heat exchanger, a mechanism for oxygenation (membrane oxygenation or surface diffusion), appropriate monitoring of perfusate temperatures and pressures, and an effective perfusion fluid. However, compared with hypometabolic flush preservation, the method is more complex, costly and may be liable to complication (by technical failure). Nevertheless, HMP techniques, when appropriately applied, still provide the maximum extension of successful storage of most organs, as will be discussed below.

Stable and readily available sterile perfusate solutions were needed to ensure good perfusion characteristics during HMP, and to provide solutes which improved preservation of organ physiology. Blood was found to be unsuitable because cold blood exhibits abnormal rheology and greater viscosity, which impart sluggish flow and may lead to micro-emboli within the organ (Humphries et al. 1946a, b). Simple balanced electrolyte solutions mimicking the extracellular milieu, such as Ringer's-lactate solution or saline, were shown to be unsatisfactory hypothermic organ-preserving solutions which exacerbated interstitial and intra-cellular oedema. Belzer and others looked towards plasma-like solutions containing albumin, dextrans, starch or other colloids to provide oncotic support and to prevent interstitial oedema. Eventually, Belzer's group developed a synthetic solution based on hydroxy-ethyl starch as colloid (Southard and Belzer 1995) and gluconate as major anion, which is known as Belzers KPS solution and has been commercially available since the late 1990s.

The beneficial effects of HMP are multiple and complex, but one fact which was recognised early in kidney HMP (Pegg et al. 1981; Southard et al. 1983; Southard et al. 1984), and has been shown consistently in other organs such as liver or heart (Fuller et al. 1990; Lockett et al. 1995; Dutkowski et al. 2006), is that oxidative energy production via mitochondrial electron transport is sustained at these hypothermic temperatures (around 10°C) during HMP. In fact, at these temperatures, aerobic metabolism can be sustained by a quite modest degree of oxygenation and at reasonably low perfusate flow rates. For example, recent work in liver HMP supports the use of normal air oxygen tension in the perfusate rather than extreme oxygenation (T' Hart et al. 2005). In a recent study on HMP in canine kidneys damaged by 60 or 75 min of warm ischaemia, surface oxygenation of the perfusate with 100% oxygen gas was applied. Good recovery of function after transplantation was noted (Lindell et al. 2005), whilst the starting perfusate oxygen tension was around 470 mm Hg – intermediate between room air and full oxygen saturation. There have been occasional reports on the value of oxygen carriers, such as perfluorocarbons or modified heamoglobins, added during HMP (Tamaki et al. 1987), and these were also used in low-flow situations. The aerobically derived ATP energy (Changani et al. 1997) is readily measurable and can be used to fuel a variety of homeostatic mechanisms for control of the intracellular environment, such as maintenance of ionic balances and pH control, and avoidance of collapse (in both a structural and functional sense) of the mitochondrial compartment (Polyak et al. 1999). Since mitochondria are central players in signalling damage events in the cell, leading to destruction via apoptosis or/and necrosis during cold preservation via a cascade of events, which together can be viewed as "death signalling", the value of HMP can be appreciated. Indeed, cardiac HMP has been reported to reduce apoptosis (Peltz et al. 2005). In situations (such as prior warm ischaemia in non heart-beating donors), the organ will have already moved some way down the path towards the cell death signalling cascade before the start of cooling, whilst resumption of mitochondrial activity can reverse some of these steps. For example, outcomes can be manipulated by pharmacological means, under conditions simulating HMP, to avoid the potentially catastrophic event of opening of the mitochondrial transition pore (McAnulty 1998). Other experimental work in liver HMP has demonstrated a positive influence of prostacyclin on tissue energy balance during the perfusion period (Changani et al. 1998). However, much remains to be understood about HMP and the molecular signals which can reverse the "death signalling", or where exactly "reversible" and "irreversible" injury diverge. It also might be predicted that the complicated effects of rewarming or "reperfusion injury" in organs following cold preservation may be less injurious following HMP, because the cells can better cope with the oxygen paradox if the principle oxygen-handling organelles (mitochondria) are themselves already in good condition. For example, experiments on livers preserved by HMP (which have maintained a good ATP balance at the end of HMP) showed lower oxidative stress during rewarming than did those organs stored by simple cold flush (Dutkowski et al. 1999).

Other attributes for HMP have been linked with positive effects on organ microcirculation, especially when prior warm ischaemia has been encountered (Hansen et al. 1997). There is a generally observed reduction in vascular resistance in organs perfused with optimal HMP protocols. The combination of improved microcirculation and cell energy status has been shown to result in improved survival in rat livers damaged by prior warm ischaemia and then transplanted after storage by HMP (Lee et al. 2003), compared with simple cold-stored organs. The crucial balance is to achieve good perfusion throughout the organ mass without injuring the microcirculatory bed (see below). In this respect, the optimal choice of colloid has remained an important research focus and the Amsterdam group have recently proposed a switch to polyethylene glycol as the major colloid for liver HMP (Bessems et al. 2005a, b). Also, the ability of HMP to regulate acid–base balance and stabilise intracellular pH has recently been revisited, with the proposal (Baicu et al. 2006) that raising the HEPES buffer concentration (from 10 mM in the standard Belzer Kidney Preservation Solution to 35 mM in a modified lactobionate/gluconate solution termed UHK) can provide tight pH control in a porcine kidney model. In contrast, colloid-free Bretschneider's HTK solution has been successfully applied to renal HMP, as long as perfusion pressures were kept at a low level (Minor et al. 2005).

Other potential applications for HMP are as reliable methods for providing cytoprotective or immuno-modulating drugs during the organ preservation period. The work of Polyak and colleagues showed that supply of prostaglandin E1 during HMP resulted in impressive results for kidney preservation (Polyak et al. 1999). However, not all cytoprotective drugs appear to behave as predicted during HMP (Bartels-Stringer et al. 2005), and little information exists on molecular signalling, pharmaco-kinetics or drug-receptor interaction under the conditions of HMP. It will be important to study these areas in future to target pharmacological agents for maximum benefit. For the future, use of HMP may be one way forward to deliver targeted gene therapy to donor organs in vitro, but significant improvements in the associated knowledge and technology bases will be needed to make this a reality (Collins et al. 2000).

Throughout the 1980s and 1990s, the subsequent expansion of clinical renal transplant activity led to a re-evaluation of many aspects of the services, including rationalisation of organ procurement and sharing between centres. The perceived high cost of HMP and the relatively difficult shipping constraints of the equipment of the period provided one stimulus to move towards cold flush preservation. The cold perfusates required for HMP were also inconvenient to prepare on a routine basis (at least, before the arrival of Belzer's KPS solution), because unstable blood components were often used. In addition, there were several studies which indicated that little or no additional benefit could be obtained using HMP (with what we now recognise as sub-optimal perfusates), compared with the standard hypometabolic flush (Halloran and Aprile 1987; Merion et al. 1991). Only a minority of centres in the United States and Europe sustained HMP programmes into the 1990s, supported by their earlier experiences, and the continuing indications that there were generally lower delayed graft function rates when using HMP (Alijani et al. 1985; Williams 1986; Merion et al. 1991; Salahudeen et al. 2000). On balance, these improvements were not considered of sufficient import to drive a consensus acceptance for HMP across the transplant community. However, during the same time period, several other factors were beginning to impact on organ procurement policies. More donors were being considered for multi-organ procurement in comparison with kidney-alone retrieval. Factors such as the greater need for organs as clinical activity increased, allied with changes in donor demographics (for e.g. improved road safety reduced the numbers of donors arising from road traffic accidents in many countries), exacerbated organ shortages and increased pressures to consider older, less fit candidates for organ donation (Bennett et al. 1993). In these situations, HMP offered a potential mechanism to evaluate some physiological functions of the procured kidneys, for example by measuring vascular resistance during the ex vivo period. Also, delayed graft function was recognised as a significant problem using such "marginal" donors, and preservation by HMP still continued to demonstrate good early function in such kidneys (Tesi et al. 1994). Another important factor was the increased commercial availability of the synthetic gluconate-based perfusate (also known as Belzer's KPS – see above) from the early 1990s. Arguments began to be made to support use of HMP not only on a clinical basis, but also on cost of the service as a whole (Wight et al. 2003), because reducing dialysis requirements and hospital stay following use of HMP after transplanting marginal kidneys could more than off-set perfusion costs (Henry et al. 1988; Johnson et al. 1990; Rosenthal et al. 1991). This case was strengthened when consideration was given to the small but significant number of kidneys assessed during HMP and discarded as having bad predicted outcome, where both monetary service and health costs to the individual recipient patient can be high if a non-functioning organ is transplanted. Some centres with a long experience of HMP continued to put high value on use of the technique not only for marginal donors, but also for traditional cadaveric kidneys (Light 1998; Merion et al. 1991).

More recently there has been a significant resurgence of interest in HMP techniques because, with the acute international shortage of suitable organs for transplantation, moves have been made in many countries to access organs from donors previously considered not to be suitable. These include donors who have cardiac instability and eventually cardiac death, the so-called donation after cardiac death (DCD) scenario. The efforts have been significantly boosted by the development of a new generation of HMP machines, such as the Lifeport Kidney

Transporter (Organ Recovery Systems), which are easily transportable and manufactured from sterile disposable components. This has led to the recent and first ever reported major multi-centre prospective trial of HMP vs. static cold preservation in renal organ preservation. The trial demonstrated a clear outcome benefit in both delayed graft function and organ graft survival at 1 year (Moers et al. 2009), and many more centres are moving to HMP with these new machines for kidney preservation in selected donor groups.

14.8 New Areas of Research in Organ Preservation

There are several new areas of research and re-evaluation of previous ideas, fuelled in part by the need to provide increased numbers of donor organs.

14.8.1 Towards a Single Preservation Solution

Because of the nature of modern multi-organ procurement operations (with up to six different organs potentially accessible from a single donor patient), the development of a single preservation solution, capable of successfully preserving all abdominal and cardiothoracic organs, would aid in logistics and reduce costs. Up until now, different solutions have been used for example, for heart or lung preservation from those used for livers or kidneys, with an inevitable wastage during the operation. To this end, solutions such as Celsior, with a physiological ion balance but containing the recognised cytoprotective agents such as lactobionate and glutathione (Faenza et al. 2001), may be one way forward.

14.8.2 Novel Modulation of the Hypometabolic State

Benefits may also be possible by further controlling hypometabolism (see Sect. 6 above). The principles of metabolic regulation of hypometabolism are highly conserved across phylogeny, and thus it seems likely that there are still new ideas to be developed in this area by continuing to reassess both natural and laboratory-applied cold tolerance (Storey 2004). For example, by applying a novel reversible inhibition of oxidative phosphorylation using gaseous hydrogen sulphide (H_2S), it has been possible to induce states of profound hypometabolism in several model organisms (Blackstone et al. 2005; Blackstone and Roth 2007). Because H_2S is a specific, potent, and reversible inhibitor of mitochondrial complex IV (cytochrome c oxidase), the terminal enzyme complex in the electron transport chain, it has been hypothesised that it could reduce metabolic rate and thus core body temperature in mammals; however, there remain controversies with this agent, which include the ability to translate experimental results to clinical applications, and the known toxicity arising from H_2S gas itself (Savolainen et al. 1980) which would dictate the need for a very fine and robust control of H_2S concentrations in the clinic for the sake of safety for both the patient and the staff.

14.8.3 A Role for Other Bioactive Gases

Improved cold preservation by the generation of bioactive gases, like Carbon Monoxide (CO) or Nitric Oxide (NO), is another recent area of focus. To reduce cold preservation-related I/R injury, restoration of adequate blood flow after transplantation is essential to maintain the microvasculature and to provide oxygen and nutrients to the recovering tissues. In respect of vascular physiology at the level of the microcirculation, carbon monoxide and nitric oxide are two endogenously produced gases that can act as second messenger molecules. Heme oxygenases and nitric oxide synthases are the in vivo enzyme systems responsible for generating carbon monoxide and nitric oxide, respectively. Experimental studies of delivering pharmacologically derived agents such as NO and CO have demonstrated protective effects against I/R injuries after cold preservation (Sandouka et al. 2006; Pizzaro et al. 2009). Both gases share similar properties, such as the ability to activate soluble guanylate cyclase to increase cyclic GMP. Administration of NO donors or CO in the form of soluble carbon monoxide-releasing molecules (CORMs) also results in enhanced function in kidneys and heart after cold storage (Vardanian et al. 2008).

Clinically, difficulties remain in safe and effective administration of these agents in clinical practice. However, these CO and/or NO delivery strategies might provide a unique opportunity to treat as "organ-targeted" therapy, the organs likely to suffer cold preservation-induced I/R before the recipient operation. A soluble form of CO in UW preservation solution (as might be provided by a CORM) would minimise the concerns about possible toxicity induced by in vivo CO inhalation (Sandouka et al. 2006). In addition, NO is itself a free radical and this is critical in areas of injury such as organ toxicity as well as in regulation of blood flow. Much still remains to be learnt about the pharmacology of these agents in the setting of organ preservation.

14.8.4 Oxygen Supply at Hypothermia

The interplay between oxygen, its link to energy metabolism and cold in organ preservation have been discussed above in Sect. 7. Both the composition of the preservation solution and the nature and concentration of dissolved gases may influence the efficacy of the preservation. Up until now, great efforts have been made to improve the composition and quality of preservation solutions (as discussed in Sect. 6), but little attention has been placed on defining the exact role of dissolved oxygen during hypothermic perfusion-preservation. The key question in the renewed interest in HMP for organ preservation (see Sect. 7) is the level of oxygen supply needed for optimal support. It has been reported that the excessive delivery of an oxygenated solution during HMP may increase reactive oxygen species (ROS) due to the high partial pressure of oxygen in the solution, depletion of natural antioxidants during HMP, and shear stress from the perfusion on endothelial cells (Silacci et al. 2001). These problems, however, do require substantial attention and should be considered as a "double-edged sword": oxygen is necessary for aerobic energy metabolism, but it could also lead to an increase in ROS, resulting in damage to cellular membranes and other cell components (Rauen and de Groot 2004; van der Plaatz et al. 2004). In this respect, there are several questions to be answered concerning the methodology involved in HMP.

- How much oxygen may be delivered during the HMP?
- Are there relationships between perfusion flow, the preservation temperature and the tissue oxygen consumption?
- Should the hypothermic perfusion be continuous or intermittent, pulsatile or not?

To explore these questions, it is important therefore to re-examine the effect of temperature on the respiration and on the diffusion of oxygen in tissues as such as liver. Historical evidence in the literature provides some background clues, and these will be briefly reviewed here. Borgmann (1974) deduced an equation that expresses the rate relationship for a complex system such as tissue respiration:

$$A = CT\mathrm{e}^{-\mu/RT},$$

where A is the tissue respiration rate, C is an empirical constant for any given tissue, T is the absolute temperature, R is the gas constant in Cal mol^{-1} T^{-1} and μ is a complex temperature coefficient, which represent a function of the energies of activation of all of the enzyme systems participating on the whole respiration process. This equation is appropriate to describe the respiration of isolated cell suspensions at PO_2 levels where diffusion profiles are not limiting the respiration. In relation to organ preservation, several authors have demonstrated that the cell respiration is related to temperature and does approximate to Arhenius-type kinetics (Farr and Fuhrman 1965; Fuhrman 1956) on cooling. The plot of ln (respiration rate) as a function of $1/T$ allows the determination of the Arrhenius coefficients which can be used to estimate the activation free energies of the respiration process. In later work, Southard et al. (1983) compared the effect of temperatures on the rate of a membrane-bound enzyme catalysed reactions and ADP-stimulated respiration in mitochondria from dog, human, pig and rabbit kidney. They found that decreasing the temperature diminished the rate of mitochondrial oxygen consumption in each species studied.

A similar phenomenon occurs for intact organs. Experiments by Fujita and co-workers during continuous perfusion of the liver at temperatures varying from 5 to 37°C demonstrated that the amount of oxygen that is consumed in the reduced metabolic state in the cold is a logarithmic function of temperature. These investigators (Fujita et al. 1993a, b) determined the temperature (T)-dependent oxygen consumption (VO_2) expressed by the equation:

$$VO_2 = 0.21 \times 10^{0.029T}\ \mu\text{mol/min/g liver}.$$

The relationship between oxygen consumption and temperature is logarithmic, and whilst oxygen consumption decreases, it may still be measurable at temperatures of 0–4°C. In fact, on a theoretical basis, at 4°C liver metabolism would still require 0.27 μmol/min/g.liver of oxygen, which suggests that a specified level of oxygen supply during hypothermic liver perfusion may be required.

To further investigate this, we have recently measured the oxygen consumption of rat livers perfused with the preservation solution Custodiol at both 10 and 5°C, having determined previously the oxygen solubility in the preservation solution at these temperatures. The livers were perfused at constant pressure (77 ± 15 mmH_2O – about 50% of the normothermic portal pressure) in a device designed to measure the oxygen consumption by the liver during hypothermic perfusion (Rodriguez et al. 2008). As shown in Table 14.2, after 2 h of perfusion at low temperatures, the livers at hypothermia possessed a reduced respiratory activity compared with control normothermic values, and this accounts for less than 5% of the values measured during perfusion with Krebs–Henseleit solution at 37°C. However, there was in the region of a 50% higher oxygen consumption moving from 5°C to 10°C, so it is important to record and consider the temperature at which HMP will be carried out, in order to achieve good aerobic metabolism.

In relation to this, another point to be considered is the effect of temperature upon oxygen diffusion in tissues. A previous study (McCabe and Maguire 2005) examined the effect of temperature on the oxygen partial pressure profiles inside respiring tissues of different architecture (two models of respiring sheets and spheres). They determined that the limiting thickness or limiting radius of the oxygenated tissue will increase or decrease corresponding to the applied temperature. It is unknown at present how this will affect HMP.

The crucial, unanswered question which we are currently unable to answer is: at what point would oxygen deprivation become significant during HMP? We need therefore a better understanding how cooling affects integrated cell metabolism and

Table 14.2 Oxygen extraction and consumption indices after normothermic and hypothermic perfusion of rat livers

Perfusion solution	Temperature (°C)	$[O_2]_{input}$ (nmol O_2 min^{-1} g $liver^{-1}$)	O_2 Extraction	VO_2 (nmol O_2 min^{-1} g $liver^{-1}$)
Krebs–Henseleit ($n = 3$)	37.2 ± 0.3	1340 ± 160	0.86 ± 0.03	1150 ± 10
Custodiol® ($n = 3$)	10.8 ± 0.6	128 ± 20	0.45 ± 0.03	57 ± 4
	4.6 ± 0.8	130 ± 40	0.36 ± 0.01	38 ± 7

Livers were perfused with normal Krebs–Henseleit at 37°C. At hypothermia, Custodiol was used – CUSTODIOL® HTK is a solution indicated for perfusion and flushing of donor kidneys, liver, pancreas and heart prior

in particular the metabolism of oxygen to determine the appropriate methodology to supply "essential" levels of oxygen during cold preservation and in particular during HMP.

14.9 Summary

The discussions in this chapter have highlighted how far we have come in being able to implement preservation based on cooling of organs and tissues in clinical therapy, but also, the wide gaps in our fundamental knowledge. Worldwide, we are able to offer a hypothermic organ preservation as a clinical service, but this is by no means optimal, since we still incur significant levels of organ dysfunction during storage, which ultimately impedes further progress in transplant therapy. The methods have been developed empirically, against a background awareness that Nature, when challenged from an evolutionary perspective to deal with similar problems of cold and hypoxia, has evolved a much more sophisticated "survival strategy". The explosive interest in molecular sciences in the recent past now gives us opportunities to revisit the problems of organ preservation, to better understand Nature's "winning hand" against hypothermic injury in tolerant species and to identify targets for translation into improved organ and tissue preservation.

Acknowledgements Part of this collaborative work was facilitated by the UNESCO Chair in Cryobiology.

References

Alijani MR, Cutler JA, Del Valle C, Morres D, Fawzy A, Pechan B, Helfrich G (1985) Single donor cold storage versus machine perfusion in cadaver kidney preservation. Transplantation 40:659–661

Andrews M (2007) Advances in molecular biology of hibernation in mammals. BioEssays 29:431–440

Arrhenius SA (1915) Quantitative laws in biological chemistry. G Bell & Sons, London

Baicu SC, Taylor MJ, Brockbank KG (2006) The role of preservation solution on acid-base regulation during machine perfusion of kidneys. Clin Transplant 20:113–121

Bartels-Stringer M, Wetzels JF, Wouterse AC, Steenbergen E, Russel FG, Kramers C (2005) Iron chelators do not reduce cold-induced cell injury in the isolated perfused rat kidney model. Nephrol Dial Transplant 20:2646–2653

Belzer FO, May R, Berry M, Lee JC (1970) Short term preservation of porcine livers. J Surg Res 10:55–61

Bennett LE, Glascock R, Breen T, Ellison M, Daily OP (1993) Organ donation in the United States: 1988 through 1992. Clin Transpl 5:85–93

Bessems M, Doorschodt BM, van Vliet AK, van Gulik TM (2005a) Improved machine perfusion preservation of the non-heart-beating donor rat liver using polysol: a new machine perfusion preservation solution. Liver Transpl 11:1379–1388

Bessems M, Doorschodt BM, van Marle J, Vreeling H, Meijer AJ, van Gulik TM (2005b) Improved rat liver preservation by hypothermic continuous machine perfusion using polysol, a new, enriched preservation solution. Liver Transpl 11:539–546

Bickford RG, Winton FR (1937) The influence of temperature on the isolated kidney of the dog. J Physiol (Lond) 89:198–219
Bickler P, Buck L (2007) Hypoxia tolerance in reptiles, amphibians, and fishes: life with variable oxygen availability. Annu Rev Physiol 69:145–170
Blackstone E, Roth MB (2007) Suspended animation-like state protects mice from lethal hypoxia. Shock 27:370–372
Blackstone E, Morrison M, Roth MB (2005) H_2S induces a suspended animation-like state in mice. Science 308:518
Borgmann U (1974) A theoretical description of biological rates. J Theor Biol 45:171–182
Boyle R (1683) New experiments and observations touching cold. R. Davis, London
Breton S, Brown D (1998) Cold-induced microtubule disruption and relocalization of membrane proteins in kidney epithelial cells. J Am Soc Nephrol 9:155–166
Brodie T (1903) The perfusion of surviving organs. J Physiol (Lond) 29:266–275
Brooks SP, Storey K (1994) cGMP-stimulated protein kinase phosphorylates protein kinase in an anoxia-tolerant marine mollusc. J Comp Physiol B 160:309–316
Calne RY, Pegg DE, Pryse-Davies J, Brown FL (1963) Renal preservation by ice-cooling: an experimental study relating to kidney transplantation from cadavers. Br Med J 2:651–655
Carrel A, Lindbergh CA (1935) The culture of whole organs. Science 81:621–623
Carrel A, Lindbergh CA (1938) The culture of organs. Hoeber, NewYork
Casillas-Ramírez A, Mosbah IB, Ramalho F, Roselló-Catafau J, Peralta C (2006) Past and future approaches to ischemia-reperfusion lesion associated with liver transplantation. Life Sci 79:1181–1194
Changani KK, Fuller BJ, Bryant DJ, Bell JD, Ala-Korpela M, Taylor-Robinson SD, Moore DP, Davidson BR (1997) Non-invasive assessment of ATP regeneration potential of the preserved donor liver. A 31P MRS study in pig liver. J Hepatol 26:336–342
Changani KK, Davidson BR, Bell J, Taylor-Robinson S, Fuller B (1998) Enhanced energy metabolism at hypothermia following addition of a prostacyclin derivative in porcine liver. CryoLetters 19:131–140
Churchill TA, Cheetham KM, Fuller BJ (1994) Glycolysis and energy metabolism in rat liver during warm and cold ischemia: evidence of an activation of the regulatory enzyme phosphofructokinase. Cryobiology 31:441–452
Collins GM, Bravo-Shugarman M, Terasaki PI (1969) Kidney preservation for transportation. Initial perfusion and 30 hours' ice storage. Lancet 2:1219–1222
Collins L, Sawyer GJ, Zhang XH, Gustafsson K, Fabre JW (2000) In vitro investigation of factors important for the delivery of an integrin-targeted nonviral DNA vector in organ transplantation. Transplantation 69:1168–1176
Crowe JH, Hoekstra FA, Crowe LM, Anchordoguy TJ, Drobnis E (1989) Lipid phase transitions measured in intact cells with Fourier transform infrared spectroscopy. Cryobiology 26:76–84
Daemen MA, de Vries B, Buurman WA (2003) Apoptosis and inflammation in renal reperfusion injury. Transplantation 73:1693–1700
Dutkowski P, Schönfeld S, Heinrich T, Watzka M, Winkelbach V, Krysiak M, Odermatt B, Junginger T (1999) Reduced oxidative stress during acellular reperfusion of the rat liver after hypothermic oscillating perfusion. Transplantation 68:44–50
Dutkowski P, Graf R, Clavien PA (2006) Rescue of the cold preserved rat liver by hypothermic oxygenated machine perfusion. Am J Transplant 6:903–912
Ellermann J, Gewiese B, David H, Stiller D, Plötz M, Wolf KJ, Lüning M (1992) Electron microscopic and 31-P NMR studies of ischemic injured rat livers during preservation and reflow. Exp Toxicol Pathol 44:245–253
Faenza A, Catena F, Nardo B, Montalti R, Capocasale E, Busi N, Boggi U, Vistoli F, Di Naro A, Albertazzi A, Mosca F, Cavallari A (2001) Kidney preservation with university of Wisconsin and Celsior solution: a prospective multicenter randomized study. Transplantation 72:1274–1277
Farr DA, Fuhrman FA (1965) Role of diffusion of oxygen in the respiration of tissues at different temperatures. J Appl Physiol 20:637–646

Feltkamp CA, van der Waerden AW (1982) Low temperature -induced displacement of cholesterol and intramembrane particles in nuclear membranes of mouse leukemia cells. Cell Biol Int Rep 6:137–145

Fuhrman FA (1956) Oxygen consumption of mammalian tissues at reduced temperatures. Proc Natl Acad Sci USA 451:50–51

Fujita S, Hamamoto I, Nakamura K, Tanaka K, Ozawa K (1993a) Evaluation of oxygen necessity during hypothermic liver perfusion. Nippon Geka Hokan 62:228–240

Fujita S, Hamamoto I, Nakamura K, Tanaka K, Ozawa K (1993b) Isolated perfusion of rat livers: effect of temperature on O_2 consumption, enzyme release, energy store, and morphology. Nippon Geka Hokan 62:58–70

Fuller BJ (1999) Organ preservation: the profit and loss account of using hypothermia to maintain viability. Transplant Rev 13:55–66

Fuller BJ (2003) Gene expression in response to low temperatures in mammalian cells: a review of current ideas. CryoLetters 24:95–102

Fuller BJ, So P-W (2001) Hepatic uptake of solutes from the preservation solution during hypothermic storage: a (1)H NMR study in rat liver. Cryobiology 42:307–313

Fuller BJ, Marley S, Green CJ (1985) Gluconeogenesis in stored kidneys from normal and cold-acclimated rats. CryoLetters 6:91–98

Fuller BJ, Gower J, Green CJ (1988) Free radical damage and organ preservation: fact or fiction? A review of the interrelationship between oxidative stress and physiological ion disbalance. Cryobiology 25:377–393

Fuller BJ, Busza AL, Proctor E (1990) Possible resuscitation of liver function by hypothermic reperfusion in vitro after prolonged (24-hour) cold preservation – a 31P NMR study. Transplantation 50:511–513

Halloran PF, Aprile M (1987) A randomised prospective trial of cold-storage versus pulsatile perfusion for cadaveric kidney preservation. Transplantation 43:827–831

Hansen TN, D'Alessandro A, Southard JH (1997) Reduced renal vascular injury following warm ischemia and preservation by hypothermic machine perfusion. Transplant Proc 29: 3577–3579

Heilbrunn LV (1937) An outline of general physiology. WB Saunders, Philadelphia, USA

Henry ML, Sommer B, Ferguson R (1988) M. Improved immediate function of renal allografts with Belzer perfusate. Transplantation 45:73–75

Hobbs KE, Hunt A, Palmer D, Badrick F, Morris A, Mitra S, Peacock J, Immelman E, Riddell AG (1968) Hypothermic low flow liver perfusion as a means of porcine hepatic storage for 6 hours. Br J Surg 55:696–703

Hochachka P (1986) Defense strategies against hypoxia and hypothermia. Science 231:234–241

Hochachka PW, Buck LT, Doll CJ, Land SC (1996) Unifying theory of hypoxia tolerance: molecular/metabolic defense and rescue mechanisms for surviving oxygen lack. Proc Natl Acad Sci USA 93:9493–9498

Holden C, Storey K (1997) Second messenger and c-AMP dependent protein kinase responses to dehydration and anoxia stress in frogs. J Comp Physiol B 167:305–312

Hume DM, Merrill J, Miller BF, Thorn GW (1955) Experiences with renal homotransplantation in the human: report of nine cases. J Clin Invest 34:327–382

Humphries AL, Russell R, Christopher P, Goodrich S, Stoddard L, Moretz WH (1964a) Succesful reimplantation of twenty-four-hour stored kidney to nephrectomized dog. Ann N Y Acad Sci 120:496–505

Humphries AL, Russell R, Gregory J, Carter R, Moretz WH (1964b) Hypothermic perfusion of the canine kidney for 48 hours followed by reimplantation. Am Surg 30:748–752

Humphries AL Jr, Russell R, Stoddard L, Moretz WH (1968a) Successful five-day kidney preservation. Perfusion with hypothermic, plasma. Invest Urol 5:609–618

Humphries AL, Russell R, Stoddard LD, Moretz WH (1968b) Perfusion of dog kidneys with cooled medium 199 followed by auto or allotransplantation. In: Norman JC (ed) Organ perfusion and preservation. Appleton Century Crofts, NewYork, pp 13–26

Jassem W, Heaton ND (2004) The role of mitochondria in ischemia/reperfusion injury in organ transplantation. Kidney Int 66:514–517

Jassem W, Fuggle SV, Rela M, Koo DD, Heaton ND (2002) The role of mitochondria in ischemia/reperfusion injury. Transplantation 73:493–499

Johansson BW (1996) The hibernator heart – nature's model of resistance to ventricular fibrillation. Cardiovasc Res 31:826–832

Johnson CP, Roza A, Adams MB (1990) Local procurement with pulsatile perfusion gives excellent results and minimises initial cost associated with renal transplantation. Transplant Proc 22:385

Keeler R, Swinney J, Taylor R, Uldall MB (1966) The problem of renal preservation. Br J Urol 38:653–655

Kupiec-Weglinski JW, Busuttil RW (2005) Ischemia and reperfusion injury in liver transplantation. Transplant Proc 37:1653–1656

Lee CY, Jain S, Duncan H, Zhang JX, Jones J, Southard JH, Clemens M (2003) Survival transplantation of preserved non-heart-beating donor rat livers. Transplantation 76:1432–1436

Lemaster JJ, Thurman RG (1993) Hypoxia and reperfusion injury to liver. Prog Liver Dis 11:85–114

Light JA (1998) A 25 year history of kidney transplantation in the Washington Hospital Centre. In: Cecka D, Terasaki PI (eds) Chapter 13, Clinical Transplants 1998, UCLA Tissue Typing Laboratory, pp 159–168

Lindell SL, Compagnon P, Mangino MJ, Southard JH (2005) UW solution for hypothermic machine perfusion of warm ischemic kidneys. Transplantation 79:1358–1361

Locke FS, Rosenheim O (1907) Contributions to the physiology of the isolated heart: the consumption of dextrose by mammalian cardiac muscle. J Physiol (Lond) 36:205–220

Lockett CJ, Busza AL, Toffa SK, Proctor E, Fuller BJ (1995) Resuscitation of cardiac energy metabolism in the rabbit heart by brief hypothermic reperfusion after preservation studied by 31P NMR spectroscopy. Transpl Int 8:8–12

Lopez-Neblina F, Toledo AH, Toledo-Pereyra LH (2005) Molecular biology of apoptosis in ischemia and reperfusion. J Invest Surg 18:335–350

Lyman CP (1982) The hibernating state. In: Willis JS, Malan A, Wang LCH (eds) Hibernation and torpor in mammals and birds. Academic Press, NewYork, pp 12–53

Lyman CP, Willis JS, Malan A, Wang LCH (1982) Hibernation and torpor in mammals and birds. Academic Press, New York

McAnulty JF (1998) The effect of calcium on hypothermia-facilitated resuscitation of warm ischemic kidney tissue slices: a role for the mitochondrial permeability transition pore? Cryobiology 36:12–19

McAnulty JF, Foley JD, Reid TW, Heath TD, Waller KR, Murphy CJ (2004) Suppression of cold ischemic injury in stored kidneys by the antimicrobial peptide bactenecin. Cryobiology 49:230–240

McCabe M, Maguire DJ (2005) Effects of temperature on oxygen transport in sheets and spheres of respiring tissues. Adv Exp Med Biol 566:105–109

Merion RM, Oh H, Port F, Toledo-Pereyra LH, Turcotte JG (1991) A prospective controlled trial of cold storage versus machine perfusion preservation in cadaveric renal transplantation. Transplantation 50:230–233

Milsom W, Beth Zimmer M, Harris B (1999) Regulation of cardiac rhythm in hibernating mammals. Comp Biochem Physiol 124:383–391

Minor T, Sitzia M, Dombrowski F (2005) Kidney transplantation from non-heart-beating donors after oxygenated low-flow machine perfusion preservation with histidine-tryptophan-ketoglutarate solution. Transpl Int 17:707–712

Moers C, Smits JM, Maathuis MH, Treckmann J, van Gelder F, Napieralski BP, van Kasterop-Kutz M, van der Heide JJ, Squifflet JP, van Heurn E, Kirste GR, Rahmel A, Leuvenink HG, Paul A, Pirenne J, Ploeg RJ (2009) Machine perfusion or cold storage in deceased-donor kidney transplantation. N Engl J Med 360:7–19

NHS Blood and Transplant (2008) Transplant activity in the UK: 2007–2008 http://www.organ-donation.nhs.uk/ukt/

NHS Blood and Transplant Annual Activity Report 2008–2009, http://www.organdonation.nhs.uk/ukt/
Palmer JF (1825) The works of John Hunter FRS with notes. Longman, Reese, Orme, Brwon, Green & Longman, London
Pegg DE (1978) An approach to hypothermic renal preservation. Cryobiology 15:1–17
Pegg D, Wusteman MC, Foreman J (1981) Metabolism of normal and ischemically injured rabbit kidneys during perfusion for 48 hours at 10 C. Transplantation 32:437–443
Peltz M, He TT, Adams GA 4th, Koshy S, Burgess SC, Chao RY, Meyer DM, Jessen ME (2005) Perfusion preservation maintains myocardial ATP levels and reduces apoptosis in an ex vivo rat heart transplantation model. Surgery 138:795–805
Pinsky DJ, Yan SF, Lawson C, Naka Y, Chen JX, Connolly ES Jr, Stern DM (1995) Hypoxia and modification of the endothelium: implications for regulation of vascular homeostatic properties. Semin Cell Biol 6:283–294
Piper HM, Meuter K, Schäfer C (2003) Cellular mechanisms of ischemia-reperfusion injury. Ann Thorac Surg 75:S644–S648
Pizzaro MD, Rodriguez JV, Mamprin M, Fuller B, Mann B, Motterlini M, Guibert E (2009) Protective effects of a carbon monoxide-releasing molecule (CORM-3) during hepatic cold preservation. Cryobiology 58:248–256
Polyak MM, Arrington BO, Stubenbord WT, Kapur S, Kinkhabwala M (1999) Prostaglandin E1 influences pulsatile preservation characteristics and early graft function in expanded criteria donor kidneys. J Surg Res 85:17–25
Rauen U, de Groot H (2002) Mammalian cell injury induced by hypothermia – the emerging role for reactive oxygen species. Biol Chem 383:477–488
Rauen U, de Groot H (2004) New insights into the cellular and molecular mechanisms of cold storage injury. J Investig Med 52:299–309
Rodriguez JV, Federico MB, Pizarro MD, Guibert EE, Scandizzi AL (2008) A device to measure oxygen consumption during the hypothermic perfusion of the liver. V Latin American Congress of Artificial Organs and Biomaterials, Ouro Preto, Brasil, N504
Rosenthal J, Danovitch G, Wilkinson A, Ettenger RB (1991) The high cost of delayed graft function in cadaveric renal transplantation. Transplantation 51:1115–1118
Ruiz J, Lillihie RC (1974) Pancreas. In: Karow A, Abouna G, Humphries AL (eds) Organ preservation for transplantation. Little Brown & Co, Boston, pp 372–382
Salahudeen AK, Huang H, Patel P, Jenkins JK (2000) Mechanism and prevention of cold storage-induced human renal tubular cell injury. Transplantation 70:1424–1431
Salehi P, Bigam DL, Ewaschuk JB, Madsen KL, Sigurdson GT, Jewell LD, Churchill TA (2008) Alleviating intestinal ischemia-reperfusion injury in an in vivo large animal model: developing an organ-specific preservation solution. Transplantation 85:878–884
Sandouka A, Fuller BJ, Mann BE, Green CJ, Foresti R, Motterlini R (2006) Treatment with CO-RMs during cold storage improves renal function at reperfusion. Kidney Int 69:239–247
Savolainen H, Tenhunen R, Elovaara E, Tossavainen A (1980) Cumulative biochemical effects of repeated subclinical hydrogen sulfide intoxication in mouse brain. Int Arch Occup Environ Health 46:87–92
Schloerb PR, Waldorf R, Welsh JS (1959) The protective effect of kidney hypothermia on total renal ischemia. Surg Gynecol Obstet 109:561–565
Silacci P, Desgeorges A, Mazzolai L, Chambaz C, Hayoz D (2001) Flow pulsatility is a critical determinant of oxidative stress in endothelial cells. Hypertension 38:1162–1166
Southard J, Belzer FO (1995) Organ preservation. Annu Rev Med 46:235–247
Southard JH, Van der Laan NC, Lutz M, Pavlock GS, Belzer JP, Belzer FO (1983) Comparison of the effect of temperature on kidney cortex mitochondria from rabbit, dog, pig, and human: Arrhenius plots of ADP-stimulated respiration. Cryobiology 20:395–400
Southard JH, Lutz M, Ametani M, Belzer FO (1984) Stimulation of ATP synthesis in hypothermically perfused dog kidneys by adenosine and PO4. Cryobiology 21:13–19
Starzl TE (2000) History of clinical transplantation. World J Surg 24:759–782

Starzl TE, Hakala TR, Shaw BW Jr, Hardesty RL, Rosenthal TJ, Griffith BP, Iwatsuki S, Bahnson HT (1984) A flexible procedure for multiple cadaveric organ procurement. Surg Gynecol Obstet 158:223–230

Stefanovich P, Ezzell RM, Sheehan SJ, Tompkins RG, Yarmush ML, Toner M (1996) Effects of hypothermia on the function, membrane integrity, and cytoskeletal structure of hepatocytes. Cryobiology 32:389–403

Stickel DL, Chen KM, Robinson R (1968) The effects of organ perfusion and other factors on the immediate function of canaine renal autografts. In: Norman JC (ed) Organ perfusion for preservation. Appleton-Century-Crofts, NewYork, pp 61–69

Storey K (2002) Life in the slow lane: molecular mechanisms of aestivation. Comp Biochem Physiol 133A:733–754

Storey K (2004) Cold ischemic organ preservation: lessons from natural systems. JInvestMed 52:315–322

Storey K, Storey J (1990) Facultative metabolic rate depression: molecular regulation and biochemical adaptation in anaerobiosis, hibernation and estivation. Q Rev Biol 65:145–174

Storey K, Storey J (2004a) Strategies for exploration of freeze responsive gene expression: advances in vertebrate freeze tolerance. Cryobiology 48:134–145

Storey K, Storey J (2004b) Metabolic rate depression in animals: transcriptional and translational controls. Biol Rev Camb Philos Soc 79:207–233

Storey K, Storey J (2007) Tribute to P. L. Lutz: putting life on 'pause' – molecular regulation of hypometabolism. J Exp Biol 210:1700–1714

Storey KB, Heldmaier G, Rider MH (2010) *Mammalian Hibernation: Physiology, Cell Signaling, and Gene Controls on Metabolic Rate Depression* XX–XX

T' Hart NA, van der Plaats A, Faber A, Leuvenink HG, Olinga P, Wiersema-Buist J, Verkerke GJ, Rakhorst G, Ploeg RJ (2005) Oxygenation during hypothermic rat liver preservation: an in vitro slice study to demonstrate beneficial or toxic oxygenation effects. Liver Transpl 11:1403–1411

Tamaki T, Kamada N, Wight D, Pegg D (1987) Successful 48-hour preservation of the rat liver by continuous hypothermic perfusion with haemaccel-isotonic citrate solution. Transplantation 43:468–471

Teoh NC, Farrell GC (2003) Hepatic ischemia reperfusion injury: pathogenic mechanisms and basis for hepatoprotection. J Gastroenterol Hepatol 18:891–902

Tesi R, Elkhammas E, Davies E, Henry M (1994) Pulsatile kidney perfusion for evaluation of high-risk kidney donors safely expands the donor pool. Transplant Proc 26:241–242

van der Plaatz A, 'T Hart NA, Verkerke GJ, Leuvenink HG, Ploeg RJ, Rakhorst G (2004) The Groningen hypothermic liver perfusion pump: functional evaluation of a new machine perfusion system. Ann Biomed Eng 32:623–631

Vardanian A, Busuttil R, Kupiec-Weglinski J (2008) Molecular mediators of liver ischemia and reperfusion injury: a brief review. Mol Med 14:337–345

Wight J, Chilcott J, Holmes M, Brewer N (2003) The clinical cost effectiveness of pulsatile machine perfusion versus cold storage of kidneys for transplantation retreived from heart-beating and non-heart beating donors. Health Technol Assess 7:1–94

Williams GM (1986) Kidney preservation and early function in the cyclosporine era: an overview. Transplant Proc 18(Suppl 1):38–40

Willis JS, Ellory JC, Becker JH (1978) Na-K pump and Na-K-ATPase: disparity of their temperature sensitivity. Am J Physiol 235:C159–C167

Zimmermann F, Diettz H, Kohler CH, Kilian N, Kosterhon J, Scholz R (1982) Effects of hypothermia on anabolic and catabolic processes and on oxygen consumption in perfused rat livers. In: Pegg DE, Jacobsen IA, Halasz N (eds) Organ preservation: basic and applied aspects. MTP Press, Lancaster, pp 121–126

Index